TENSOR CALCULUS

MATHEMATICAL EXPOSITIONS

Volumes Already Published

No. 1: *The Foundations of Geometry* – G. de B. Robinson
No. 2: *Non-Euclidean Geometry* – H. S. M. Coxeter
No. 3: *The Theory of Potential and Spherical Harmonics* – W. J. Sternberg and T. L. Smith
No. 4: *The Variational Principles of Mechanics*–Cornelius Lanczos
No. 5: *Tensor Calculus* – J. L. Synge and A. E. Schild

In Press

No. 6: *The Theory of Functions of a Real Variable* – R. L. Jeffery

There are many books dealing in an individual way with elementary aspects of Algebra, Geometry, or Analysis. In recent years various advanced topics have been treated exhaustively, but there is need in English of books which emphasize fundamental principles while presenting the material in a less elaborate manner. A series of books, published under the auspices of the University of Toronto and bearing the title "Mathematical Expositions," represents an attempt to meet this need. It will be the first concern of each author to take into account the natural background of his subject and to present it in a readable manner.

MATHEMATICAL EXPOSITIONS, No. 5

TENSOR CALCULUS

by

J. L. SYNGE
Professor of Mathematics
Dublin Institute for Advanced Studies

and

A. SCHILD
Assistant Professor of Mathematics
Carnegie Institute of Technology

UNIVERSITY OF TORONTO PRESS
TORONTO, 1949

LONDON:
GEOFFREY CUMBERLEGE
OXFORD UNIVERSITY PRESS

PRINTED IN CANADA

PREFACE

A MATHEMATICIAN unacquainted with tensor calculus is at a serious disadvantage in several fields of pure and applied mathematics. He is cut off from the study of Riemannian geometry and the general theory of relativity. Even in Euclidean geometry and Newtonian mechanics (particularly the mechanics of continua) he is compelled to work in notations which lack the compactness of tensor calculus.

This book is intended as a general brief introduction to tensor calculus, without claim to be exhaustive in any particular direction. There is no attempt to be historical or to assign credit to the originators of the various lines of development of the subject. A bibliography at the end gives the leading texts to which the reader may turn to trace the development of tensor calculus or to go more deeply into some of the topics. As treatments of tensor calculus directed towards relativity are comparatively numerous, we have excluded relativity almost completely, and emphasized the applications to classical mathematical physics. However, by using a metric which may be indefinite, we have given an adequate basis for applications to relativity.

Each chapter ends with a summary of the most important formulae and a set of exercises; there are also exercises scattered through the text. A number of the exercises appeared on examination papers at the University of Toronto, and our thanks are due to the University of Toronto Press for permission to use them.

The book has grown out of lectures delivered over a number of years by one of us (J.L.S.) at the University of Toronto, the Ohio State University, the Carnegie Institute of Technology. Many suggestions from those who attended the lectures have been incorporated into the book. Our special thanks are due to Professors G. E. Albert, B. A. Griffith, E. J. Mickle, M. Wyman, and Mr. C. W. Johnson, each of whom has read the manuscript in whole or in part and made valuable suggestions for its improvement.

J. L. SYNGE
A. SCHILD

CONTENTS

I. SPACES AND TENSORS

II. BASIC OPERATIONS IN RIEMANNIAN SPACE

III. CURVATURE OF SPACE

VII. RELATIVE TENSORS, IDEAS OF VOLUME, GREEN-STOKES THEOREMS

VIII. NON-RIEMANNIAN SPACES

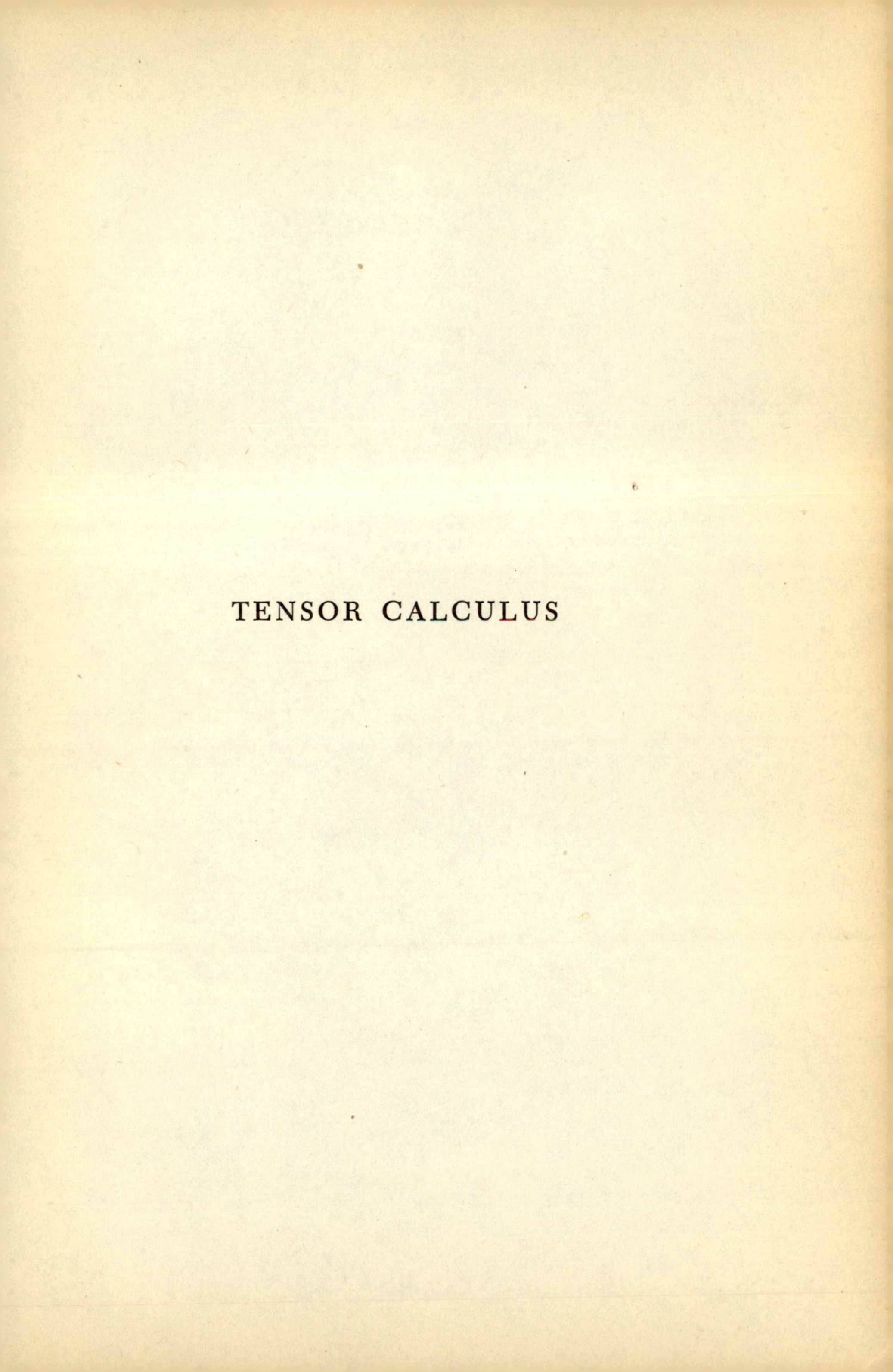

TENSOR CALCULUS

CHAPTER I

SPACES AND TENSORS

1.1. The generalized idea of a space. In dealing with two real variables (the pressure and volume of a gas, for example), it is a common practice to use a geometrical representation. The variables are represented by the Cartesian coordinates of a point in a plane. If we have to deal with three variables, a point in ordinary Euclidean space of three dimensions may be used. The advantages of such geometrical representation are too well known to require emphasis. The analytic aspect of the problem assists us with the geometry and *vice versa*.

When the number of variables exceeds three, the geometrical representation presents some difficulty, for we require a space of more than three dimensions. Although such a space need not be regarded as having an actual physical existence, it is an extremely valuable concept, because the language of geometry may be employed with reference to it. With due caution, we may even draw diagrams in this "space," or rather we may imagine multidimensional diagrams projected on to a two-dimensional sheet of paper; after all, this is what we do in the case of a diagram of a three-dimensional figure.

Suppose we are dealing with N real variables $x^1, x^2, \ldots, x^N$. For reasons which will appear later, it is best to write the numerical labels as superscripts rather than as subscripts. This may seem to be a dangerous notation on account of possible confusion with powers, but this danger does not turn out to be serious.

We call a set of values of $x^1, x^2, \ldots, x^N$ a *point*. The variables $x^1, x^2, \ldots, x^N$ are called *coordinates*. The totality of points corresponding to all values of the coordinates within certain ranges constitute a *space of N dimensions*. Other words, such as *hyperspace*, *manifold*, or *variety* are also used to avoid confusion with the familiar meaning of the word "space." The ranges of the coordinates may be from $-\infty$ to $+\infty$, or they may be restricted. A space of N dimensions is referred to by a symbol such as V_N.

Excellent examples of generalized spaces are given by dynamical systems consisting of particles and rigid bodies. Suppose we have a bar which can slide on a plane. Its position (or *configuration*) may be fixed by assigning the Cartesian coordinates x, y of one end and the angle θ which the bar makes with a fixed direction. Here the space of configurations is of three dimensions and the ranges of the coordinates are

$$-\infty < x < +\infty, \quad -\infty < y < +\infty, \quad 0 \leqslant \theta < 2\pi.$$

Exercise. How many dimensions has the configuration-space of a rigid body free to move in ordinary space? Assign coordinates and give their ranges.

It will be most convenient in our general developments to discuss a space with an unspecified number of dimensions N, where $N \geqslant 2$. It is a remarkable feature of the tensor calculus that no essential simplification is obtained by taking a small value of N; a space of two million dimensions is as easy to discuss (in its general aspects) as a space of two dimensions. Nevertheless the cases $N = 2$, $N = 3$, and $N = 4$ are of particular interest: $N = 2$ gives us results in the intrinsic geometry of an ordinary surface; $N = 3$ gives us results in the geometry of ordinary space; $N = 4$ gives us results in the space-time of relativity.

The development of the geometry of V_N is a game which must be played with adroitness. We take the familiar words of geometry and try to give them meanings in V_N. But we must of course remember that N might be 3 and V_N might be our familiar Euclidean space of three dimensions. Therefore, to avoid confusion, we must be careful to frame our definitions so that, in this particular case, these definitions agree with the familiar ones.

A curve is defined as the totality of points given by the equations

1.101. $$x^r = f^r(u) \qquad (r = 1, 2, \ldots, N).$$

Here u is a parameter and f^r are N functions.

Next we consider the totality of points given by

1.102. $$x^r = f^r(u^1, u^2, \ldots, u^M) \qquad (r = 1, 2, \ldots, N),$$

where the u's are parameters and $M < N$. This totality of points may be called V_M, a *subspace* of V_N. There are two cases of special interest, namely $M = 2$ and $M = N - 1$. Either of these might be called a *surface*, because if $N = 3$ they both coincide with the familiar concept of "surface." It seems, however, that V_{N-1} has the better right to be called a surface, because it has (for any N) the fundamental property of a surface in ordinary space, viz. it divides the neighbouring portion of space into two parts. To see this, we eliminate the parameters from 1.102. Since $M = N - 1$, the number of parameters is one less than the number of equations, and so elimination gives just one equation:

1.103. $$F(x^1, x^2, \ldots, x^N) = 0.$$

The adjacent portion of V_N is divided into two parts for which respectively F is positive and negative. V_{N-1} is often called a *hypersurface* in V_N.

Other familiar geometrical ideas will be extended to V_N as the occasion arises.

Exercise. The parametric equations of a hypersurface in V_N are

$$\begin{aligned}
x^1 &= a \cos u^1, \\
x^2 &= a \sin u^1 \cos u^2, \\
x^3 &= a \sin u^1 \sin u^2 \cos u^3, \\
&\cdots\cdots \\
x^{N-1} &= a \sin u^1 \sin u^2 \sin u^3 \ldots \sin u^{N-2} \cos u^{N-1}, \\
x^N &= a \sin u^1 \sin u^2 \sin u^3 \ldots \sin u^{N-2} \sin u^{N-1},
\end{aligned}$$

where a is a constant. Find the single equation of the hypersurface in the form 1.103, and determine whether the points $(\frac{1}{2}a, 0, 0, \ldots . 0)$, $(0, 0, 0, \ldots 0, 2a)$ lie on the same or opposite sides of the hypersurface.

Exercise. Let U_2 and W_2 be two subspaces of V_N. Show that if $N = 3$ they will in general intersect in a curve; if $N = 4$ they will in general intersect in a finite number of points; and if $N > 4$ they will not in general intersect at all.

1.2. Transformation of coordinates. Summation convention. It is a basic principle of tensor calculus that we should not tie ourselves down to any one system of coordinates. We seek statements which are true, not for one system of coordinates, but for all.

Let us suppose that in a V_N there is a system of coordinates $x^1, x^2, \ldots, x^N$. Let us write down equations

1.201. $$x'^r = f^r(x^1, x^2, \ldots, x^N) \qquad (r = 1, 2, \ldots, N),$$

where the f's are single valued continuous differentiable functions for certain ranges of $x^1, x^2, \ldots, x^N$. These equations assign to any point $x^1, x^2, \ldots, x^N$ a new set of coordinates $x'^1, x'^2, \ldots, x'^N$. The Jacobian of the transformation is

1.202. $$J' = \begin{vmatrix} \dfrac{\partial x'^1}{\partial x^1} & \cdots\cdots & \dfrac{\partial x'^1}{\partial x^N} \\ & \cdots\cdots & \\ \dfrac{\partial x'^N}{\partial x^1} & \cdots\cdots & \dfrac{\partial x'^N}{\partial x^N} \end{vmatrix},$$

or, in a briefer notation,

1.203. $$J' = \left| \frac{\partial x'^r}{\partial x^s} \right|,$$

the ranges $r, s = 1, 2, \ldots, N$ being understood. We shall suppose that the Jacobian does not vanish. Then, as is well known from the theory of implicit functions, the equations 1.201 may be solved to read

1.204. $$x^r = g^r(x'^1, x'^2, \ldots, x'^N) \quad (r = 1, 2, \ldots, N).$$

Differentiation of 1.201 gives

1.205. $$dx'^r = \sum_{s=1}^{N} \frac{\partial x'^r}{\partial x^s} dx^s \quad (r = 1, 2, \ldots, N).$$

Thus the transformation of the differentials of the coordinates is a linear homogeneous transformation, the coefficients being functions of position in V_N. We shall return to this transformation presently, but first let us introduce two notational conventions which will save us an enormous amount of writing.

Range Convention. *When a small Latin suffix (superscript or subscript) occurs unrepeated in a term, it is understood to take all the values* 1, 2, . . . , N, *where* N *is the number of dimensions of the space.*

Summation Convention. *When a small Latin suffix is repeated in a term, summation with respect to that suffix is understood, the range of summation being* 1, 2, . . . , N.

It will be noticed that the reference is to small Latin suffixes only. Some other range (to be specified later) will be understood for small Greek suffixes, while if the suffix is a capital letter no range or summation will be understood.

To see the economy of this notation, we observe that 1.205 is completely expressed by writing

1.206. $$dx'^r = \frac{\partial x'^r}{\partial x^s} dx^s.$$

Repeated suffixes are often referred to as "dummies" since, due to the implied summation, any such pair may be replaced by any other pair of repeated suffixes without changing the expression. We have, for example,

$$a_{rs}b^s = a_{rk}b^k.$$

This device of changing dummies is often employed as a useful manipulative trick for simplifying expressions.

In order to avoid confusion we make it a general rule that the same suffix must never be repeated more than twice in any single term or product. If this cannot be avoided, the summation convention should be suspended and all sums should be indicated explicitly.

Exercise. Show that

$$(a_{rst} + a_{str} + a_{srt})x^r x^s x^t = 3a_{rst}x^r x^s x^t.$$

Exercise. If $\phi = a_{rs}x^r x^s$, show that

$$\frac{\partial \phi}{\partial x^r} = (a_{rs} + a_{sr})\, x^s, \quad \frac{\partial^2 \phi}{\partial x^r \partial x^s} = a_{rs} + a_{sr}.$$

Simplify these expressions in the case where $a_{rs} = a_{sr}$.

Let us introduce a symbol δ^r_s called the *Kronecker delta*; it is defined by

1.207.
$$\begin{aligned} \delta^r_s &= 1 \text{ if } r = s, \\ \delta^r_s &= 0 \text{ if } r \neq s. \end{aligned}$$

Exercise. Prove the relations

$$\begin{aligned} \delta^r_s\, a^s{}_{ij} &= a^r{}_{ij}, \\ \delta^r_s\, b_{rij} &= b_{sij}. \end{aligned}$$

It is evident that $\partial x^r / \partial x^s = \delta^r_s$, or equivalently

1.208.
$$\frac{\partial x^r}{\partial x'^n}\frac{\partial x'^n}{\partial x^s} = \delta^r_s.$$

From this we may derive an identity which will be useful later. Partial differentiation with respect to x^p gives (since the Kronecker delta is constant)

1.209.
$$\frac{\partial^2 x^r}{\partial x'^m \partial x'^n}\frac{\partial x'^m}{\partial x^p}\frac{\partial x'^n}{\partial x^s} + \frac{\partial x^r}{\partial x'^n}\frac{\partial^2 x'^n}{\partial x^p \partial x^s} = 0.$$

If we multiply across by $\dfrac{\partial x'^q}{\partial x^r}$, we get

1.210.
$$\frac{\partial^2 x'^q}{\partial x^p \partial x^s} + \frac{\partial^2 x^r}{\partial x'^m \partial x'^n}\frac{\partial x'^m}{\partial x^p}\frac{\partial x'^n}{\partial x^s}\frac{\partial x'^q}{\partial x^r} = 0.$$

We may of course interchange primed and unprimed symbols in the above equations.

Exercise. If A^r_s are the elements of a determinant A, and B^r_s the elements of a determinant B, show that the element of the product determinant AB is $A^r_n B^n_s$. Hence show that the product of the two Jacobians

$$J = \left| \frac{\partial x^r}{\partial x'^s} \right|, \quad J' = \left| \frac{\partial x'^r}{\partial x^s} \right|$$

is unity.

1.3. Contravariant vectors and tensors. Invariants. Consider a point P with coordinates x^r and a neighbouring point Q with coordinates $x^r + dx^r$. These two points define an infinitesimal displacement or *vector* $\overrightarrow{PQ}$; for the given coordinate system this vector is described by the quantities dx^r, which may be called the *components* of this vector in the given coordinate system. The vector dx^r is not to be regarded as "free," but as associated with (or attached to) the point P with coordinates x^r.

Let us still think of the same two points, but use a different coordinate system x'^r. In this coordinate system the components of the vector $\overrightarrow{PQ}$ are dx'^r; these quantities are connected with the components in the coordinate system x^r by the equation

$$\textbf{1.301.} \qquad dx'^r = \frac{\partial x'^r}{\partial x^s} dx^s,$$

as in 1.206. If we keep the point P fixed, but vary Q in the neighbourhood of P, the coefficients $\partial x'^r/\partial x^s$ remain constant. In fact, under these conditions, the transformation 1.301 is a linear homogeneous (or affine) transformation.

The vector is to be considered as having an absolute meaning, but the numbers which describe it depend on the coordinate system employed. The infinitesimal displacement is the prototype of a class of *geometrical objects* which are called *contravariant vectors*. The word "contravariant" is used to distinguish these objects from "covariant" vectors, which will be introduced in 1.4. The definition of a contravariant vector is as follows:

A set of quantities T^r, associated with a point P, are said to be the components of a contravariant vector if they transform, on change of coordinates, according to the equation

1.302. $$T'^r = T^s \frac{\partial x'^r}{\partial x^s},$$

where the partial derivatives are evaluated at P. Thus an infinitesimal displacement is a particular example of a contravariant vector. It should be noted that there is no general restriction that the components of a contravariant vector should be infinitesimal. For a curve, given by the equations 1.101, the derivatives dx^r/du are the components of a finite contravariant vector. It is called a *tangent vector* to the curve.

Any *infinitesimal* contravariant vector T^r may be represented geometrically by an infinitesimal displacement. We have merely to write

1.303. $$dx^r = T^r.$$

If we use a different coordinate system x'^r, and write

1.304. $$dx'^r = T'^r,$$

we get an infinitesimal displacement. The whole point of the argument is that these two equations define the same displacement, provided T^r are the components of a contravariant vector. If T^r and T'^r were two sets of quantities connected by a transformation which was not of the form 1.302, but something different, say, $T'^r = T^s \partial x^s / \partial x'^r$, then the connection between the dx^r of 1.303 and the dx'^r of 1.304 would not be the transformation 1.301 which connects the components of a single infinitesimal displacement in the two coordinate systems. In that case, dx^r and dx'^r would not represent the same infinitesimal displacement in V_N.

This point has been stressed because it is very useful to have geometrical representations of geometrical objects, in order that we may use the intuitions we have developed in ordinary geometry. But it is not always easy to do this. Although we can do it for an infinitesimal contravariant vector, we cannot do it so completely for a finite contravariant vector. This may appear strange to the physicist who is accustomed to represent a finite vector by a finite directed segment in space. This representation does not work in the general type of space we have in mind at present.

Exercise. Show that a finite contravariant vector determines the ratios of the components of an infinitesimal displacement. (Consider the transformation of the equation $dx^r = \theta T^r$, where θ is an arbitrary infinitesimal factor which does not change under the transformation. Alternatively, show that the equations $T^r dx^s - T^s dx^r = 0$ remain true when we transform the coordinates.)

We now proceed to define geometrical objects of the contravariant class, more complicated in character than the contravariant vector. We set down the definition:

A set of quantities T^{rs} are said to be the components of a contravariant tensor of the second order if they transform according to the equation

1.305.
$$T'^{rs} = T^{mn} \frac{\partial x'^r}{\partial x^m} \frac{\partial x'^s}{\partial x^n}.$$

It is immediately obvious that if U^r and V^r are two contravariant vectors, then the product $U^r V^s$ is a contravariant tensor of the second order.

The definitions of tensors of the third, fourth, or higher orders will at once suggest themselves, and it is unnecessary to write them down here. But going in the opposite direction, we notice that a contravariant vector is a contravariant tensor of the first order, and this suggests that there should be a contravariant tensor of zero order, a single quantity, transforming according to the identical relation

$$T' = T. \tag{1.306}$$

Such a quantity is called an *invariant*; its value is independent of the coordinate system used.

Exercise. Write down the equation of transformation, analogous to 1.305, of a contravariant tensor of the third order. Solve the equation so as to express the unprimed components in terms of the primed components.

1.4. Covariant vectors and tensors. Mixed tensors. Let ϕ be an invariant function of the coordinates. Then

$$\frac{\partial \phi}{\partial x'^r} = \frac{\partial \phi}{\partial x^s} \frac{\partial x^s}{\partial x'^r}. \tag{1.401}$$

This law of transformation is rather like 1.301, but the partial derivative involving the two sets of coordinates is the other way up. Just as the infinitesimal displacement was the prototype of the general contravariant vector, so the partial derivative of an invariant is the prototype of the general *covariant* vector. We define it as follows:

A set of quantities T_r are said to be the components of a covariant vector if they transform according to the equation

$$T'_r = T_s \frac{\partial x^s}{\partial x'^r}. \tag{1.402}$$

It is a well-established convention that suffixes indicating contravariant character are placed as superscripts, and those indicating covariant character as subscripts. It was to satisfy this convention that the coordinates were written x^r rather than x_r, although of course it is only the differentials of the coordinates, and not the coordinates themselves, that have tensor character.

There is no difficulty in defining covariant tensors of various orders, the covariant vector being a tensor of the first order. Thus for a tensor of the second order we make this definition:

A set of quantities T_{rs} are said to be the components of a covariant tensor of the second order if they transform according to the equation

1.403. $$T'_{rs} = T_{mn} \frac{\partial x^m}{\partial x'^r} \frac{\partial x^n}{\partial x'^s}.$$

Having set down definitions of contravariant and covariant tensors, definitions of *mixed tensors* suggest themselves. Suppose a set of quantities T^r_{st} transform according to

1.404. $$T'^r_{st} = T^m_{np} \frac{\partial x'^r}{\partial x^m} \frac{\partial x^n}{\partial x'^s} \frac{\partial x^p}{\partial x'^t}.$$

We would say that they are the components of a mixed tensor of the third order, with one contravariant and two covariant suffixes.

In adopting for the Kronecker delta the notation δ^r_s we anticipated the fact that it has mixed tensor character. Let us now prove this, i.e. let us show that

1.405. $$\delta'^r_s = \delta^m_n \frac{\partial x'^r}{\partial x^m} \frac{\partial x^n}{\partial x'^s},$$

where δ'^r_s is unity if $r = s$ and zero if $r \neq s$. Holding m fixed temporarily, and summing with respect to n, we get no contribution unless $n = m$. Hence the right-hand side of 1.405 reduces to

$$\frac{\partial x'^r}{\partial x^m} \frac{\partial x^m}{\partial x'^s},$$

and this is equal to δ^r_s; thus the truth of 1.405 is established.

The importance of tensors in mathematical physics and geometry rests on the fact that a tensor equation is true in all coordinate systems, if true in one. This follows from the fact that the tensor transformations are linear and homogeneous. Suppose, for example, that we are given that $T_{rs} = 0$; it is an immediate consequence of 1.403 that $T'_{rs} = 0$ also. More generally, if we are given

1.406. $$A_{rs} = B_{rs},$$
then

1.407. $$A'_{rs} = B'_{rs}.$$

Let us now consider what may be called the *transitivity* of tensor character. We shall speak of covariant tensors of the second order for simplicity, but the results hold quite generally. Let there be three coordinate systems: x^r, x'^r, x''^r. Suppose that a set of quantities T_{rs} transform tensorially when we pass from the first to the second set of coordinates, and also when we pass from the second to the third. These two transformations combine to form a "product" transformation from the first set of coordinates to the third. If this transformation is tensorial, then we say that the tensor character is *transitive*. To establish this transitivity, we have to prove the following statement:

Given that

1.408. $$T'_{rs} = T_{mn} \frac{\partial x^m}{\partial x'^r} \frac{\partial x^n}{\partial x'^s},$$

and

1.409. $$T''_{rs} = T'_{mn} \frac{\partial x'^m}{\partial x''^r} \frac{\partial x'^n}{\partial x''^s},$$

then

1.410. $$T''_{rs} = T_{mn} \frac{\partial x^m}{\partial x''^r} \frac{\partial x^n}{\partial x''^s}.$$

This is easy to show, and is left as an exercise.

A tensor may be given at a single point of the space V_N, or it may be given along a curve, or throughout a subspace, or throughout V_N itself. In the last three cases we refer to a *tensor field*, if we wish to emphasize the fact that the tensor is given throughout a continuum.

Exercise. For a transformation from one set of rectangular Cartesian coordinates to another in Euclidean 3-space, show that the law of transformation of a contravariant vector is precisely the same as that of a covariant vector. Can this statement be extended to cover tensors of higher orders?

1.5. Addition, multiplication, and contraction of tensors. Two tensors of the same order and type may be added together to give another tensor. Suppose, for example, that A^r_{st} and B^r_{st} are tensors, and that the quantities C^r_{st} are defined by

1.501. $$C^r_{st} = A^r_{st} + B^r_{st}.$$

Then it is easy to prove that C^r_{st} are the components of a tensor.

A set of quantities A_{rs} (not necessarily components of a tensor) is said to be *symmetric* if

1.502. $$A_{rs} = A_{sr},$$

and *skew-symmetric* or *antisymmetric* if

1.503. $$A_{rs} = -A_{sr}.$$

If the quantities have tensor character, the property of symmetry (or of skew-symmetry) is conserved under transformation of coordinates. This follows from the fact that

$$A_{rs} - A_{sr}, \quad A_{rs} + A_{sr}$$

are themselves tensors, and so vanish in all coordinate systems, if they vanish in one.

These remarks about symmetry apply equally to contravariant tensors, the subscripts being replaced by superscripts. They do not apply to a mixed tensor A^r_s; the relationship $A^r_s = A^s_r$ does not in general carry over from one coordinate system to another.

The definitions of symmetry and skew-symmetry may be extended to more complicated tensors. We say that a tensor is symmetric with respect to a pair of suffixes (both superscripts or both subscripts) if the value of the component is unchanged on interchanging these suffixes. It is skew-symmetric if interchange of suffixes leads to a change of sign without change of absolute value.

The following result is of considerable importance in the application of tensor calculus to physics: *Any tensor of the second order (covariant or contravariant) may be expressed as the sum of a symmetric tensor and a skew-symmetric tensor.* This is easy to prove. Let us take a contravariant tensor A^{rs} for illustration. We have merely to write

1.504. $$A^{rs} = \tfrac{1}{2}(A^{rs} + A^{sr}) + \tfrac{1}{2}(A^{rs} - A^{sr});$$

each of the two terms on the right is a tensor—the first is symmetric and the second skew-symmetric.

Exercise. In a space of four dimensions, the tensor A_{rst} is skew-symmetric in the last pair of suffixes. Show that only 24 of the 64 components may be chosen arbitrarily. If the further condition

$$A_{rst} + A_{str} + A_{trs} = 0$$

is imposed, show that only 20 components may be chosen arbitrarily.

Exercise. If A^{rs} is skew-symmetric and B_{rs} symmetric, prove that

$$A^{rs}B_{rs} = 0.$$

Hence show that the quadratic form $a_{ij}x^i x^j$ is unchanged if a_{ij} is replaced by its symmetric part.

Let us now consider the multiplication of tensors. In adding or subtracting tensors we use only tensors of a single type, and add components with the same literal suffixes, although these need not occur in the same order. This is not the case in multiplication. The only restriction here is that we never multiply two components with the same literal suffix at the same level in each. (This general rule may be broken in the case of Cartesian tensors, to be discussed in chapter IV, but this exception is unimportant.) These restrictive rules on addition and multiplication are introduced in order that the results of the operations of addition and multiplication may themselves be tensors.

To multiply, we may take two tensors of different types and different literal suffixes, and simply write them in juxtaposition. Thus, suppose A_{rs} and B^m_n are tensors of the types indicated. If we write

1.505. $$C^m_{rsn} = A_{rs}B^m_n,$$

then these quantities are the components of a tensor of the type indicated. This follows immediately from the formulae of tensor transformation.

Such a product as 1.505, in which all the suffixes are different from one another, is called an *outer product.* The *inner product* is obtained from the outer product by the process of *contraction*, which we shall now explain.

Consider a tensor with both contravariant and covariant suffixes, such as T^m_{nrs}. Consider the quantities T^m_{mrs}, in which there is of course summation with respect to m in accordance with the summation convention. What is its tensor character, if any? We have

1.506. $$T'^m_{nrs} = T^p_{qtu}\frac{\partial x'^m}{\partial x^p}\frac{\partial x^q}{\partial x'^n}\frac{\partial x^t}{\partial x'^r}\frac{\partial x^u}{\partial x'^s}.$$

Putting $n = m$, and using 1.208, we obtain

1.507. $$T'^m_{mrs} = T^p_{ptu}\frac{\partial x^t}{\partial x'^r}\frac{\partial x^u}{\partial x'^s}.$$

In fact, the tensor character is that of B_{rs}, covariant of the second order. We have here *contracted* with respect to the suffixes m, n (one above and one below), with the result that these suffixes become dummy suffixes of summation and no longer imply any tensor character.

The general rule is as follows: *Given a mixed tensor, if we contract by writing the same letter as a superscript and as a subscript, the result has the tensor character indicated by the remaining suffixes.*

Applying contraction to the outer product 1.505, we get the inner products

$$A_{ms}B^m_n,\ A_{rm}B^m_n,$$

each of which is a covariant tensor of the second order.

The process of contraction cannot be applied to suffixes at the same level. Of course, there is nothing to stop us writing down the expression A^m_{nrr}; but it has not tensor character, and so is of minor interest, since our object is to deal (as far as possible) only with tensors.

Exercise. What are the values (in a space of N dimensions) of the following contractions formed from the Kronecker delta?

$$\delta^m_m, \quad \delta^m_n\delta^n_m, \quad \delta^m_n\delta^n_r\delta^r_m.$$

1.6. Tests for tensor character. The direct test for the tensor character of a set of quantities is this: see whether the components obey the law of tensor transformation when the coordinates are changed. However, it is sometimes much more convenient to proceed indirectly as follows.

Suppose that A_r is a set of quantities which we wish to test for tensor character. Let X^r be the components of an *arbitrary* contravariant tensor of the first order. We shall now prove that *if the inner product* A_rX^r *is an invariant, then* A_r *are the components of a covariant tensor of the first order.* We have, by the given invariance,

1.601. $$A_rX^r = A'_rX'^r,$$

and, by the law of tensor transformation,

1.602. $$X'^r = X^s\frac{\partial x'^r}{\partial x^s}.$$

Substituting this in the right-hand side of 1.601, rearranging, and making a simple change in notation, we have

1.603. $$\left(A_s - A'_r\frac{\partial x'^r}{\partial x^s}\right)X^s = 0.$$

Since the quantities X^s are arbitrary, the quantity inside the parentheses vanishes; this establishes the tensor character of A_r by 1.402.

The above example is illustrative of the indirect test for tensor character. The test is by no means confined to the case of a tensor of the first order, nor is it necessary that an invariant should be formed. It is, however, essential that there should enter into the test some quantities which are arbitrary and are known to have tensor character.

Exercise. If X^r, Y^r are arbitrary contravariant vectors and $a_{rs}X^rY^s$ is an invariant, then a_{rs} are the components of a covariant tensor of the second order.

Exercise. If X_{rs} is an arbitrary covariant tensor of the second order, and $A_r^{mn}X_{mn}$ is a covariant vector, then A_r^{mn} has the mixed tensor character indicated by the positions of its suffixes.

The following case is of some importance. Suppose that a_{rs} is a set of quantities whose tensor character is under investigation. Let X^r be an arbitrary contravariant vector. Suppose we are given that $a_{rs}X^rX^s$ is an invariant. What can we tell about the tensor character of a_{rs}?

We have

1.604.
$$a_{rs}X^rX^s = a'_{rs}X'^rX'^s = a'_{rs}X^mX^n \frac{\partial x'^r}{\partial x^m}\frac{\partial x'^s}{\partial x^n},$$

and so

1.605.
$$\left(a_{mn} - a'_{rs}\frac{\partial x'^r}{\partial x^m}\frac{\partial x'^s}{\partial x^n}\right)X^mX^n = 0.$$

This quadratic form vanishes for arbitrary X^r, but we cannot jump to the conclusion that the quantity inside the parentheses vanishes. We must remember that in a form $b_{mn}X^mX^n$ the coefficient of the product X^1X^2 is mixed up with the coefficient of X^2X^1; it is, in fact, $b_{12} + b_{21}$. Thus we can deduce from 1.605 only that

1.606.
$$a_{mn} + a_{nm} = a'_{rs}\frac{\partial x'^r}{\partial x^m}\frac{\partial x'^s}{\partial x^n} + a'_{rs}\frac{\partial x'^r}{\partial x^n}\frac{\partial x'^s}{\partial x^m}.$$

The trick now is to interchange the dummies r, s in the last term; this gives

$$1.607. \quad a_{mn} + a_{nm} = (a'_{rs} + a'_{sr}) \frac{\partial x'^r}{\partial x^m} \frac{\partial x'^s}{\partial x^n}.$$

This establishes the tensor character of $a_{mn} + a_{nm}$. If we are given that a_{mn} and a'_{mn} are symmetric, the tensor character of a_{mn} follows, and we obtain this result: *If $a_{rs}X^rX^s$ is invariant, X^r being an arbitrary contravariant vector and a_{rs} being symmetric in all coordinate systems, then a_{rs} are the components of a covariant tensor of the second order.*

1.7. Compressed notation. The range convention and the summation convention, introduced in 1.2, save a great deal of unnecessary writing. But still more can be done to improve the symbolism. In the present section we shall discuss a "compressed notation." This notation certainly simplifies the proofs of some results, but it is questionable whether it is advisable to adopt it as standard notation. On the whole, it has seemed best to introduce the present notation as a sample of what can be done in the way of smoother notation, but to revert, in the subsequent parts of the book, to the notation which we have used up to the present.

Suppose we have a space of N dimensions. Let x^r be a system of coordinates, small Latin suffixes having the range $1, 2, \ldots, N$. Let x^ρ be another system of coordinates, small Greek suffixes also taking the range $1, 2, \ldots, N$. At first sight this appears to be an impossible notation, violating the fundamental rule that one mathematical symbol shall not denote two different quantities at the same time. We ask: What does x^1 mean? To which of the two systems of coordinates does it belong? The answer is: Never write x^1, but $(x^r)_{r=1}$ or $(x^\rho)_{\rho=1}$ according to whether you wish to denote the first or the second coordinate system. This is clumsy, but does not spoil the notation for general arguments, in which we do not require to give numerical values to the suffixes. As long as the suffix remains literal, the fact that it is Latin or Greek tells us which coordinate system is involved.

We now denote partial differentiation by ∂, so that

1.701. $$\partial_r = \frac{\partial}{\partial x^r}, \quad \partial_\rho = \frac{\partial}{\partial x^\rho}.$$

Further, let us write

1.702. $$X^r_s = \partial_s x^r,\ X^r_\sigma = \partial_\sigma x^r,\ X^\rho_s = \partial_s x^\rho,\ X^\rho_\sigma = \partial_\sigma x^\rho,$$

so that obviously

1.703. $$X^r_s = \delta^r_s, \quad X^r_\rho X^\rho_s = \delta^r_s.$$

The following are tensor transformations expressed in this notation:

$$T^r = T^\sigma X^r_\sigma, \qquad T^\rho = T^s X^\rho_s,$$

$$T_r = T_\sigma X^\sigma_r, \qquad T_\rho = T_s X^s_\rho,$$

1.704. $$T_{rs} = T_{\rho\sigma} X^\rho_r X^\sigma_s,$$

$$T^{r_1 \dots r_m}_{s_1 \dots s_n} = T^{\rho_1 \dots \rho_m}_{\sigma_1 \dots \sigma_n} X^{r_1}_{\rho_1} \dots X^{r_m}_{\rho_m} X^{\sigma_1}_{s_1} \dots X^{\sigma_n}_{s_n}.$$

The last line in 1.704 is the general formula of tensor transformation.

It is convenient to use the following notation for second derivatives:

1.705. $$\partial_r X^\rho_s = \partial_s X^\rho_r = \frac{\partial^2 x^\rho}{\partial x^r \partial x^s} = X^\rho_{rs}.$$

In this notation the equation 1.210 reads

1.706. $$X^\mu_{rs} + X^m_{\rho\sigma} X^\rho_r X^\sigma_s X^\mu_m = 0.$$

Exercise. If A_{rs} is a skew-symmetric covariant tensor, prove that B_{rst}, defined as

1.707. $$B_{rst} = \partial_r A_{st} + \partial_s A_{tr} + \partial_t A_{rs},$$

is a covariant tensor, and that it is skew-symmetric in all pairs of suffixes.

When calculations become complicated, notational devices become of real importance as labour-saving devices, and to keep the bulk of formulae under control. It sometimes happens that more than two coordinate systems are involved. There are two methods of handling such situations.

The first plan is to break up the Latin alphabet into groups such as (abcde), (fghij), (klmno), . . . , and assign a group to each coordinate system. Thus, coordinates for the first coordinate system would be denoted by $x^a, \ldots, x^e$, for the second $x^f, \ldots, x^j$, and so on. Using X as the base letter for partial derivatives, we would then have formulae such as the following:

1.708. $$X^a_c = \delta^a_c, \qquad X^a_f = \frac{\partial x^a}{\partial x^f}, \qquad X^a_f X^f_k = X^a_k,$$

$$T^{fg} = T^{ab} X^f_a X^g_b, \qquad T^{kl} = T^{ab} X^k_a X^l_b.$$

The second plan is to use one alphabet, but to put a sign on the suffix indicative of the coordinate system involved. Thus, coordinates for the first coordinate system would be denoted by $x^r, x^m, x^n, \ldots$, for the second by $x^{r'}, x^{m'}, x^{n'}, \ldots$, for the third by $x^{r''}, x^{m''}, x^{n''}, \ldots$, and we would have formulae such as the following:

1.709. $$X^r_m = \delta^r_m, \qquad X^r_{m'} = \frac{\partial x^r}{\partial x^{m'}}, \qquad X^r_{m'} X^{m'}_{n''} = X^r_{n''},$$

$$T^{r'm'} = T^{np} X^{r'}_n X^{m'}_p, \qquad T^{r''m''} = T^{np} X^{r''}_n X^{m''}_p.$$

SUMMARY I

Contravariant tensor:

$$T'^{mn} = T^{rs} \frac{\partial x'^m}{\partial x^r} \frac{\partial x'^n}{\partial x^s}.$$

Covariant tensor:

$$T'_{mn} = T_{rs} \frac{\partial x^r}{\partial x'^m} \frac{\partial x^s}{\partial x'^n}.$$

Mixed tensor:

$$T'^m_n = T^r_s \frac{\partial x'^m}{\partial x^r} \frac{\partial x^s}{\partial x'^n}.$$

Invariant:

$$T' = T.$$

Kronecker delta:

$$\delta^r_s \begin{cases} = 1 \text{ if } r = s, \\ = 0 \text{ if } r \neq s. \end{cases}$$

EXERCISES I

1. In a V_4 there are two 2-spaces with equations

$$x^r = f^r(u^1, u^2), \quad x^r = g^r(u^3, u^4).$$

Prove that if these 2-spaces have a curve of intersection, then the determinantal equation

$$\left| \frac{\partial x^r}{\partial u^s} \right| = 0$$

is satisfied along this curve.

2. In Euclidean space of three dimensions, write down the equations of transformation between rectangular Cartesian coordinates x, y, z, and spherical polar coordinates r, θ, ϕ. Find the Jacobian of the transformation. Where is it zero or infinite?

3. If X, Y, Z are the components of a contravariant vector for rectangular Cartesian coordinates in Euclidean 3-space, find its components for spherical polar coordinates.

4. In a space of three dimensions, how many different expressions are represented by the product $A^m_{np} B^{pq}_{rs} C^s_{tu}$? How many terms occur in each such expression, when written out explicitly?

5. If A is an invariant in V_N, are the second derivatives $\dfrac{\partial^2 A}{\partial x^r \partial x^s}$ the components of a tensor?

6. Suppose that in a V_2 the components of a contravariant tensor field T^{mn} in a coordinate system x^r are

$$T^{11} = 1 \qquad T^{12} = 0.$$
$$T^{21} = 0 \qquad T^{22} = 1.$$

Find the components T'^{mn} in a coordinate system x'^{r}, where

$$x'^1 = (x^1)^2, \quad x'^2 = (x^2)^2.$$

Write down the values of these components in particular at the point $x^1 = 1$, $x^2 = 0$.

7. Given that if T_{mnrs} is a covariant tensor, and

$$T_{mnrs} + T_{mnsr} = 0$$

in a coordinate system x^p, establish directly that

$$T'_{mnrs} + T'_{mnsr} = 0$$

in any other coordinate system x'^q.

8. Prove that if A_r is a covariant vector, then

$$\frac{\partial A_r}{\partial x^s} - \frac{\partial A_s}{\partial x^r}$$

is a skew-symmetric covariant tensor of the second order. (Use the notation of 1.7.)

9. Let x^r, $\bar{x}^r$, y^r, $\bar{y}^r$ be four systems of coordinates. Examine the tensor character of $\partial x^r/\partial y^s$ with respect to the following transformations: (i) A transformation $x^r = f^r(\bar{x}^1, \ldots, \bar{x}^N)$ with y^r unchanged; (ii) a transformation $y^r = g^r(\bar{y}^1, \ldots, \bar{y}^N)$ with x^r unchanged.

10. If x^r, y^r, z^r are three systems of coordinates, prove the following rule for the multiplication of Jacobians:

$$\left|\frac{\partial x^m}{\partial y^n}\right| \cdot \left|\frac{\partial y^r}{\partial z^s}\right| = \left|\frac{\partial x^t}{\partial z^u}\right|.$$

11. Prove that with respect to transformations

$$x'^r = C_{rs}x^s,$$

where the coefficients are constants satisfying

$$C_{mr}C_{ms} = \delta^r_s,$$

contravariant and covariant vectors have the same formula of transformation:

$$A'^r = C_{rs}A^s, \quad A'_r = C_{rs}A_s.$$

12. Prove that

$$\frac{\partial}{\partial x^r} \ln \left| \frac{\partial y^m}{\partial x^n} \right| = \frac{\partial^2 y^m}{\partial x^r \partial x^n} \frac{\partial x^n}{\partial y^m}.$$

13. Consider the quantities dx^r/dt for a particle moving in a plane. If x^r are rectangular Cartesian coordinates, are these quantities the components of a contravariant or covariant vector with respect to rotation of the axes? Are they components of a vector with respect to transformation to any curvilinear coordinates (e.g. polar coordinates)?

14. Consider the questions raised in No. 13 for the acceleration (d^2x^r/dt^2).

15. It is well known that the equation of an ellipse may be written

$$ax^2 + 2hxy + by^2 = 1.$$

What is the tensor character of a, h, b with respect to transformation to any Cartesian coordinates (rectangular or oblique) in the plane?

16. Matter is distributed in a plane and A, B, H are the moments and product of inertia with respect to rectangular axes Oxy in the plane. Examine the tensor character of the set of quantities A, B, H under rotation of the axes. What notation would you suggest for moments and product of inertia in order to exhibit the tensor character? What simple invariant can be formed from A, B, H?

17. Given a tensor S_{mnr} skew-symmetric in the first two suffixes, find a tensor f_{mnr} skew-symmetric in the last two suffixes and satisfying the relation

$$-f_{mnr} + f_{nmr} = S_{mnr}.$$

[Answer: $f_{mnr} = \frac{1}{2}(-S_{mnr} - S_{rmn} + S_{nrm})$].

CHAPTER II

BASIC OPERATIONS IN RIEMANNIAN SPACE

2.1. The metric tensor and the line element. We shall lead up to the concept of a Riemannian space by first discussing properties of curvilinear coordinates in the familiar Euclidean space of three dimensions. Suppose that y^1, y^2, y^3 are rectangular Cartesian coordinates. Then the square of the distance between adjacent points is

2.101. $$ds^2 = (dy^1)^2 + (dy^2)^2 + (dy^3)^2.$$

Let x^1, x^2, x^3 be any system of curvilinear coordinates (e.g. cylindrical or spherical polar coordinates). Then the y's are functions of the x's, and the dy's are linear homogeneous functions of the dx's. When we substitute these linear functions in 2.101, we get a homogeneous quadratic expression in the dx's. This may be written

2.102. $$ds^2 = a_{mn} dx^m dx^n,$$

where the coefficients a_{mn} are functions of the x's. Since the a_{mn} do not occur separately, but only in the combinations $(a_{mn} + a_{nm})$, there is no loss of generality in taking a_{mn} symmetric, so that $a_{mn} = a_{nm}$.

No matter what curvilinear coordinates are used, the distance between two given points has the same value, i.e. ds (or ds^2) is an invariant. If we keep one of the two points fixed and allow the other to vary arbitrarily in its neighbourhood, then dx^r is an arbitrary contravariant vector. It follows from 1.6 that a_{mn} is a covariant tensor of the second order. It is called the *metric tensor*, or *fundamental tensor* of space.

Exercise. Take polar coordinates r, θ in a plane. Draw the infinitesimal triangle with vertices at the points (r, θ), $(r+dr, \theta)$ $(r, \theta + d\theta)$. Evaluate the square on the hypotenuse of this infinitesimal triangle, and so obtain the metric tensor for the plane for the coordinates (r, θ).

Exercise. Show that if $x^1 = r$, $x^2 = \theta$, $x^3 = \phi$, in the usual notation for spherical polar coordinates, then

$$a_{11} = 1,\ a_{22} = r^2,\ a_{33} = r^2 \sin^2 \theta,$$

and the other components vanish.

Suppose now that we draw a surface in Euclidean 3-space. According to the method of Gauss, we may write its equations in the form

2.103. $$y^1 = f^1(x^1, x^2),\ y^2 = f^2(x^1, x^2),\ y^3 = f^3(x^1, x^2),$$

where x^1, x^2 are curvilinear coordinates on the surface. The square of the distance between two adjacent points on the surface is again given by 2.101, and we may use 2.103 to transform this expression into a homogeneous quadratic expression in the dx's. This expression may be written in the form 2.102, but the range of the suffixes is now only 1,2, and not 1, 2, 3 as before. It follows that a_{mn} is again a covariant tensor; it is the metric tensor of the surface, which is itself a space of two dimensions.

Exercise. Starting from 2.103, show that

$$a_{mn} = \frac{\partial y^1}{\partial x^m}\frac{\partial y^1}{\partial x^n} + \frac{\partial y^2}{\partial x^m}\frac{\partial y^2}{\partial x^n} + \frac{\partial y^3}{\partial x^m}\frac{\partial y^3}{\partial x^n},$$

and calculate these quantities for a sphere, taking as curvilinear coordinates on the sphere $x^1 = y^1$, $x^2 = y^2$.

The differential expression which represents ds^2 may be called the *metric form* or *fundamental form* of the space under consideration. It may also be called the square of the *line element*.

What has been explained in the preceding argument is basic in the application of tensor calculus to classical geometry and classical mathematical physics. But we have in mind the extension of these ideas to spaces of higher dimensionality. Therefore we shall not pause here to develop the immediate implications of what we have done, but rather use it as a source of suggestion for generalization. As a basis for generalization, let us summarize our result as follows:

In ordinary space, or on a surface in that space, the square of the line element is a homogeneous quadratic form, and the coefficients of that form, when written symmetrically, are the components of a covariant tensor of the second order.

We now pass to a general space of N dimensions, discussed in 1.1. Consider two adjacent points P, Q in it. Does there exist something which may be called the *distance* between P and Q? Although the basic ideas of tensor calculus originated with Riemann, he himself was a little confused on this essential question. He apparently thought that the concept of distance was intrinsic in a space. We know now that this is not the case. We can develop a logically consistent theory of a non-metrical space, in which the concept of distance never enters. If there is to be a measure of distance in a space of N dimensions, it is something that we must put in for ourselves. The question now before us is this: How shall we define distance in V_N to satisfy the following criteria?

(i) The definition should give a comparatively simple geometry;

(ii) The definition should agree with the ordinary definition in the particular cases where the space is ordinary Euclidean space, or a surface in that space.

These criteria are best satisfied by using the italicized statement above as the basis of definition. But a little caution is necessary if we are to make our definition wide enough to include the space-time of relativity. Thus, without introducing the word "distance," we lay down the following definition of a Riemannian space:

A space V_N is said to be Riemannian if there is given in it a metric (or fundamental) covariant tensor of the second order, which is symmetric.

If we denote this tensor by a_{mn}, we may write down a *metric* (or *fundamental*) *form*

2.104. $$\Phi = a_{mn}dx^m dx^n.$$

This form is, of course, invariant.

We might proceed to define the "distance" between adjacent points by means of the equation $ds^2 = \Phi$. However, with a view to relativity, we must admit the possibility of an important difference between the form 2.104 and the form we encountered in Euclidean space. In Euclidean space the form 2.102 is *positive-definite*; this means that it is positive unless all the differentials vanish. In other words, the distance between two points vanishes only if the points are brought into coincidence.

We shall not impose on the form Φ the condition that it shall be positive-definite. We shall admit the possibility of an *indefinite* form, such as $\Phi = (dx^1)^2 - (dx^2)^2$, which vanishes if $dx^1 = dx^2$. Then for some displacements dx^r the form Φ may be positive and for others it may be zero or negative. If $\Phi = 0$, for dx^r not all zero, the displacement is called a *null* displacement. For any displacement dx^r which is not null, there exists an *indicator* ϵ, chosen equal to $+1$, or -1, so as to make $\epsilon\Phi$ positive. We may use this indicator to overcome a difficulty in the definition of distance arising from the indefiniteness of the form. We define the *length* of the displacement dx^r (or the *distance* between its end points) to be ds, where

2.105. $$ds^2 = \epsilon\Phi = \epsilon a_{mn}dx^m dx^n, \quad ds > 0.$$

We define the length of a null displacement to be zero. Thus, in a Riemannian space with an indefinite metric form, two points may be at zero distance from one another without being coincident

It is most important to note that Riemannian geometry is built up on the concept of the distance between two neighbouring points, rather than on the concept of finite distance.

2.2. The conjugate tensor. Lowering and raising suffixes. From the covariant metric tensor a_{mn} we can obtain another tensor, also of the second order, but contravariant. Consider the determinant

2.201.
$$a = |a_{mn}| = \begin{vmatrix} a_{11} & a_{12} & \cdots & a_{1N} \\ a_{21} & a_{22} & \cdots & a_{2N} \\ \cdots & \cdots & \cdots & \cdots \\ a_{N1} & a_{N2} & \cdots & a_{NN} \end{vmatrix}.$$

We shall suppose, here and throughout, that a is not zero. Let Δ^{mn} be the cofactor of a_{mn} in this determinant, so that

2.202.
$$a_{mr}\Delta^{ms} = a_{rm}\Delta^{sm} = \delta_r^s a.$$

This follows from the ordinary rules for developing a determinant. Now let us define a^{mn} by the equation

2.203.
$$a^{mn} = \frac{\Delta^{mn}}{a}.$$

It follows from 2.202 that

2.204.
$$a_{mr}a^{ms} = \delta_r^s,$$

and, similarly,

2.205.
$$a_{rm}a^{sm} = \delta_r^s.$$

Since a_{mn} is symmetric, it is obvious that a^{mn} is symmetric also.

Note that 2.204, or 2.205, might be regarded as a *definition* of a^{mn}, since either of these sets of equations determines these quantities uniquely. Let us take 2.204 and multiply both sides by Δ^{kr}; then

2.206.
$$a\delta_m^k a^{ms} = \delta_r^s \Delta^{kr}$$

by 2.202. Thus

2.207.
$$a^{ks} = \frac{\Delta^{ks}}{a},$$

which is the same as 2.203. We obtain the same result from 2.205 on multiplication by Δ^{rk}. This proves our assertion.

We are now ready to investigate the tensor character of a^{mn}. For this purpose we introduce a contravariant tensor $\bar{a}^{mn}$ which in one particular system of coordinates x^r, say, coincides with a^{mn}. Then

2.208. $$a_{mr}\bar{a}^{ms} = \delta^s_r.$$

This is obviously a tensor equation and therefore holds in any other system of coordinates x'^r; thus

2.209. $$a'_{mr}\bar{a}'^{ms} = \delta^s_r.$$

Comparing this with

2.210. $$a'_{mr}a'^{ms} = \delta^s_r,$$

which, as we have seen, determines a'^{mn} uniquely, we have $a'^{mn} = \bar{a}'^{mn}$. Thus a^{mn} and $\bar{a}^{mn}$ coincide in all coordinate systems. The latter being a tensor by definition, it follows that *a^{mn} is a contravariant tensor of the second order*. It is said to be *conjugate* to a_{mn}, the fundamental tensor.

Exercise. Show that if $a_{mn} = 0$ for $m \neq n$, then

$$a^{11} = \frac{1}{a^{11}}, \quad a^{22} = \frac{1}{a^{22}}, \ldots, a^{12} = 0, \ldots$$

Exercise. Find the components of a^{mn} for spherical polar coordinates in Euclidean 3-space.

Having now at our disposal the covariant fundamental tensor and its contravariant conjugate, we are able to introduce the processes known as the *lowering* and *raising* of suffixes. We shall in future refrain from writing a subscript and a superscript on the same vertical line; in vacant spaces we shall write dots, thus: $T^{mn}_{\cdot\cdot rs}$.

Take a tensor $T^m_{\cdot rs}$, and write

2.211. $$S_{nrs} = a_{nm}T^m_{\cdot rs};$$

this has the tensor character indicated. The tensor S has been generated from the tensor T by lowering a suffix. We may raise a suffix by means of a^{mn}:

2.212. $$U^m_{\cdot rs} = a^{mn}S_{nrs}.$$

It is easy to see that this tensor U is precisely the original tensor T. This suggests that in the processes of lowering or raising suffixes we should retain the same principal letter. Thus we write 2.211 in the form

2.213. $$T_{nrs} = a_{nm} T^{m}_{.rs}.$$

Exercise. Find the mixed metric tensor $a_{m}^{.n}$, obtained from a_{mn} by raising the second subscript.

Before we introduced the metric tensor of Riemannian space we recognized a tensor as a geometrical object—a thing which had different representations in different coordinate systems, but at the same time an existence of its own. The tensors T^{mn} and T_{mn} were entirely unrelated; one was contravariant and the other covariant, and there was no connection between the one and the other. The use of the same basic letter T in both implied no relationship. But in a Riemannian space these two tensors are essentially the same geometrical object; if we know the components of one, we can obtain the components of the other. In most of the physical applications of tensor calculus, the space is Riemannian; a physical object (e.g. stress in an elastic body) is represented by a tensor, and we can suit our convenience as to whether we express that tensor in contravariant, covariant, or mixed form.

We shall now establish a useful formula for the derivative of the determinant a. Let us forget the assumed symmetry of a_{mn}, so that a_{mn} and a_{nm} are regarded as independent quantities. Then a is a function of the N^2 quantities a_{mn}, and from the expansion of the determinant it is evident that

2.214. $$\frac{\partial a}{\partial a_{mn}} = a a^{mn},$$

or

2.215. $$\frac{\partial}{\partial a_{mn}} \ln a = a^{mn}.$$

Hence

2.216. $$\frac{\partial}{\partial x^r} \ln a = \left(\frac{\partial}{\partial a_{mn}} \ln a \right) \frac{\partial a_{mn}}{\partial x^r} = a^{mn} \frac{\partial a_{mn}}{\partial x^r},$$

which is the formula required. If a is negative, the same equation holds with $-a$ written for a.

Exercise. Prove that $a_{mn} a^{mn} = N$.

2.3. Magnitude of a vector. Angle between vectors. To find the magnitude of a vector in ordinary space, we square the components, add, and take the square root. This simple plan does not work in a general space for a very good reason: the result obtained by this process is not invariant, i.e., it depends on the coordinate system employed.

If the space has no metric, there is no way of defining the magnitude of a vector. But if a Riemannian metric tensor a_{mn} is given, the definition is easy. *The magnitude X of a contravariant vector X^r is the positive real quantity satisfying*

2.301. $$X^2 = \epsilon a_{mn} X^m X^n,$$

where ϵ is the indicator of X^r (see 2.1). For a covariant vector we take, instead of 2.301,

2.302. $$X^2 = \epsilon a^{mn} X_m X_n.$$

Obviously there will be null vectors, i.e. vectors of zero magnitude, if the metric form is indefinite. We note that the definition 2.105 defines ds as the magnitude of the infinitesimal vector dx^r.

Exercise. Show that in Euclidean 3-space with rectangular Cartesian coordinates, the definition 2.301 coincides with the usual definition of the magnitude of a vector.

Suppose we are given a curve with equations

2.303. $$x^r = x^r(u),$$

where u is a parameter. If we write $p^r = dx^r/du$, an infinitesimal displacement along the curve is

2.304. $$dx^r = p^r(u) du.$$

The length of this displacement is

2.305. $$ds = [\epsilon a_{mn} p^m p^n]^{\frac{1}{2}} du,$$

where ϵ is the indicator of dx^r; the length of the curve from $u = u_1$ to $u = u_2$ is

2.306.
$$s = \int_{u_1}^{u_2} [\epsilon a_{mn} p^m p^n]^{\frac{1}{2}} \, du.$$

Exercise. A curve in Euclidean 3-space has the equations

$$x^1 = a \cos u, \; x^2 = a \sin u, \; x^3 = bu,$$

where x^1, x^2, x^3 are rectangular Cartesian coordinates, u is a parameter, and a, b are positive constants. Find the length of this curve between the points $u = 0$ and $u = 2\pi$.

What is the length of the curve with these same equations, between the same values of u, if the metric form of the 3-space is

$$(dx^1)^2 + (dx^2)^2 - (dx^3)^2?$$

Consider your result for the cases where a is greater than, equal to, and less than b.

Unless we are dealing with a null curve, for which $s = 0$, we may take s as parameter along the curve. The finite vector $p^r = dx^r/ds$ has the same direction as the infinitesimal displacement dx^r along the curve, i.e., it is a *tangent vector*. Moreover, its magnitude is unity, since

2.307.
$$\epsilon a_{mn} dx^m dx^n = ds^2,$$

and so

2.308.
$$\epsilon a_{mn} \frac{dx^m}{ds} \frac{dx^n}{ds} = 1.$$

Any vector with unit magnitude is called a *unit vector*. Thus dx^r/ds is the *unit tangent vector* to the curve.

Let us now consider the *angle* between two curves, or, what is the same thing, the angle between two unit vectors, tangent to the curves. Difficulties arise if we attempt to define angle in a space with an indefinite line element. Accordingly we shall confine ourselves here to a space with a positive-definite line element. We might write down a formal definition of angle, but it is more interesting to develop it as a natural generalization from familiar concepts.

Consider, in Euclidean 3-space, two curves issuing from a point A. Let B and C be points on them, one on each; join B and C by a curve (Fig. 1). We shall consider a limit as B and C

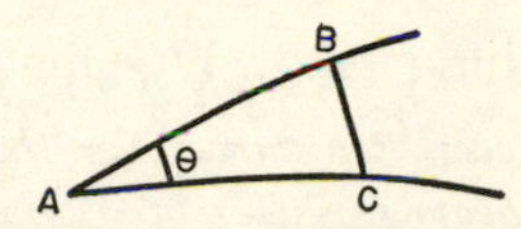

FIG. 1. Angle between curves in Riemannian space.

both tend to A; it will be assumed that the curve joining B and C maintains a finite curvature during this process. It is then easy to see that the angle θ between the curves satisfies the equation

2.309. $$\cos\theta = \lim \frac{AB^2 + AC^2 - BC^2}{2\,AB.AC},$$

where AB, AC, BC indicate arc-lengths.

The above equation is set up in such a way that it can be taken over into Riemannian N-space as a *definition* of angle. Two things remain to be done: first, to turn 2.309 into a usable formula; secondly, to show that the angle θ defined in this way is real.

Suppose that Fig. 1 now refers to Riemannian N-space. Let the coordinates be as follows:

$$\begin{aligned} &A \;\ldots\ldots\ldots\; x^r, \\ &B \;\ldots\ldots\ldots\; x^r + \xi^r, \\ &C \;\ldots\ldots\ldots\; x^r + \eta^r. \end{aligned}$$

The principal parts of the squares of the small arcs are

$$\begin{aligned} AB^2 &= a_{mn}\xi^m\xi^n, \\ AC^2 &= a_{mn}\eta^m\eta^n, \\ BC^2 &= a_{mn}(\xi^m - \eta^m)(\xi^n - \eta^n). \end{aligned}$$

Since we are interested only in principal parts, it is not necessary to allow for the change in the metric tensor on passing from A to B or C; we consider a_{mn} evaluated at A. The formula 2.309 gives

2.310. $$\cos\theta = \lim \frac{a_{mn}\xi^m\eta^n}{AB.AC}.$$

The unit tangent vectors to the two curves at A are

2.311. $$X^r = \lim \xi^r/AB, \quad Y^r = \lim \eta^r/AC.$$

Hence *the angle θ between two curves (or the angle between any two unit vectors at a point) satisfies*

2.312. $$\cos\theta = a_{mn}X^mY^n.$$

This equation determines a unique angle θ in the range $(0, \pi)$, provided the right hand side does not exceed unity in absolute value. We shall now show that this condition is fulfilled in a space with positive-definite metric form.

From the assumed positive-definite character, we have

2.313. $$a_{mn}(X^m + kY^m)(X^n + kY^n) \geqslant 0;$$

here X^r and Y^r are any unit vectors, and k any real number. Multiplying out, we get

2.314. $$k^2 + 2ka_{mn}X^mY^n + 1 \geqslant 0$$

or

2.315. $$(k + a_{mn}X^mY^n)^2 + [1 - (a_{mn}X^mY^n)^2] \geqslant 0.$$

Since this holds for arbitrary k, and k may be chosen to make the first term vanish, it follows that

2.316. $$|a_{mn}X^mY^n| \leqslant 1.$$

This proves the required result.

In a space with an indefinite metric form, 2.312 may be used as a formal definition of angle, but the definition is not of much use since the angle turns out to be imaginary in some cases. However, whether the metric form is definite or indefinite, we adopt as definition of *perpendicularity* or *orthogonality* of vectors X^r, Y^r the condition

2.317. $$a_{mn}X^mY^n = 0.$$

Exercise. Show that the small angle between unit vectors X^r and $X^r + dX^r$ (these increments being infinitesimal) is given by

$$\theta^2 = a_{mn} dX^m dX^n.$$

2.4. Geodesics and geodesic null lines. Christoffel symbols. In Euclidean 3-space a straight line is usually regarded as a basic concept. But if we wish to build up Euclidean geometry from the Euclidean line element, there is no difficulty in *defining* a straight line; it is the shortest curve between its end points.

We carry over this definition into Riemannian N-space in a modified form. The modification is suggested by considering a surface in Euclidean 3-space. In general, this surface contains no straight lines; nevertheless, there are on it certain curves analogous to the straight lines of space. They are curves of stationary length, or geodesics (great circles on a sphere). This idea of stationary length, rather than shortest length, is what we carry over into Riemannian space as a basis of definition: *A geodesic is a curve whose length has a stationary value with respect to arbitrary small variations of the curve, the end points being held fixed.*

In the notation of the calculus of variations, a geodesic joining points A and B satisfies the variational condition

2.401. $$\delta \int_A^B ds = 0.$$

Our next task is to find the differential equations of a geodesic, using the technique of the calculus of variations. We shall use an argument valid whether the metric form is definite or indefinite.

Consider the equations

2.402. $$x^r = x^r(u, v).$$

If we hold v fixed and let u vary, we get a curve. Thus 2.402 represents a singly infinite family of curves, v being constant along each curve.

A singly infinite family of curves joining common end points A, B may be represented by equations of the form 2.402. There is no loss of generality in supposing the para-

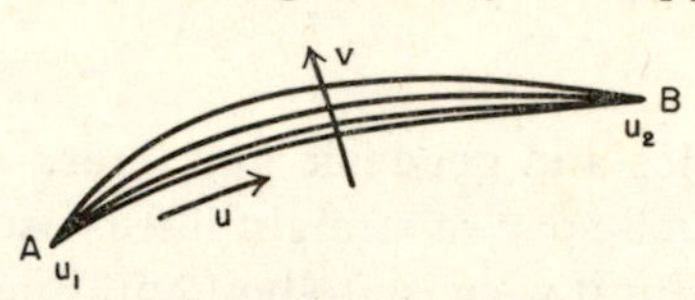

FIG. 2. Family of curves with common end points.

meter u so chosen that u has the same value (u_1) for all the curves at A, and the same value (u_2) for all the curves at B (Fig. 2). The length of any curve of this family is

2.403.
$$L = \int_A^B ds = \int_{u_1}^{u_2} (\epsilon a_{mn} p^m p^n)^{\frac{1}{2}} \, du,$$

where $p^r = \partial x^r / \partial u$ is a tangent to the curve and ϵ its indicator. We shall suppose that all the curves have the same indicator. For shortness let us write

2.404.
$$w = a_{mn} p^m p^n.$$

Then 2.403 reads

2.405.
$$L = \int_{u_1}^{u_2} (\epsilon w)^{\frac{1}{2}} \, du.$$

This length is a function of v, and its derivative is

2.406.
$$\frac{dL}{dv} = \int_{u_1}^{u_2} \frac{\partial}{\partial v} (\epsilon w)^{\frac{1}{2}} \, du.$$

Now w is a function of the x's and the p's; hence

2.407.
$$\frac{\partial}{\partial v} (\epsilon w)^{\frac{1}{2}} = \frac{\partial}{\partial x^r} (\epsilon w)^{\frac{1}{2}} \frac{\partial x^r}{\partial v} + \frac{\partial}{\partial p^r} (\epsilon w)^{\frac{1}{2}} \frac{\partial p^r}{\partial v}.$$

But

2.408.
$$\frac{\partial p^r}{\partial v} = \frac{\partial}{\partial v} \frac{\partial x^r}{\partial u} = \frac{\partial}{\partial u} \frac{\partial x^r}{\partial v},$$

and so, if we substitute from 2.407 in 2.406 and integrate by parts, we get

2.409. $$\frac{dL}{dv} = \left[\frac{\partial}{\partial p^r}(\epsilon w)^{\frac{1}{2}} \frac{\partial x^r}{\partial v}\right]_{u_1}^{u_2} - \int_{u_1}^{u_2} \left(\frac{\partial}{\partial u}\frac{\partial}{\partial p^r}(\epsilon w)^{\frac{1}{2}} - \frac{\partial}{\partial x^r}(\epsilon w)^{\frac{1}{2}}\right)\frac{\partial x^r}{\partial v}\, du\,.$$

We may also express this result in terms of infinitesimals. The change in length δL when we pass from a curve v to a neighbouring curve $v + \delta v$ is

2.410. $$\delta L = \frac{dL}{dv}\delta v = \left[\frac{\partial}{\partial p^r}(\epsilon w)^{\frac{1}{2}}\, \delta x^r\right]_{u_1}^{u_2} - \int_{u_1}^{u_2} \left(\frac{\partial}{\partial u}\frac{\partial}{\partial p^r}(\epsilon w)^{\frac{1}{2}} - \frac{\partial}{\partial x^r}(\epsilon w)^{\frac{1}{2}}\right)\delta x^r du,$$

where

2.411. $$\delta x^r = \frac{\partial x^r}{\partial v}\delta v.$$

This last differential is the increment in x^r on passing from a point on the curve v to the point on the curve $v + \delta v$ with the same value of u. Since the points A and B are fixed, $\delta x^r = 0$ there, and so the first term on the right hand side of 2.410 disappears; thus we have

2.412. $$\delta L = -\int_{u_1}^{u_2} \left(\frac{d}{du}\frac{\partial}{\partial p^r}(\epsilon w)^{\frac{1}{2}} - \frac{\partial}{\partial x^r}(\epsilon w)^{\frac{1}{2}}\right)\delta x^r du;$$

we have replaced $\partial/\partial u$ by d/du, since there is now no chance of confusion.

If the particular curxe v is a geodesic, the integral in 2.412 vanishes for variations δx^r which are arbitrary, except at the end points (where they are zero). It follows from the fundamental lemma of the calculus of variations* that the equations

*R. Courant, *Differential and Integral Calculus* (Blackie, London and Glasgow, 1936), II, p. 499.

2.413. $$\frac{d}{du}\frac{\partial}{\partial p^r}(\epsilon w)^{\frac{1}{2}} - \frac{\partial}{\partial x^r}(\epsilon w)^{\frac{1}{2}} = 0$$

are satisfied at all points on a geodesic. These are called the "Euler equations" of the variational problem. It is easy to see that they may be written

2.414. $$\frac{d}{du}\frac{\partial w}{\partial p^r} - \frac{\partial w}{\partial x^r} = \frac{1}{2w}\frac{dw}{du}\frac{\partial w}{\partial p^r}.$$

So far the parameter u has been arbitrary. Let us now choose it equal to the arc-length s along the geodesic, so that

2.415. $$u = s, \quad p^r = \frac{dx^r}{ds}, \quad w = a_{mn}\, p^m p^n = \epsilon, \quad \frac{dw}{du} = 0.$$

The differential equations of a geodesic now read

2.416. $$\frac{d}{ds}\frac{\partial w}{\partial p^r} - \frac{\partial w}{\partial x^r} = 0.$$

To obtain a more explicit form, we substitute for w. This gives

2.417. $$\frac{d}{ds}(2a_{rm}p^m) - \frac{\partial a_{mn}}{\partial x^r}p^m p^n = 0,$$

or

2.418. $$a_{rm}\frac{dp^m}{ds} + \frac{\partial a_{rm}}{\partial x^n}p^m p^n - \tfrac{1}{2}\frac{\partial a_{mn}}{\partial x^r}p^m p^n = 0.$$

By a mere rearrangement of dummy suffixes, we have identically

2.419. $$\frac{\partial a_{rm}}{\partial x^n}p^m p^n = \tfrac{1}{2}\left(\frac{\partial a_{rm}}{\partial x^n} + \frac{\partial a_{rn}}{\partial x^m}\right)p^m p^n,$$

and so the equations of a geodesic may be written

2.420. $$a_{rm}\frac{dp^m}{ds} + [mn, r]\, p^m p^n = 0,$$

where

2.421.
$$[mn, r] = \tfrac{1}{2}\left(\frac{\partial a_{rm}}{\partial x^n} + \frac{\partial a_{rn}}{\partial x^m} - \frac{\partial a_{mn}}{\partial x^r}\right).$$

This expression $[mn, r]$ is called the *Christoffel symbol of the first kind.*

The *Christoffel symbol of the second kind* is defined as

2.422.
$$\left\{ {r \atop mn} \right\} = a^{rs}\,[mn, s].$$

If we multiply 2.420 by a^{rs}, we obtain another form for the equations of a geodesic:

2.423.
$$\frac{dp^r}{ds} + \left\{ {r \atop mn} \right\} p^m p^n = 0.$$

In more explicit form, this reads

2.424.
$$\frac{d^2x^r}{ds^2} + \left\{ {r \atop mn} \right\} \frac{dx^m}{ds}\frac{dx^n}{ds} = 0.$$

We have now found the differential equations of a geodesic in three different forms, as shown in equations 2.416, 2.420, and 2.423 or 2.424. In the discussion we have tacitly assumed that the curve under consideration has not a null direction at any point on it; if it had, we would have $ds = 0$, and the equations would have become meaningless. But, as we shall see in equation 2.445, by taking a different approach, we can define curves analogous to geodesics, but with $ds = 0$.

As regards the amount of information necessary to determine a geodesic, we note that the equations 2.424 are differential equations of the second order. A solution $x^r(s)$ is determined uniquely if we are given initial values of x^r and dx^r/ds. In geometrical language, this means that a geodesic is determined if we are given a point on it and the direction of the tangent at that point; in this respect, as well as in its stationary property, a geodesic resembles a straight line in Euclidean 3-space.

Exercise. Prove the following identities:

2.425. $[mn,r] = [nm, r], \quad [rm, n] + [rn, m] = \partial a_{mn}/\partial x^r.$

Exercise. Prove that

2.426.
$$[mn, r] = a_{rs}\left\{ {s \atop mn} \right\}.$$

In 2.414 we obtained the differential equation of a geodesic in terms of an arbitrary parameter u. This equation may be reduced to a form analogous to 2.424, or we may obtain the result directly from 2.424 by transforming the independent variable from s to u. In this way we get

2.427.
$$\frac{d^2x^r}{du^2} + \left\{ {r \atop mn} \right\} \frac{dx^m}{du}\frac{dx^n}{du} = \lambda \frac{dx^r}{du},$$

with

2.428.
$$\lambda = -\frac{d^2u}{ds^2} \bigg/ \left(\frac{du}{ds}\right)^2.$$

Since the right-hand side of 2.427 is a vector, so is the left-hand side.* We may therefore say that on a geodesic, no matter what parameter u is used, the vector

2.429.
$$\frac{d^2x^r}{du^2} + \left\{ {r \atop mn} \right\} \frac{dx^m}{du}\frac{dx^n}{du}$$

is codirectional with (or opposed to) the tangent vector dx^r/du. Conversely, if we are given that along a curve C the vector 2.429 is codirectional with (or opposed to) the vector dx^r/du, and if further dx^r/du is not a null vector, then C must be a geodesic. This may be proved without difficulty by starting with 2.427, in which λ is a known function of u, and defining s by the relation

2.430.
$$s = \int_{u_0}^{u} \left(exp \int_{v_0}^{v} \lambda(w)dw \right) dv,$$

u_0, v_0 being constants. No matter what values these constants have 2.424 is satisfied, and by adjusting the constant v_0 we can ensure that $a_{mn}(dx^m/ds)(dx^n/ds) = \pm 1$ along C, so that s is actually the arc length.

*This tells us that it is a vector when calculated for a geodesic. In 2.437 we establish its vector character for any curve.

The fact that we can put the equations of a geodesic into the form 2.427, where $\lambda(u)$ is an arbitrary function, enables us to discuss the possibility of drawing a geodesic through two given points. Let the points be $x^r = a^r$ and $x^r = b^r$. Let us choose $u = x^N$, and assign to Greek suffixes the range 1, 2, . . . , $N - 1$. Then the last of 2.427 gives

$$\lambda = \left\{ \begin{matrix} N \\ \mu\nu \end{matrix} \right\} \frac{dx^\mu}{dx^N}\frac{dx^\nu}{dx^N} + 2\left\{ \begin{matrix} N \\ \mu N \end{matrix} \right\} \frac{dx^\mu}{dx^N} + \left\{ \begin{matrix} N \\ NN \end{matrix} \right\},$$

and the other equations read

$$\frac{d^2x^\rho}{(dx^N)^2} = \lambda\frac{dx^\rho}{dx^N} - \left\{ \begin{matrix} \rho \\ \mu\nu \end{matrix} \right\} \frac{dx^\mu}{dx^N}\frac{dx^\nu}{dx^N} - 2\left\{ \begin{matrix} \rho \\ \mu N \end{matrix} \right\} \frac{dx^\mu}{dx^N} - \left\{ \begin{matrix} \rho \\ NN \end{matrix} \right\}.$$

Here we have $(N - 1)$ ordinary differential equations, each of the second order, and their solution will contain $2(N - 1)$ constants of integration. The conditions that we should have $x^\rho = a^\rho$ when $x^N = a^N$, and $x^\rho = b^\rho$ when $x^N = b^N$ are $2(N - 1)$ in number. Hence we may form the general conclusion that it is possible to draw at least one geodesic through two given points, but of course this argument is suggestive rather than convincing. We can also approach the question through the variational principle, seeking the curve of shortest length connecting the given points. However, difficulties occur here in the case of an indefinite metric, since any two points can be joined by an indefinite number of curves of zero length. We shall not pursue further the question of the existence of a geodesic joining two given points. When we have occasion to use this construction, we shall assume that it can be done.

The importance of stationary principles, such as that used in defining a geodesic, was recognized long before the tensor calculus was invented. Such a principle is *invariant*, in the sense that no particular coordinate system is mentioned in stating the principle. Hence the differential equations obtained from the principle must share this invariant character. If the equations 2.416 are satisfied by a certain curve for one coordinate system, then they must be satisfied for all coordinate systems. This suggests (but does not, of course, prove) that the left-hand side of 2.416 has tensor character. Let us examine this question in a slightly more general form.

Consider a curve $x^r = x^r(u)$, where u is some parameter, not necessarily the arc-length. Write $p^r = dx^r/du$, and $w = a_{mn}p^m p^n$, where a_{mn} is the metric tensor. Define f_r by

2.431. $$2f_r = \frac{d}{du}\left(\frac{\partial w}{\partial p^r}\right) - \frac{\partial w}{\partial x^r}.$$

We ask: Is f_r a tensor?

We know that p^r is a contravariant vector, and so

2.432. $$p^r = p'^s \frac{\partial x^r}{\partial x'^s}.$$

Hence, if we regard p^r as a function of p'^m and x'^m, we have

2.433. $$\frac{\partial p^r}{\partial p'^s} = \frac{\partial x^r}{\partial x'^s}.$$

Then

2.434. $$\frac{\partial w}{\partial p'^r} = \frac{\partial w}{\partial p^s}\frac{\partial p^s}{\partial p'^r} = \frac{\partial w}{\partial p^s}\frac{\partial x^s}{\partial x'^r},$$

and

2.435. $$\frac{d}{du}\left(\frac{\partial w}{\partial p'^r}\right) = \frac{d}{du}\left(\frac{\partial w}{\partial p^s}\right)\frac{\partial x^s}{\partial x'^r} + \frac{\partial w}{\partial p^s}\frac{\partial^2 x^s}{\partial x'^m \partial x'^r}\frac{dx'^m}{\partial u}.$$

Further

2.436. $$\frac{\partial w}{\partial x'^r} = \frac{\partial w}{\partial x^s}\frac{\partial x^s}{\partial x'^r} + \frac{\partial w}{\partial p^s}\frac{\partial p^s}{\partial x'^r}.$$

To evaluate $\partial p^s/\partial x'^r$ we use 2.432, with a change of suffixes. Then subtraction of 2.436 from 2.435 gives

2.437. $$2\left(f'_r - f_s\frac{\partial x^s}{\partial x'^r}\right)$$
$$= \frac{\partial w}{\partial p^s}\frac{\partial^2 x^s}{dx'^m \partial x'^r}\frac{dx'^m}{du} - \frac{\partial w}{\partial p^s}p'^m\frac{\partial^2 x^s}{\partial x'^r \partial x'^m} = 0.$$

Hence f_r *is a covariant vector.*

The transition from 2.416 to 2.420, 2.423, and 2.424 is a formal process which holds equally well when an arbitrary parameter u is employed instead of s. Hence

2.438. $$f_r \equiv a_{rm}\frac{d^2x^m}{du^2} + [mn, r]\frac{dx^m}{du}\frac{dx^n}{du}$$

is a covariant vector, and

2.439. $$f^r \equiv \frac{d^2x^r}{du^2} + \left\{ {r \atop mn} \right\}\frac{dx^m}{du}\frac{dx^n}{du}$$

is a contravariant vector.

It may be noted that, in proving the tensor character of f_r defined by 2.431, no use was made of the actual form of w. All we needed to know about w was that it was an invariant function of the x's and the p's.

Let us now establish an important property of the differential equations

2.440. $$\frac{d}{du}\left(\frac{\partial w}{\partial p^r}\right) - \frac{\partial w}{\partial x^r} = 0.$$

Multiplication by p^r gives (since $p^r = dx^r/du$)

2.441. $$\frac{d}{du}\left(p^r\frac{\partial w}{\partial p^r}\right) - \frac{dp^r}{du}\frac{\partial w}{\partial p^r} - \frac{dx^r}{du}\frac{\partial w}{\partial x^r} = 0.$$

The last two terms together give dw/du, and so 2.441 may be written

2.442. $$\frac{d}{du}\left(p^r\frac{\partial w}{\partial p^r} - w\right) = 0;$$

hence

2.443. $$p^r\frac{\partial w}{\partial p^r} - w = \text{constant}.$$

This is a *first integral* of the equations 2.440. If we now put $w = a_{mn}p^m p^n$, we get

2.444. $$w \equiv a_{mn}p^m p^n = \text{constant}.$$

If we put $u = s$ in 2.440, we get the equations of a geodesic; then, as we already know, the constant in 2.444 is ϵ, the indicator of the direction of the geodesic ($\epsilon = \pm 1$). But if we start a curve with a null direction, so that $w = 0$, and if this curve obeys the differential equations 2.440, then the constant in 2.444 has the value zero; the curve has a null direction at each of its points. Such a curve is a *geodesic null line.* Its differential equations read

$$\text{2.445.} \qquad \frac{d^2x^r}{du^2} + \left\{ {r \atop mn} \right\} \frac{dx^m}{du} \frac{dx^n}{du} = 0,$$

with the particular first integral

$$\text{2.446.} \qquad a_{mn} \frac{dx^m}{du} \frac{dx^n}{du} = 0.$$

To sum up: *A geodesic null line is a curve which, for some parameter u, satisfies the differential equations* 2.440, *or equivalently* 2.445, *with the particular first integral* 2.446.

If we change the parameter from u to v, where v is some function of u, the equations 2.445 and 2.446 become respectively

$$\text{2.447.} \qquad \frac{d^2x^r}{dv^2} + \left\{ {r \atop mn} \right\} \frac{dx^m}{dv} \frac{dx^n}{dv} = \lambda \frac{dx^r}{dv},$$

with

$$\lambda = -\frac{d^2v}{du^2} \bigg/ \left(\frac{dv}{du}\right)^2,$$

and

$$\text{2.448.} \qquad a_{mn} \frac{dx^m}{dv} \frac{dx^n}{dv} = 0.$$

By suitable choice of the parameter v, λ can be made any preassigned function of v. Hence, sufficient conditions that a curve be a geodesic null line are that the quantities

$$\text{2.449.} \qquad \frac{d^2x^r}{dv^2} + \left\{ {r \atop mn} \right\} \frac{dx^m}{dv} \frac{dx^n}{dv}$$

be proportional to dx^r/dv and that 2.448 hold.

Exercise. The class of all parameters u, for which the equations of a geodesic null line assume the simple form 2.445, are obtained from any one such parameter by the linear transformation

$$\bar{u} = au + b,$$

a and b being arbitrary constants.

Like a geodesic, a geodesic null line is determined by a point on it, and the direction of its tangent there; but of course this direction must be a null direction. The geodesic null line is important in relativity; it represents the history of a light pulse in space-time.

Exercise. Consider a 3-space wtih coordinates x, y, z, and a metric form $\Phi = (dx)^2 + (dy)^2 - (dz)^2$. Prove that the geodesic null lines may be represented by the equations

$$x = au + a', \quad y = bu + b', \quad z = cu + c',$$

where u is a parameter and a, a', b, b', c, c' are constants which are arbitrary except for the relation $a^2 + b^2 - c^2 = 0$.

2.5. Derivatives of tensors. We saw in 1.4 that the partial derivative of an invariant with respect to one of the coordinates is a covariant vector. One might think that the partial derivative of any tensor is itself a tensor. That is not so (cf. Exercises I, No. 5). But by adding certain terms to the derivative we obtain a tensor. This is a very important idea in tensor calculus, and we shall devote the present section to it. First we shall see how the Christoffel symbols transform.

In 2.4 we saw that f^r, where

2.501. $$f^r = \frac{d^2x^r}{du^2} + \begin{Bmatrix} r \\ mn \end{Bmatrix} \frac{dx^m}{du}\frac{dx^n}{du},$$

is a contravariant tensor. From this fact it is easy to deduce the law of transformation of the Christoffel symbols. We have

2.502. $$f'^r = f^s \frac{\partial x'^r}{\partial x^s}.$$

Now

2.503. $$\frac{d^2x^s}{du^2} = \frac{d}{du}\left(\frac{dx^s}{dx'^m}\frac{dx'^m}{du}\right)$$
$$= \frac{\partial x^s}{\partial x'^m}\frac{d^2x'^m}{du^2} + \frac{\partial^2 x^s}{\partial x'^m \partial x'^n}\frac{dx'^m}{du}\frac{dx'^n}{du},$$

and so

2.504. $$\frac{d^2x^s}{du^2}\frac{\partial x'^r}{\partial x^s} = \frac{d^2x'^r}{du^2} + \frac{\partial x'^r}{\partial x^s}\frac{\partial^2 x^s}{\partial x'^m \partial x'^n}\frac{dx'^m}{du}\frac{dx'^n}{du}.$$

Thus 2.502 gives

2.505. $$A^r_{mn}\frac{dx'^m}{du}\frac{dx'^n}{du} = 0,$$

where we have written for brevity

2.506. $$A^r_{mn} = \left\{ {r \atop mn} \right\}' - \left\{ {s \atop pq} \right\}\frac{\partial x'^r}{\partial x^s}\frac{\partial x^p}{\partial x'^m}\frac{\partial x^q}{\partial x'^n} - \frac{\partial x'^r}{\partial x^s}\frac{\partial^2 x^s}{\partial x'^m \partial x'^n}.$$

Obviously $A^r_{mn} = A^r_{nm}$, and these quantities are independent of dx'^r/du. Hence $A^r_{mn} = 0$, and so we have the formula of transformation for the Christoffel symbols of the second kind.

2.507. $$\left\{ {r \atop mn} \right\}' = \left\{ {s \atop pq} \right\}\frac{\partial x'^r}{\partial x^s}\frac{\partial x^p}{\partial x'^m}\frac{\partial x^q}{\partial x'^n} + \frac{\partial x'^r}{\partial x^s}\frac{\partial^2 x^s}{\partial x'^m \partial x'^n}$$

In the notation of 1.7, this reads

2.508. $$\left\{ {\rho \atop \mu\nu} \right\} = \left\{ {r \atop mn} \right\} X^\rho_r X^m_\mu X^n_\nu + X^\rho_r X^r_{\mu\nu}.$$

Exercise. Prove that the Christoffel symbols of the first kind transform according to the equation

2.509. $$[mn, r]' = [pq, s]\frac{\partial x^p}{\partial x'^m}\frac{\partial x^q}{\partial x'^n}\frac{\partial x^s}{\partial x'^r} + a_{pq}\frac{\partial x^p}{\partial x'^r}\frac{\partial^2 x^q}{\partial x'^m \partial x'^n},$$

or, in the notation of 1.7,

2.510. $$[\mu\nu, \rho] = [mn, r] X^m_\mu X^n_\nu X^r_\rho + a_{rs} X^r_\rho X^s_{\mu\nu}.$$

It will be noticed that neither Christoffel symbol is a tensor; the first terms on the right-hand sides of the equations of transformation are those of tensor transformation, but the last terms spoil the tensor character.

We are now ready to discuss the differentiation of tensors. We shall start with a contravariant vector field T^r, defined along a curve $x^r = x^r(u)$, and prove the following result:

The absolute derivative $\delta T^r/\delta u$ of the vector T^r, defined as

2.511. $$\frac{\delta T^r}{\delta u} = \frac{dT^r}{du} + \begin{Bmatrix} r \\ mn \end{Bmatrix} T^m \frac{dx^n}{du},$$

is itself a contravariant vector.

This is easily shown by using the formulae of transformation of the vector T^r and the Christoffel symbols. We find

$$\frac{\delta T'^r}{\delta u} - \frac{\delta T^s}{\delta u}\frac{\partial x'^r}{\partial x^s} = T^m \frac{dx^n}{du}\left(\frac{\partial^2 x'^r}{\partial x^m \partial x^n} + \frac{\partial^2 x^s}{\partial x'^p \partial x'^q}\frac{\partial x'^r}{\partial x^s}\frac{\partial x'^p}{\partial x^m}\frac{\partial x'^q}{\partial x^n}\right),$$

and this vanishes by 1.210; this establishes the tensor character of $\delta T^r/\delta u$.

If the vector T^r satisfies the differential equations

2.512. $$\frac{\delta T^r}{\delta u} \equiv \frac{dT^r}{du} + \begin{Bmatrix} r \\ mn \end{Bmatrix} T^m \frac{dx^n}{du} = 0$$

along a curve, then the vector T^r is said to be propagated *parallelly* along the curve. If the space is Euclidean 3-space and the coordinates are rectangular Cartesians, the Christoffel symbols vanish, and 2.512 reduces to $dT^r/du = 0$; in this particular case, parallel propagation implies the constancy of the components, i.e. the vector passes through a sequence of parallel positions, using the word "parallel" in the ordinary sense.

Referring to 2.424, we note that *the unit tangent vector to a geodesic is propagated parallelly along it*: in symbols

2.513. $$\frac{\delta}{\delta s}\frac{dx^r}{ds} = 0.$$

The fact that there exists a tensorial derivative of a contravariant vector suggests that the same might be true for a covariant vector. This is correct, and we can easily find the form of this derivative indirectly as follows. Let T_r be a covariant vector field, defined along a curve $x^r = x^r(u)$, and let S^r be a contravariant vector which is propagated parallelly along the curve, and therefore satisfies 2.512. Then $T_r S^r$ is an invariant, and so is $\frac{d}{du}(T_r S^r)$. But

$$\text{2.514.} \qquad \frac{d}{du}(T_r S^r) = \frac{dT_r}{du} S^r + T_r \frac{dS^r}{du}$$

$$= \left(\frac{dT_r}{du} - \left\{ \begin{matrix} m \\ rn \end{matrix} \right\} T_m \frac{dx^n}{du} \right) S^r.$$

Now, at any point of the curve, S^r may be chosen arbitrarily; hence, by the test of 1.6, it follows that

$$\text{2.515.} \qquad \frac{\delta T_r}{\delta u} = \frac{dT_r}{du} - \left\{ \begin{matrix} m \\ rn \end{matrix} \right\} T_m \frac{dx^n}{du}$$

is a covariant vector. We call it the *absolute derivative* of T_r.

The equation for parallel propagation of a covariant vector is $\delta T_r / \delta u = 0$.

This method opens up the possibility of defining the absolute derivative of any tensor given along a curve. We have merely to build up an invariant by multiplying the given tensor by vectors which are propogated parallelly along the curve. Consider, for example, the tensor T_{rs}. We build up an invariant $T_{rs} S^r U^s$ and differentiate. We find at once, on using the equations of parallel propagation for S^r and U^r, that

$$\text{2.516.} \qquad \frac{\delta T_{rs}}{\delta u} = \frac{dT_{rs}}{du} - \left\{ \begin{matrix} m \\ rn \end{matrix} \right\} T_{ms} \frac{dx^n}{du} - \left\{ \begin{matrix} m \\ sn \end{matrix} \right\} T_{rm} \frac{dx^n}{du}$$

is a covariant tensor of the second order; we call it the *absolute derivative* of T_{rs}.

Applying the same method, we obtain the following definitions of the absolute derivatives of contravariant and mixed tensors of the second order:

$$\textbf{2.517.} \qquad \frac{\delta T^{rs}}{\delta u} = \frac{dT^{rs}}{du} + \begin{Bmatrix} r \\ mn \end{Bmatrix} T^{ms} \frac{dx^n}{du} + \begin{Bmatrix} s \\ mn \end{Bmatrix} T^{rm} \frac{dx^n}{du},$$

$$\textbf{2.518.} \qquad \frac{\delta T^r_{\cdot s}}{\delta u} = \frac{dT^r_{\cdot s}}{du} + \begin{Bmatrix} r \\ mn \end{Bmatrix} T^m_{\cdot s} \frac{dx^n}{du} - \begin{Bmatrix} m \\ sn \end{Bmatrix} T^r_{\cdot m} \frac{dx^n}{du}.$$

There is no difficulty at all in applying the same method to tensors of any order.

Exercise. Find the absolute derivative of $T^r{}_{st}$.

In order that the absolute derivative of a tensor may have a meaning, the tensor must be given along a curve. If it is given throughout a region of space, we may define the *covariant derivative* of a tensor in the following way. Let us start with a contravariant vector, for which the absolute derivative is as in 2.511. For any curve traversing the region in which the vector is given, we have

$$\textbf{2.519.} \qquad \frac{\delta T^r}{\delta u} = \left(\frac{\partial T^r}{\partial x^n} + \begin{Bmatrix} r \\ mn \end{Bmatrix} T^m \right) \frac{dx^n}{du}.$$

Now the left-hand side is a contravariant vector; dx^n/du is also a contravariant vector, and it is arbitrary. Hence by the tests of 1.6, the coefficient of dx^n/du is a mixed tensor. We write it

$$\textbf{2.520.} \qquad T^r{}_{|n} = \frac{\partial T^r}{\partial x^n} + \begin{Bmatrix} r \\ mn \end{Bmatrix} T^m,$$

and call it the covariant derivative of T^r.*

The same method may be applied to obtain, from the absolute derivative, the covariant derivative of any tensor. The plan is so obvious that we shall merely write down the results for a covariant vector and for tensors of the second order:

$$\textbf{2.521.} \qquad T_{r|n} = \frac{\partial T_r}{\partial x^n} - \begin{Bmatrix} m \\ rn \end{Bmatrix} T_m,$$

$$\textbf{2.522.} \qquad T^{rs}{}_{|n} = \frac{\partial T^{rs}}{\partial x^n} + \begin{Bmatrix} r \\ mn \end{Bmatrix} T^{ms} + \begin{Bmatrix} s \\ mn \end{Bmatrix} T^{rm},$$

*Other notations are $T^r{}_{,n}$, $T^r{}_{;n}$, $\nabla_n T^r$, and $D_n T^r$.

2.523. $$T_{rs|n} = \frac{\partial T_{rs}}{\partial x^n} - \begin{Bmatrix} m \\ rn \end{Bmatrix} T_{ms} - \begin{Bmatrix} m \\ sn \end{Bmatrix} T_{rm},$$

2.524. $$T^r_{.\,s|n} = \frac{\partial T^r_{.\,s}}{\partial x^n} + \begin{Bmatrix} r \\ mn \end{Bmatrix} T^m_{.\,s} - \begin{Bmatrix} m \\ sn \end{Bmatrix} T^r_{.\,m}.$$

The formulae for covariant derivatives are fundamental in tensor calculus, and it is useful to remember them. This is easy, if we observe the way in which they are built up. The formula for the covariant derivative of a general (mixed) tensor of any order may be split into three parts:

(1) The partial derivative of the tensor.

(2) A sum of terms, each prefixed with a plus sign and corresponding to one of the contravariant suffixes of the tensor. This suffix is taken off the original tensor and put into the upper line of the Christoffel symbol occurring in the term. It is replaced in the tensor by a dummy, which is also inserted in the lower line of the Christoffel symbol. The vacant space in the Christoffel symbol is filled with the suffix of the x with respect to which we differentiate.

(3) A sum of terms, each prefixed with a minus sign and corresponding to one of the covariant suffixes of the tensor. These terms are formed in very much the same way as those in (2), the guiding principle being that of taking a suffix from the tensor and putting it into the Christoffel symbol at the same level.

Here is the formula for the covariant derivative of the most general tensor:

2.525. $$T^{r_1 \ldots r_m}_{s_1 \ldots s_n|p} = \frac{\partial}{\partial x^p} T^{r_1 \ldots r_m}_{s_1 \ldots s_n} + \begin{Bmatrix} r_1 \\ qp \end{Bmatrix} T^{qr_2 \ldots r_m}_{s_1 \ldots s_n} + \ldots$$
$$+ \begin{Bmatrix} r_m \\ pq \end{Bmatrix} T^{r_1 \ldots r_{m-1}q}_{s_1 \ldots s_n} - \begin{Bmatrix} q \\ s_1 p \end{Bmatrix} T^{r_1 \ldots r_m}_{qs_2 \ldots s_n} - \ldots$$
$$- \begin{Bmatrix} q \\ s_n p \end{Bmatrix} T^{r_1 \ldots r_m}_{s_1 \ldots s_{n-1}q}.$$

A very important special case of 2.523 is that in which $T_{rs} = a_{rs}$, the metric tensor. We find

2.526.
$$a_{rs|t} = \frac{\partial a_{rs}}{\partial x^t} - \begin{Bmatrix} m \\ rt \end{Bmatrix} a_{ms} - \begin{Bmatrix} m \\ st \end{Bmatrix} a_{rm}$$
$$= \frac{\partial a_{rs}}{\partial x^t} - [rt,s] - [st,r]$$
$$= 0.$$

The covariant derivative of the metric tensor vanishes.

Exercise. Prove that

2.527.
$$\delta^r_{s|t} = 0, \; a^{rs}{}_{|t} = 0.$$

For the sake of completeness we define the absolute and covariant derivatives of an invariant to be the ordinary and partial derivatives respectively. If T is an invariant,

2.528.
$$\frac{\delta T}{\delta u} = \frac{dT}{du}, \; T_{|r} = \frac{\partial T}{\partial x^r}.$$

An important difference between the covariant derivative and the partial derivative should be noticed. Suppose we have a tensor T_{rs}. Consider the quantities
$$T_{12|r}, \; \frac{\partial T_{12}}{\partial x^r}.$$
The partial derivative can be calculated if we know the component T_{12} as a function of the coordinates. But we cannot calculate the covariant derivative unless we know *all* the components. Thus, partial differentiation is an operation which is applied to a single quantity; covariant differentiation is an operation which is applied to a whole set of quantities.

Absolute and covariant differentiation obey the following basic laws of elementary calculus: (i) the derivative of a sum is the sum of the derivatives. (ii) The derivative of a product UV is the sum of two products: the product of U and the derivative of V, and the product of V and the derivative of U.

The first is obvious, from the definitions of absolute and covariant derivatives. The second is not so immediate. But there is no difficulty (only length of writing) in giving a straightforward proof. We shall not give such a proof here, because the next section will provide us with a new approach which cuts out the need for long calculations.

The rule for differentiating a product, together with the fact that the covariant derivative of the metric tensor vanishes, implies that the order of the operations of lowering (or raising) a suffix and of differentiating can be reversed. For example,

2.529. $$T_{r.m|t}^{.n} = (a_{rs}T_{..m}^{sn})_{|t} = a_{rs|t}T_{..m}^{sn} + a_{rs}T_{..m|t}^{sn} = a_{rs}T_{..m|t}^{sn}.$$

This means that it is immaterial whether we differentiate the covariant, contravariant or one of the mixed representations of a tensor.

Exercise. Prove that

2.530. $$\frac{d}{ds}(a_{mn}\lambda^m\lambda^n) = 2a_{mn}\lambda^m\frac{\delta\lambda^n}{\delta s},$$

where λ^r is any vector field given along a curve for which s is the arc length.

Exercise. Without assuming (ii) above, prove that

2.531. $$(T^rS_s)_{|n} = T^r{}_{|n}S_s + T^rS_{s|n}.$$

As a general rule, Christoffel symbols are clumsy to handle in explicit calculations, and we avoid their use whenever possible. Thus, the vector f_r is easier to compute from 2.431 than from 2.438 or 2.439. We shall illustrate this by making explicit calculations for a Euclidean 3-space with spherical polar coordinates

$$x^1 = r,\ x^2 = \theta,\ x^3 = \phi.$$

The metric is

2.532. $$ds^2 = (dx^1)^2 + (x^1dx^2)^2 + (x^1 \sin x^2\, dx^3)^2,$$

and consequently the metric tensor is

2.533. $$a_{11} = 1,\ a_{22} = (x^1)^2,\ a_{33} = (x^1 \sin x^2)^2,$$
$$a_{23} = a_{31} = a_{12} = 0.$$

Then the function w occurring in 2.431 is

2.534. $$w = (p^1)^2 + (x^1 p^2)^2 + (x^1 \sin x^2\ p^3)^2,$$

and so 2.431 read explicitly

2.535. $$f_1 = \frac{d}{du}(p^1) - x^1(p^2)^2 - x^1(\sin x^2\ p^3)^2,$$
$$f_2 = \frac{d}{du}((x^1)^2 p^2) - \sin x^2 \cos x^2\ (x^1 p^3)^2,$$
$$f_3 = \frac{d}{du}((x^1 \sin x^2)^2 p^3),$$

where $p^1 = dx^1/du$, $p^2 = dx^2/du$, $p^3 = dx^3/du$. After performing the differentiations, we have

2.536. $$f_1 = \frac{d^2x^1}{du^2} - x^1\left(\frac{dx^2}{du}\right)^2 - x^1(\sin x^2)^2\left(\frac{dx^3}{du}\right)^2,$$
$$f_2 = (x^1)^2\frac{d^2x^2}{du^2} - (x^1)^2 \sin x^2 \cos x^2\left(\frac{dx^3}{du}\right)^2 + 2x^1\frac{dx^1}{du}\frac{dx^2}{du},$$
$$f_3 = (x^1 \sin x^2)^2\frac{d^2x^3}{du^2} + 2(x^1)^2 \sin x^2 \cos x^2\frac{dx^2}{du}\frac{dx^3}{du} + 2x^1(\sin x^2)^2\frac{dx^3}{du}\frac{dx^1}{du}.$$

Comparing these expressions with 2.438, and noting that the coefficient of $(dx^1/du)(dx^2/du)$ in f_2, say, is $[12, 2] + [21, 2] = 2\,[12, 2]$, we immediately read off the Christoffel symbols of the first kind. The complete table of all non-vanishing symbols of the first kind is as follows:

2.537.
$$[mn, 1] : [22, 1] = -x^1, \qquad [33, 1] = -x^1(\sin x^2)^2;$$
$$[mn, 2] : [33, 2] = -(x^1)^2 \sin x^2 \cos x^2, \qquad [12, 2] = x^1;$$
$$[mn, 3] : [23, 3] = (x^1)^2 \sin x^2 \cos x^2, \qquad [31, 3] = x^1(\sin x^2)^2.$$

Noting that

2.538.

$$a^{11} = 1,\ a^{22} = \frac{1}{(x^1)^2},\ a^{33} = \frac{1}{(x^1 \sin x^2)^2},$$

$$a^{23} = a^{31} = a^{12} = 0,$$

we can derive from 2.537 the Christoffel symbols of the second kind. However, we shall illustrate a more direct method of obtaining them. This method is useful since we often require the $\left\{ {r \atop mn} \right\}$ but not the $[mn, r]$, as, for instance, in the computation of the curvature tensors which are discussed in the following chapter.

Since $f^m = a^{mn} f_n$, we can immediately obtain the f^m from 2.536. In the special, but important, case of an "orthogonal" metric ($a_{mn} = 0$ when $m \neq n$), the f^m are obtained by dividing the corresponding expressions for the f_m by the coefficient of the second order derivative. Thus we have

2.539.

$$f^1 = f_1 = \frac{d^2x^1}{du^2} - x^1\left(\frac{dx^2}{du}\right)^2 - x^1(\sin x^2)^2\left(\frac{dx^3}{du}\right)^2,$$

$$f^2 = \frac{f_2}{(x^1)^2} = \frac{d^2x^2}{du^2} - \sin x^2 \cos x^2\left(\frac{dx^3}{du}\right)^2 + \frac{2}{x^1}\frac{dx^1}{du}\frac{dx^2}{du},$$

$$f^3 = \frac{f_3}{(x^1 \sin x^2)^2} = \frac{d^2x^3}{du^2} + 2 \cot x^2 \frac{dx^2}{du}\frac{dx^3}{du} + \frac{2}{x^1}\frac{dx^3}{du}\frac{dx^1}{du}.$$

Comparing these expressions with 2.439, we read off all non-vanishing Christoffel symbols of the second kind:

2.540.

$$\left\{ {1 \atop mn} \right\}: \left\{ {1 \atop 22} \right\} = -x^1, \qquad \left\{ {1 \atop 33} \right\} = -x^1(\sin x^2)^2;$$

$$\left\{ {2 \atop mn} \right\}: \left\{ {2 \atop 33} \right\} = -\sin x^2 \cos x^2, \qquad \left\{ {2 \atop 12} \right\} = \frac{1}{x^1};$$

$$\left\{ {3 \atop mn} \right\}: \left\{ {3 \atop 23} \right\} = \cot x^2, \qquad \left\{ {3 \atop 31} \right\} = \frac{1}{x^1}.$$

Exercise. Compute the Christoffel symbols in 2.540 directly from the definitions 2.421 and 2.422. Check that all Christoffel symbols not shown explicitly in 2.540 vanish.

Returning to a general Riemannian space, let us note a short cut for calculating the contracted Christoffel symbol $\left\{ {n \atop rn} \right\}$. Using 2.216, we have

2.541.
$$\begin{aligned} 2\left\{ {n \atop rn} \right\} &= 2a^{nm}[rn, m] \\ &= a^{nm}\left(\frac{\partial a_{rm}}{\partial x^n} + \frac{\partial a_{mn}}{\partial x^r} - \frac{\partial a_{rn}}{\partial x^m}\right) \\ &= a^{mn}\frac{\partial a_{mn}}{\partial x^r} \\ &= \frac{\partial}{\partial x^r}\ln a, \end{aligned}$$

or

2.542.
$$\left\{ {n \atop rn} \right\} = \frac{\partial}{\partial x^r}\ln\sqrt{a} = \frac{1}{\sqrt{a}}\frac{\partial}{\partial x^r}\sqrt{a},$$

assuming that the determinant a is positive. If a is negative, 2.542 holds with a replaced by $-a$.

Exercise. Show that for the spherical polar metric 2.532, we have $\ln\sqrt{a} = 2\ln x^1 + \ln\sin x^2$, and

2.543.
$$\left\{ {n \atop 1n} \right\} = \frac{2}{x^1}, \quad \left\{ {n \atop 2n} \right\} = \cot x^2, \quad \left\{ {n \atop 3n} \right\} = 0.$$

The result 2.542 leads to a useful formula for the "divergence" $T^n{}_{|n}$ of a vector T^r. By 2.520 we have

2.544.
$$\begin{aligned} T^n{}_{|n} &= \frac{\partial T^n}{\partial x^n} + \left\{ {n \atop mn} \right\} T^m \\ &= \frac{\partial T^n}{\partial x^n} + \left(\frac{1}{\sqrt{a}}\frac{\partial}{\partial x^m}\sqrt{a}\right) T^m, \end{aligned}$$

or

2.545. $$T^n{}_{|n} = \frac{1}{\sqrt{a}} \frac{\partial}{\partial x^n} (\sqrt{a} T^n).$$

Here we have assumed a positive; if a is negative we replace it by $-a$.

Exercise. Show that for spherical polar coordinates

2.546. $$T^n{}_{|n} = \frac{1}{r^2} \frac{\partial}{\partial r} (r^2 T^1) + \frac{1}{\sin\theta} \frac{\partial}{\partial \theta} (\sin\theta T^2) + \frac{\partial}{\partial \phi} T^3.$$

Obtain a similar expression for the "Laplacian" ΔV of an invariant V defined by

2.547. $$\Delta V = \left(a^{nm} \frac{\partial V}{\partial x^m} \right)_{|n}.$$

2.6. Special coordinate systems. Tensor calculus gives us a symbolism which avoids reference to particular coordinate systems. This "democratic principle" is, in fact, the idea underlying the whole subject. Nevertheless there are occasions when special coordinate systems prove very useful. To think of Euclidean 3-space for a moment, the formalism of tensor calculus applies to the most general curvilinear coordinate system; but there are many occasions when it is much simpler to work with rectangular Cartesians. In a general Riemannian V_N there exists no system of coordinates as simple as rectangular Cartesians. But there are several systems with certain simplifying properties, and these we shall now discuss.

First we shall consider *local Cartesians*. The terms in the general form $\Phi = a_{mn} dx^m dx^n$ which contain dx^1 are

$$a_{11}(dx^1)^2 + 2a_{12} dx^1 dx^2 + 2a_{13} dx^1 dx^3 + \ldots + 2a_{1N} dx^1 dx^N.$$

Let us assume that a_{11} is not zero.* Then this expression differs from

$$a_{11} \left[dx^1 + \frac{a_{12}}{a_{11}} dx^2 + \ldots + \frac{a_{1N}}{a_{11}} dx^N \right]^2$$

*See Appendix A.

by certain terms which contain $dx^2, \ldots, dx^N$, but do not contain dx^1. Thus we may write

$$\text{2.601.} \qquad \Phi = \epsilon_1 \Psi_1^2 + \Phi_1,$$

where $\epsilon_1 = \pm 1$, to make $\epsilon_1 a_{11}$ positive,

$$\text{2.602.} \qquad \Psi_1 = \sqrt{\epsilon_1 a_{11}} \left[dx^1 + \frac{a_{12}}{a_{11}} dx^2 + \ldots + \frac{a_{1N}}{a_{11}} dx^N \right],$$

and Φ_1 is a homogeneous quadratic form in $dx^2, \ldots, dx^N$, the coefficients being functions of $x^1, x^2, \ldots, x^N$.

Applying the same process to Φ_1, and so on, we finally get

$$\text{2.603.} \qquad \Phi = \epsilon_1 \Psi_1^2 + \epsilon_2 \Psi_2^2 + \ldots + \epsilon_N \Psi_N^2,$$

where each ϵ is ± 1 and each Ψ is a differential form of the first degree, as in 2.602; we may write

$$\text{2.604.} \qquad \Psi_m = b_{mn} dx^n.$$

In the general case, it is impossible to integrate these differential forms; we cannot obtain a set of coordinates y_m such that the differentials of these coordinates are given by 2.604.

But let us fasten our attention on a point O with coordinates a^r; let us write

$$\text{2.605.} \qquad y_m = (b_{mn})_O (x^n - a^n),$$

the subscript O indicating evaluation at the point O. Then, by 2.603 and 2.604, we have at O

$$\text{2.606.} \qquad \Phi = \epsilon_1 dy_1^2 + \epsilon_2 dy_2^2 + \ldots + \epsilon_N dy_N^2.$$

To sum up: *It is possible to choose coordinates so that the metric form reduces to* 2.606 *at any one assigned point of space. Such* coordinates are called *local Cartesians.**

The next special coordinates to be discussed are *Riemannian coordinates.* Let x^r be a general coordinate system and let a^r be the coordinates of a point O. Consider the family of geodesics drawn out from O; each geodesic satisfies the differential equation

*See Appendix A.

2.607. $$\frac{d^2x^r}{ds^2} + \begin{Bmatrix} r \\ mn \end{Bmatrix} \frac{dx^m}{ds} \frac{dx^n}{ds} = 0.$$

Let p^r be the unit tangent vector at O to one of these geodesics. Then by 2.607, we have at O

$$\frac{d^2x^r}{ds^2} = -\begin{Bmatrix} r \\ mn \end{Bmatrix} p^m p^n,$$

2.608. $$\frac{d^3x^r}{ds^3} = A^r_{.mns} p^m p^n p^s,$$

where

2.609. $$A^r_{.mns} = -\frac{\partial}{\partial x^s}\begin{Bmatrix} r \\ mn \end{Bmatrix} + 2\begin{Bmatrix} r \\ sp \end{Bmatrix}\begin{Bmatrix} p \\ mn \end{Bmatrix}.$$

There are similar but more complicated expressions for the higher derivatives. Consider a point P on the geodesic at a distance s from O. Its coordinates may be written in the form of power series in s:

2.610. $$x^r = a^r + sp^r - \tfrac{1}{2} s^2 \begin{Bmatrix} r \\ mn \end{Bmatrix} p^m p^n + \tfrac{1}{6} s^3 A^r_{.mns} p^m p^n p^s + \ldots .$$

The coefficients are of course evaluated at O.

We define the Riemannian coordinates of P to be

2.611. $$x'^r = sp^r,$$

where s is the arc length OP and p^r is the unit tangent vector at O to the geodesic OP. The Riemannian coordinates of O are therefore $x'^r = 0$.

The first thing to show regarding Riemannian coordinates is that they form a regular system in the neighbourhood of O, i.e. we are to show that the Jacobian $\left|\frac{\partial x^r}{\partial x'^s}\right|$ is neither zero nor infinite. Substitution from 2.611 in 2.610 gives

2.612. $$x^r = a^r + x'^r - \tfrac{1}{2}\begin{Bmatrix} r \\ mn \end{Bmatrix} x'^m x'^n + \tfrac{1}{6} A^r_{.mns} x'^m x'^n x'^s + \ldots,$$

and so

2.613.
$$\frac{\partial x^r}{\partial x'^s} = \delta^r_s - \left\{ {r \atop ms} \right\} x'^m + \ldots.$$

Thus at O the Jacobian is

2.614.
$$\left| \frac{\partial x^r}{\partial x'^s} \right| = \left| \delta^r_s \right| = 1,$$

and in the neighbourhood of O it is neither zero nor infinite.

We shall now prove the fundamental property of Riemannian coordinates: *At the origin of Riemannian coordinates, the Christoffel symbols of both kinds and the first-order partial derivatives of the metric tensor all vanish.*

Consider the geodesics drawn through O, the origin of Riemannian coordinates. Along each of them the equation

2.615.
$$\frac{d^2 x'^r}{ds^2} + \left\{ {r \atop mn} \right\}' \frac{dx'^m}{ds} \frac{dx'^n}{ds} = 0$$

is satisfied. Here the primed Christoffel symbol is calculated for the metric tensor a'_{rs} corresponding to the Riemannian coordinate system x'^r. Now substitute from 2.611 in 2.615, remembering that p^r is constant as we pass along the geodesic since it represents a quantity calculated at the fixed point O. It follows that

2.616.
$$\left\{ {r \atop mn} \right\}' p^m p^n = 0$$

along the geodesic. Therefore at O, where the ratios of the p's are arbitrary (corresponding to the arbitrary direction of the geodesic there), we have

2.617.
$$\left\{ {r \atop mn} \right\}' = 0.$$

Hence, by 2.426 and 2.425, we have

2.618.
$$[mn, r]' = 0, \quad \frac{\partial a'_{mn}}{\partial x'^r} = 0$$

at O. This establishes the result italicized above. Note that 2.617 and 2.618 hold only at the point O, and not elsewhere.

We recall from 2.5 that the absolute and covariant derivatives differ from their leading terms (ordinary or partial derivatives) only by the addition of certain terms each of which has a Christoffel symbol as a factor. It follows immediately that *at the origin of Riemannian coordinates absolute and covariant derivatives reduce to ordinary and partial derivatives.*

This is a very useful result. Let us use it to prove the statement regarding the absolute and covariant derivatives of products, made (but not proved) in 2.5 (p. 53). One particular case will serve to show the method.

Consider the statement

2.619. $$(T^{rs} S_{rm})_{|t} = T^{rs}{}_{|t} S_{rm} + T^{rs} S_{rm|t}.$$

At present, we do not know whether it is true or not. However, we recognize that each side is a tensor, and the equation must be true in all coordinate systems, if it is true in one. Let us fix our attention on some arbitrary point O. Take Riemannian coordinates with O as origin. At O, the tentative equation 2.619 becomes a statement connecting partial derivatives instead of covariant derivatives. We know this statement to be true from elementary calculus. Therefore the general statement 2.619 is true.

The above argument is typical of the way in which Riemannian coordinates may be used to avoid a great deal of tedious calculation with Christoffel symbols.

Exercise. Prove that if a pair of vectors are unit orthogonal vectors at a point on a curve, and if they are both propagated parallelly along the curve, then they remain unit orthogonal vectors along the curve.

Exercise. Given that λ^r is a unit vector field, prove that

$$\lambda^r{}_{|s}\lambda_r = 0 \text{ and } \lambda^r\lambda_{r|s} = 0.$$

Is the relation $\lambda^r{}_{|s}\lambda^s = 0$ true for a general unit vector field?

We proceed now to another special coordinate system—*normal coordinates* or *orthogonal trajectory coordinates*. Riemannian coordinates have certain analogies with Cartesian coordinates; normal coordinates have some resemblance to spherical polars, but neither analogy is to be pressed too far.

Consider a singly infinite family of surfaces, i.e. subspaces of $(N - 1)$ dimensions, in a Riemannian N-space. Their equations may be written

2.620. $$x^r = f^r(u^1, u^2, \ldots, u^{N-1}; C),$$

where the u's are parameters defining the position of a point on a particular surface, and C is a parameter which is constant over each surface. If we eliminate the u's and solve for C we get a single equation of the form

2.621. $$F(x^1, x^2, \ldots, x^N) = C.$$

This single equation represents the whole family of surfaces, the value of the constant C determining the particular surface.

For an infinitesimal displacement in one of these surfaces we have

2.622. $$\frac{\partial F}{\partial x^n} dx^n = 0.$$

If we define

2.623. $$X_n = \frac{\partial F}{\partial x^n}, \quad X^m = a^{mn} \frac{\partial F}{\partial x^n},$$

the equation 2.622 may be written

2.624. $$a_{mn} X^m dx^n = 0.$$

According to 2.317, this expresses the orthogonality of the vector X^r and *any* infinitesimal displacement in the surface. In fact, the vector X^r (defined in 2.623) is a *normal vector* to the surface.

We now seek the *orthogonal trajectories* of the family of surfaces 2.621, i.e. a family of curves cutting the family of surfaces orthogonally. At each point an infinitesimal displacement dx^r along such a curve must have the direction of X^r. Thus it is a question of solving the ordinary differential equations

2.625.
$$\frac{dx^r}{dx^N} = \frac{X^r}{X^N}$$

for $x^1, \ldots, x^{N-1}$ in terms of the remaining coordinate x^N. Under very general conditions these equations will have a solution with a sufficient number of constants of integration to give one curve through each point of space. In brief, *a singly infinite family of surfaces in Riemannian N-space possesses a family of orthogonal trajectories.*

In the above argument the coordinate system was general. We shall now see how a normal system of coordinates is defined. We start with a singly infinite family of surfaces. We define x^N to be a parameter which is constant over each surface. On *one* of the surfaces we set up a coordinate system $x^1, \ldots, x^{N-1}$. We shall use the convention that Greek suffixes have the range 1 to $N - 1$; thus we refer to these coordinates as x^ρ.

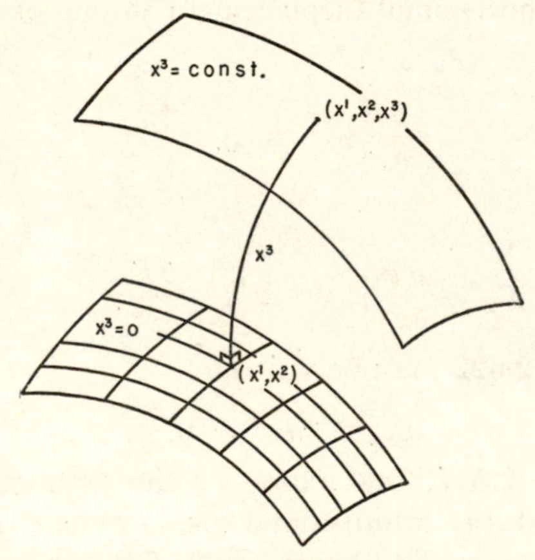

FIG. 3. Normal co-ordinates in 3-space.

The coordinates of any point P in space are now assigned in the following way. The coordinate x^N has the value belonging to the surface passing through P. We pass along the orthogonal trajectory which passes through P until we meet the surface on which x^ρ are assigned. We attach to P the values of x^ρ corresponding to the intersection of the orthogonal trajectory with this surface (Fig. 3).

Consider now two vectors at any point. One vector X^r is tangent to the orthogonal trajectory and the other Y^r is tangent to the surface of the family. Since the coordinates x^ρ remain constant as we pass along the orthogonal trajectory, we have

2.626. $$X^\rho = 0.$$

Since x^N remains constant as we move in one of the surfaces, we have

2.627. $$Y^N = 0.$$

But from the basic property of orthogonal trajectories, these two vectors are orthogonal, so that $a_{mn}X^m Y^n = 0$. By virtue of 2.626 and 2.627 this reduces to

2.628. $$a_{N\rho}X^N Y^\rho = 0.$$

Since X^N does not vanish and Y^ρ are arbitrary, we deduce that

2.629. $$a_{N\rho} = 0.$$

This is the characteristic property of normal coordinate systems. The metric form is

2.630. $$\Phi = a_{\mu\nu}dx^\mu dx^\nu + a_{NN}(dx^N)^2$$

Exercise. Deduce from 2.629 that

2.631. $$a^{N\rho} = 0, \ a^{NN} = 1/a_{NN}.$$

On any one of the surfaces (which is itself a V_{N-1}) the metric form is

2.632. $$\Phi' = a_{\mu\nu}dx^\mu dx^\nu.$$

The coefficients are functions of the coordinates x^ρ and also contain x^N as a parameter. Since the surface is a Riemannian space, it will have its own tensor technique, which is related to the tensor technique of the parent N-space but is not to be confused with it. In the tensor calculus of the surface we consider only transformations of the coordinates x^ρ, instead of the whole set x^r.

If we write down the symbol $a^{\rho\sigma}$, what do we mean by it?

Either it stands for the conjugate of the tensor $a_{\rho\sigma}$ in V_{N-1}, or it stands for some of the components of the conjugate of the tensor a_{rs} in V_N. If these two things were different from one another, the notation would be confusing. Actually they are the same, as we shall now see.

To make it quite clear what we are doing, let $a'^{\rho\sigma}$ be the conjugate of $a_{\rho\sigma}$ in V_{N-1}, and $a^{\rho\sigma}$ be some of the components of the conjugate of a_{rs} in V_N. Now, as in 2.204,

2.633. $$a_{mr}a^{ms} = \delta^s_r,$$

and if we take r and s in the range $1, \ldots, N-1$, this gives

2.634. $$a_{m\rho}a^{m\sigma} = \delta^\sigma_\rho,$$

or, by 2.629,

2.635. $$a_{\mu\rho}a^{\mu\sigma} = \delta^\sigma_\rho.$$

These equations determine $a^{\rho\sigma}$. But

2.636. $$a_{\mu\rho}a'^{\mu\sigma} = \delta^\sigma_\rho,$$

and these equations determine $a'^{\rho\sigma}$. Hence

2.637. $$a'^{\rho\sigma} = a^{\rho\sigma},$$

which is the required result.

The next question is this: Do the Christoffel symbols of V_{N-1} coincide in value with the Christoffel symbols of V_N when the indices of the latter lie in the range $1, \ldots, N-1$? It is easy to see that they do; we have (indicating the symbols for V_{N-1} with a prime)

2.638. $$[\mu\nu, \rho]' = [\mu\nu, \rho], \quad \left\{ \begin{matrix} \rho \\ \mu\nu \end{matrix} \right\}' = \left\{ \begin{matrix} \rho \\ \mu\nu \end{matrix} \right\}.$$

As for the other Christoffel symbols in V_N, we have

2.639. $$[\rho N, \sigma] = [\sigma N, \rho] = -[\rho\sigma, N] = \tfrac{1}{2}\frac{\partial a_{\rho\sigma}}{\partial x^N},$$

$$[\rho N, N] = -[NN, \rho] = \tfrac{1}{2}\frac{\partial a_{NN}}{\partial x^\rho}, \quad [NN, N] = \tfrac{1}{2}\frac{\partial a_{NN}}{\partial x^N},$$

$$\left\{ \begin{matrix} \rho \\ N\sigma \end{matrix} \right\} = \tfrac{1}{2} a^{\rho\mu} \frac{\partial a_{\mu\sigma}}{\partial x^N}, \quad \left\{ \begin{matrix} N \\ \rho\ \sigma \end{matrix} \right\} = -\frac{1}{2a_{NN}} \frac{\partial a_{\rho\sigma}}{\partial x^N},$$

$$\left\{ \begin{matrix} N \\ N\rho \end{matrix} \right\} = \frac{1}{2a_{NN}} \frac{\partial a_{NN}}{\partial x^\rho}, \quad \left\{ \begin{matrix} \rho \\ N\ N \end{matrix} \right\} = -\tfrac{1}{2} a^{\rho\mu} \frac{\partial a_{NN}}{\partial x^\mu},$$

$$\left\{ \begin{matrix} N \\ N\ N \end{matrix} \right\} = \frac{1}{2a_{NN}} \frac{\partial a_{NN}}{\partial x^N}.$$

Consider a set of quantities $T_{\alpha\beta}$ which transform according to the tensor law

2.640.
$$T'_{\alpha\beta} = T_{\mu\nu} \frac{\partial x^\mu}{\partial x'^\alpha} \frac{\partial x^\nu}{\partial x'^\beta}$$

if we transform the coordinates $x^1, x^2, \ldots, x^{N-1}$, without changing x^N. We shall refer to the set of quantities $T_{\alpha\beta}$ as the components of a *subtensor*. Since the transformation is of the form

2.641.
$$x'^\alpha = f^{(\alpha)}(x^1, \ldots, x^{N-1}), \qquad x'^N = x^N,$$

we have

2.642.
$$\frac{\partial x'^\alpha}{\partial x^N} = 0, \quad \frac{\partial x'^N}{\partial x^\alpha} = 0, \quad \frac{\partial x'^N}{\partial x^N} = 1,$$
$$\frac{\partial x^\alpha}{\partial x'^N} = 0, \quad \frac{\partial x^N}{\partial x'^\alpha} = 0, \quad \frac{\partial x^N}{\partial x'^N} = 1.$$

It follows that if we split up a tensor T_{rs} of V_N into the groups of components

$$T_{\rho\sigma},\ T_{\rho N},\ T_{N\rho},\ T_{NN},$$

then, for the transformations 2.641,

$T_{\rho\sigma}$ is a covariant subtensor of the second order,
$T_{\rho N}$, $T_{N\rho}$ are covariant subvectors,
T_{NN} is a subinvariant.

Let us look into the covariant differentiation of subtensors with respect to the metric form $a_{\mu\nu}dx^\mu dx^\nu$. We shall denote this type of differentiation with a double stroke. We shall illustrate by examining the covariant derivative of a covariant subvector. Then, remembering 2.638,

$$\textbf{2.643.} \qquad T_{\alpha\|\beta} = \frac{\partial T_\alpha}{\partial x^\beta} - \left\{ \begin{matrix} \gamma \\ \alpha\ \beta \end{matrix} \right\} T_\gamma.$$

Now

$$\textbf{2.644.} \qquad \begin{aligned} T_{\alpha|\beta} &= \frac{\partial T_\alpha}{\partial x^\beta} - \left\{ \begin{matrix} m \\ \alpha\ \beta \end{matrix} \right\} T_m \\ &= T_{\alpha\|\beta} - \left\{ \begin{matrix} N \\ \alpha\ \beta \end{matrix} \right\} T_N, \end{aligned}$$

and so, by 2.639,

$$\textbf{2.645.} \qquad \begin{aligned} T_{\alpha|\beta} &= T_{\alpha\|\beta} + \frac{1}{2a_{NN}} \frac{\partial a_{\alpha\beta}}{\partial x^N} T_N \\ &= T_{\alpha\|\beta} + \tfrac{1}{2} \frac{\partial a_{\alpha\beta}}{\partial x^N} T^N. \end{aligned}$$

We also have

$$\textbf{2.646.} \qquad \begin{aligned} T_{N|\alpha} &= \frac{\partial T_N}{\partial x^\alpha} - \left\{ \begin{matrix} m \\ N\ \alpha \end{matrix} \right\} T_m \\ &= \frac{\partial T_N}{\partial x^\alpha} - \left\{ \begin{matrix} \beta \\ N\ \alpha \end{matrix} \right\} T_\beta - \left\{ \begin{matrix} N \\ N\ \alpha \end{matrix} \right\} T_N \\ &= \frac{\partial T_N}{\partial x^\alpha} - \tfrac{1}{2} a^{\beta\mu} \frac{\partial a_{\mu\alpha}}{\partial x^N} T_\beta - \frac{1}{2a_{NN}} \frac{\partial a_{NN}}{\partial x^\alpha} T_N \\ &= \frac{\partial T_N}{\partial x^\alpha} - \tfrac{1}{2} \frac{\partial a_{\mu\alpha}}{\partial x^N} T^\mu - \tfrac{1}{2} \frac{\partial a_{NN}}{\partial x^\alpha} T^N, \end{aligned}$$

and

$$\textbf{2.647.} \qquad \begin{aligned} T_{\alpha|N} &= \frac{\partial T_\alpha}{\partial x^N} - \left\{ \begin{matrix} m \\ \alpha\ N \end{matrix} \right\} T_m \\ &= \frac{\partial T_\alpha}{\partial x^N} - \tfrac{1}{2} \frac{\partial a_{\alpha\mu}}{\partial x^N} T^\mu - \tfrac{1}{2} \frac{\partial a_{NN}}{\partial x^\alpha} T^N. \end{aligned}$$

Exercise. Show that

$$\text{2.648.} \qquad T^{\alpha}{}_{|\beta} = T^{\alpha}{}_{||\beta} + \tfrac{1}{2}a^{\alpha\mu}\frac{\partial a_{\mu\beta}}{\partial x^N}T^N,$$

$$\text{2.649.} \qquad T^{N}{}_{|\alpha} = \frac{\partial T^N}{\partial x^\alpha} - \frac{1}{2a_{NN}}\frac{\partial a_{\alpha\mu}}{\partial x^N}T^\mu + \frac{1}{2a_{NN}}\frac{\partial a_{NN}}{\partial x^\alpha}T^N,$$

$$\text{2.650.} \qquad T^{\alpha}{}_{|N} = \frac{\partial T^\alpha}{\partial x^N} + \tfrac{1}{2}\,a^{\alpha\mu}\frac{\partial a_{\mu\sigma}}{\partial x^N}T^\sigma - \tfrac{1}{2}\,a^{\alpha\mu}\frac{\partial a_{NN}}{\partial x^\mu}T^N.$$

The above relations hold for a normal coordinate system. There is a special type of normal coordinate system, namely, a *geodesic normal coordinate system.* This will be described below, but first we must establish an important property of geodesics.

Let us take a surface V_{N-1} in V_N, and draw the geodesics normal to V_{N-1}. Let us measure off along all these geodesics the same length. The points so obtained give another surface

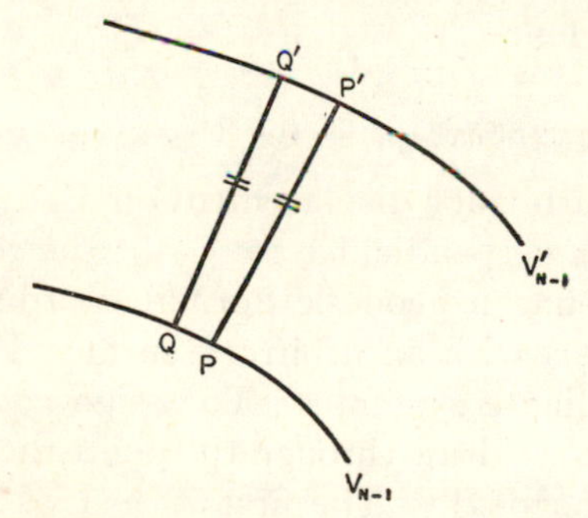

FIG. 4. Construction of geodesic normal coordinate system.

V'_{N-1} (Fig. 4). The essential fact we wish to prove is this: *All the geodesics normal to* V_{N-1} *are also normal to* V'_{N-1}.

Let P, Q be adjacent points on V_{N-1} and let P', Q' be the points where the geodesics through P, Q cut V'_{N-1}. Let us look back to the formula 2.410. It is a formula for the variation in the length of a curve when the curve receives a small displacement. When it was obtained in 2.4 we had in mind curves with common ends points, but actually this condition was not used in obtaining the formula in question. In applying the formula to the case now under consideration, we put $\delta L = 0$, since $PP' = QQ'$ by construction. Further, the integral on the right hand side vanishes because PP' is a geodesic. Thus we have

2.651. $$\left[\frac{\partial}{\partial p^n} (\epsilon w)^{\frac{1}{2}} \delta x^n \right]_{u_1}^{u_2} = 0,$$

where δx^r at $u = u_1$ represents the infinitesimal displacement PQ and δx^r at $u = u_2$ represents the infinitesimal displacement $P'Q'$. Let us take the parameter u equal to the arc length along the geodesics. Then 2.415 hold, and 2.651 reduces to

2.652. $$\left[a_{mn} p^m \, \delta x^n \right]_P^{P'} = 0.$$

Here p^m is the unit tangent vector to the geodesic. But the contribution from P vanishes, since the geodesics cut V_{N-1} orthogonally. Hence

2.653. $$(a_{mn} p^m \delta x^n)_{P'} = 0.$$

Since δx^r is an arbitrary displacement on V'_{N-1}, it follows that the vector p^r is perpendicular to V'_{N-1}; the result is proved.

We now define a geodesic normal coordinate system as follows. We start with an arbitrary surface V_{N-1} and assign over it a coordinate system x^ρ. To assign coordinates to an arbitrary point, we draw through the point the geodesic which cuts V_{N-1} orthogonally. The first $N - 1$ of the coordinates are defined to be the coordinates of the point where this geodesic meets V_{N-1}; the last cordinate x^N is the arc length of the geodesic.

It is clear that the geodesic normal coordinate system is a special case of the normal coordinate system, and so all the relations given above hold good. But there are other simplifications. As we go along a parametric line* of x^N we have

*A *parametric line* is a curve along which only one coordinate changes.

2.654. $$ds^2 = \epsilon_N a_{NN}(dx^N)^2,$$

where ϵ_N is the indicator of the parametric line. But $dx^N = ds$, by the definition of x^N. Hence

2.655. $$a_{NN} = \epsilon_N, \quad a^{NN} = \epsilon_N.$$

These equations are additional to 2.629 and 2.631. Equations 2.639 hold, of course, for a geodesic normal system, but there are some obvious simplifications since the derivatives of a_{NN} vanish.

For a geodesic normal coordinate system, the metric form is

2.656. $$\Phi = a_{\mu\nu}dx^\mu dx^\nu + \epsilon_N(dx^N)^2.$$

Exercise. Write down equations 2.643 to 2.650 for the special case of a geodesic normal coordinate system.

The special types of coordinates considered above—local Cartesians, Riemannian coordinates, normal coordinates, and geodesic normal coordinates—all exist in a general Riemannian space of N dimensions. Let us now consider *orthogonal coordinates* for which the parametric lines of the coordinates are perpendicular to one another at every point. For an infinitesimal displacement $d_{(1)}x^r$ along the parametric line of x^1, we have

$$d_{(1)}x^2 = d_{(1)}x^3 = \ldots = d_{(1)}x^N = 0,$$

and for an infinitesimal displacement $d_{(2)}x^r$ along the parametric line of x^2, we have

$$d_{(2)}x^1 = d_{(2)}x^3 = \ldots = d_{(2)}x^N = 0.$$

The condition of orthogonality

$$a_{mn}d_{(1)}x^m d_{(2)}x^n = 0$$

reduces to

$$a_{12}d_{(1)}x^1 d_{(2)}x^2 = 0,$$

and so $a_{12} = 0$. Taking into consideration the other pairs of parametric lines, we see that *the coordinate system is orthogonal if, and only if,*

2.657. $$a_{mn} = 0 \text{ for } m \neq n.$$

The corresponding metric form is

2.658. $$\Phi = \epsilon_1(h_1 dx^1)^2 + \epsilon_2(h_2 dx^2)^2 + \ldots + \epsilon_N(h_N dx^N)^2,$$

where the ϵ's are $+1$ or -1 and the h's are functions of the coordinates.

Orthogonal coordinates do not exist in a general Riemannian space of N dimensions, and consequently they cannot be used in general arguments. They do exist, however, in special types of space, and in particular in Euclidean 3-space, where polar coordinates, confocal coordinates, and other orthogonal systems are familiar. They also exist in special types of Riemannian 4-space occurring in the general theory of relativity. In such cases they have distinct advantages when detailed computations are required, since the conditions 2.657 reduce the number of components of a_{mn} from $\frac{1}{2}N(N+1)$ to N, so that in V_4 we have only 4 components instead of 10.

2.7. Frenet formulae. With any point on a twisted curve in Euclidean 3-space there is associated an orthogonal triad consisting of the tangent, principal normal, and binormal, and two numbers, the curvature and the torsion. We shall now extend these ideas to the case of a curve in a Riemannian space of N dimensions.*

Let

2.701. $$x^r = x^r(s)$$

be the equation of the curve, s being the arc length measured from some point on the curve. Then

2.702. $$\lambda^r = \frac{dx^r}{ds}$$

is the *unit tangent vector*, and we have

*We shall use an indefinite metric form, but we shall for simplicity exclude the possibility that any of the vectors encountered has a null-direction. We shall also exclude the possibility that any of the curvatures vanishes. However, these possibilities will be discussed after the general formulae have been set up.

2.703. $$a_{mn}\lambda^m\lambda^n = \epsilon,$$

where ϵ is the indicator of the direction of the tangent. Differentiation gives, by 2.530,

2.704. $$\lambda_n \frac{\delta\lambda^n}{\delta s} = 0,$$

which shows that $\delta\lambda^r/\delta s$ is perpendicular to the tangent. The unit vector $\lambda_{(1)}{}^r$ codirectional with $\delta\lambda^r/\delta s$ is called the *unit first normal*, and the magnitude of $\delta\lambda^r/\delta s$ is called the *first curvature* $\kappa_{(1)}$. Thus we have

2.705. $$\frac{\delta\lambda^r}{\delta s} = \kappa_{(1)}\lambda_{(1)}{}^r, \quad \epsilon_{(1)}\lambda_{(1)n}\lambda_{(1)}{}^n = 1,$$

where $\epsilon_{(1)}$ is the indicator of $\lambda_{(1)}{}^r$. This is the first of the Frenet formulae.

Let us now define a unit vector $\lambda_{(2)}{}^r$ and a positive invariant $\kappa_{(2)}$ by the equations

2.706. $$\frac{\delta\lambda_{(1)}{}^r}{\delta s} = \kappa_{(2)}\lambda_{(2)}{}^r - \epsilon\epsilon_{(1)}\kappa_{(1)}\lambda^r, \quad \epsilon_{(2)}\lambda_{(2)n}\lambda_{(2)}{}^n = 1,$$

where $\epsilon_{(2)}$ is the indicator of $\lambda_{(2)}{}^r$. This vector $\lambda_{(2)}{}^r$ is perpendicular to both λ^r and $\lambda_{(1)}{}^r$; for we have, by 2.706, on multiplication by λ^r, $\lambda_{(1)}{}^r$, respectively,

2.707. $$\kappa_{(2)}\lambda_{(2)}{}^n\lambda_n = \frac{\delta\lambda_{(1)}{}^n}{\delta s}\lambda_n + \epsilon_{(1)}\kappa_{(1)}, \quad \kappa_{(2)}\lambda_{(2)}{}^n\lambda_{(1)n} = 0,$$

since $\lambda_{(1)}{}^r$ is a unit vector perpendicular to λ^r. In the first of these equations we may put

2.708. $$\frac{\delta\lambda_{(1)}{}^n}{\delta s}\lambda_n = -\lambda_{(1)}{}^n\frac{\delta\lambda_n}{\delta s} = -\epsilon_{(1)}\kappa_{(1)},$$

so that the stated perpendicularities follow. The vector $\lambda_{(2)}{}^r$ is called the *unit second normal* and $\kappa_{(2)}$ is called the *second curvature*.*

Let us next define a unit vector $\lambda_{(3)}{}^r$ and a positive invariant $\kappa_{(3)}$ by the equations

*Sometimes called *torsion*.

$$2.709. \quad \frac{\delta\lambda_{(2)}{}^r}{\delta s} = \kappa_{(3)}\lambda_{(3)}{}^r - \epsilon_{(1)}\epsilon_{(2)}\kappa_{(2)}\lambda_{(1)}{}^r, \quad \epsilon_{(3)}\lambda_{(3)n}\lambda_{(3)}{}^n = 1,$$

where $\epsilon_{(3)}$ is the indicator of $\lambda_{(3)}{}^r$. It is easily proved in the same way that $\lambda_{(3)}{}^r$ is perpendicular to all three vectors λ^r, $\lambda_{(1)}{}^r$, and $\lambda_{(2)}{}^r$; it is called the *unit third normal* and $\kappa_{(3)}$ is called the *third curvature.*

Consider the sequence of formulae

$$2.710. \quad \frac{\delta\lambda_{(M-1)}{}^r}{\delta s} = \kappa_{(M)}\lambda_{(M)}{}^r - \epsilon_{(M-2)}\epsilon_{(M-1)}\kappa_{(M-1)}\lambda_{(M-2)}{}^r,$$
$$\lambda_{(0)}{}^r = \lambda^r, \ \kappa_{(0)} = 0, \ \epsilon_{(0)} = \epsilon,$$
$$\epsilon_{(M-1)}\lambda_{(M-1)}{}^n\lambda_{(M-1)n} = 1,$$
$$(M = 1, 2, \ldots)$$

in which there is no summation with respect to repeated capital letters. These formulae for $M = 1$, 2, and 3 coincide respectively with 2.705, 2.706, and 2.709. It is easily proved by mathematical induction that the whole sequence of vectors defined by 2.710 are perpendicular to the tangent and to one another. The vectors $\lambda_{(1)}{}^r, \lambda_{(2)}{}^r, \ldots$, are the unit first, second, ..., normals, and the invariants $\kappa_{(1)}, \kappa_{(2)}, \ldots$, are the first, second, ..., curvatures of the curve.

It would appear at first sight that the formula 2.710 defines an infinite sequence of vectors. This cannot be the case, however, because at a point in V_N we can draw only N mutually perpendicular vectors. Therefore 2.710 with $M = N$ cannot yield a vector $\lambda_{(N)}{}^r$ with non-zero components. In fact, we must have $\kappa_{(N)} = 0$, so that for $M = N$ 2.710 reads

$$2.711. \quad \frac{\delta\lambda_{(N-1)}{}^r}{\delta s} = -\epsilon_{(N-2)}\epsilon_{(N-1)}\kappa_{(N-1)}\lambda_{(N-2)}{}^r.$$

This equation terminates the sequence.

The complete set of Frenet formulae may be written

$$2.712. \quad \frac{\delta\lambda_{(M-1)}{}^r}{\delta s} = \kappa_{(M)}\lambda_{(M)}{}^r - \epsilon_{(M-2)}\epsilon_{(M-1)}\kappa_{(M-1)}\lambda_{(M-2)}{}^r,$$
$$\epsilon_{(M-1)}\lambda_{(M-1)}{}^n\lambda_{(M-1)n} = 1, \ (M = 1, 2, \ldots, N),$$

where we define

2.713. $$\lambda_{(0)}{}^r = \lambda^r, \quad \kappa_{(0)} = 0, \quad \kappa_{(N)} = 0.$$

Exercise. For positive definite metric forms, write out explicitly the Frenet formulae for the cases $N = 2, N = 3, N=4$.

In the above discussion it was assumed that the successive curvatures could be determined with non-zero values. But there is no reason to suppose that this will necessarily be the case. The equation 2.705 gives

2.714. $$\kappa_{(1)}{}^2 = \epsilon_{(1)}\, a_{mn} \frac{\delta\lambda^m}{\delta s} \frac{\delta\lambda^n}{\delta s}, \quad \epsilon_{(1)} = \pm 1.$$

If $\delta\lambda^r/\delta s = 0$ (which means that the curve is a geodesic), then $\kappa_{(1)} = 0$. Even though the separate components do not all vanish, we shall have $\kappa_{(1)} = 0$ if $\delta\lambda^r/\delta s$ is a null-vector. (This can occur only with an indefinite metric form.) In either case, 2.705 fails to define the normal vector $\lambda_{(1)}{}^r$, and so the procedure breaks down. If we get past the first stage with $\kappa_{(1)} > 0$, it may happen that $\kappa_{(2)} = 0$, and the procedure breaks down at the second stage. In such cases we have a truncated set of Frenet formulae. Let us illustrate by the case where $\kappa_{(1)} > 0$, $\kappa_{(2)} > 0$, $\kappa_{(3)} = 0$. Then the Frenet formulae are

2.715. $$\begin{aligned} \frac{\delta\lambda^r}{\delta s} &= \kappa_{(1)}\lambda_{(1)}{}^r, \\ \frac{\delta\lambda_{(1)}{}^r}{\delta s} &= \kappa_{(2)}\lambda_{(2)}{}^r - \epsilon\epsilon_{(1)}\kappa_{(1)}\lambda^r, \\ \frac{\delta\lambda_{(2)}{}^r}{\delta s} &= \qquad\qquad - \epsilon_{(1)}\epsilon_{(2)}\kappa_{(2)}\lambda_{(1)}{}^r. \end{aligned}$$

Other normal directions remain undefined, but we can fill in a complete set of mutually perpendicular vectors by taking at a point on the curve unit vectors $\lambda_{(3)}{}^r, \ldots, \lambda_{(N-1)}{}^r$ mutually perpendicular and perpendicular to λ^r, $\lambda_{(1)}{}^r$ and $\lambda_{(2)}{}^r$. If we subject $\lambda_{(3)}{}^r, \ldots, \lambda_{(N-1)}{}^r$ to parallel propagation along the curve, the set of $N - 1$ vectors remain unit vectors, mutually perpendicular and perpendicular to the vectors λ^r, $\lambda_{(1)}{}^r$, $\lambda_{(2)}{}^r$. The formulae 2.712 are satisfied with

$$\kappa_{(3)} = \kappa_{(4)} = \ldots = \kappa_{(N-1)} = 0.$$

Exercise. In a Euclidean space V_N, the fundamental form is given as $\Phi = dx^n dx^n$. Show that a curve which has $\kappa_{(2)} = 0$ and $\kappa_{(1)} =$ constant satisfies equations of the form

$$x^r = A^r \cos \kappa_{(1)} s + B^r \sin \kappa_{(1)} s + C^r,$$

where A^r, B^r, C^r are constants, satisfying

$$A^r A^r = B^r B^r = \frac{1}{\kappa_{(1)}{}^2}, \; A^r B^r = 0,$$

so that A^r and B^r are vectors of equal magnitude and perpendicular to one another. (This curve is a circle in the N-space.)

SUMMARY II

Metric form:

$$\Phi = a_{mn} dx^m dx^n, \; ds^2 = \epsilon \, a_{mn} dx^m dx^n.$$

Metric tensor and conjugate tensor:

$$a_{mn} = a_{nm}, \quad a^{mn} = a^{nm}, \quad a_{mr} a^{ms} = \delta^s_r.$$

Magnitude of vector:

$$X^2 = \epsilon \, a_{mn} X^m X^n.$$

Condition of orthogonality:

$$a_{mn} X^m Y^n = 0.$$

Raising and lowering suffixes:

$$X^m = a^{mn} X_n, \quad X_m = a_{mn} X^n.$$

Christoffel symbols:

$$[mn, r] = \tfrac{1}{2}\left(\frac{\partial a_{mr}}{\partial x^n} + \frac{\partial a_{nr}}{\partial x^m} - \frac{\partial a_{mn}}{\partial x^r}\right), \; \left\{ {r \atop mn} \right\} = a^{rs}[mn, s].$$

Geodesic:

$$\delta \int ds = 0,$$

$$\frac{d}{ds}\frac{\partial w}{\partial p^r} - \frac{\partial w}{\partial x^r} = 0, \; p^r = \frac{dx^r}{ds}, \; w = a_{mn} p^m p^n,$$

$$\frac{d^2x^r}{ds^2} + \left\{ {r \atop mn} \right\} \frac{dx^m}{ds}\frac{dx^n}{ds} = 0, \quad a_{mn}\frac{dx^m}{ds}\frac{dx^n}{ds} = \epsilon.$$

Geodesic null line:

$$\frac{d^2x^r}{du^2} + \left\{ {r \atop mn} \right\} \frac{dx^m}{du}\frac{dx^n}{du} = 0, \quad a_{mn}\frac{dx^m}{du}\frac{dx^n}{du} = 0.$$

Absolute derivative:

$$\frac{\delta T^r}{\delta u} = \frac{dT^r}{du} + \left\{ {r \atop mn} \right\} T^m \frac{dx^n}{du},$$

$$\frac{\delta T_r}{\delta u} = \frac{dT_r}{du} - \left\{ {m \atop rn} \right\} T_m \frac{dx^n}{du}.$$

Covariant derivative:

$$T^r{}_{|s} = \frac{\partial T^r}{\partial x^s} + \left\{ {r \atop ms} \right\} T^m, \quad T_{r|s} = \frac{\partial T_r}{\partial x^s} - \left\{ {m \atop rs} \right\} T_m,$$

$$a_{rs|t} = 0, \quad a^{rs}{}_{|t} = 0, \quad \delta^r_{s|t} = 0.$$

Local Cartesians:

$$\Phi = \epsilon_1 (dy_1)^2 + \epsilon_2 (dy_2)^2 + \ldots + \epsilon_N (dy_N)^2 \text{ at origin.}$$

Riemannian coordinates:

$$[mn, r] = 0, \quad \left\{ {r \atop mn} \right\} = 0, \quad \frac{\partial a_{mn}}{\partial x^r} = 0 \text{ at origin.}$$

Normal coordinate system:

$$\Phi = a_{\mu\nu} dx^\mu dx^\nu + a_{NN} (dx^N)^2.$$

Geodesic normal coordinate system:

$$\Phi = a_{\mu\nu} dx^\mu dx^\nu + \epsilon_N (dx^N)^2.$$

First normal $\lambda_{(1)}{}^r$ and first curvature $\kappa_{(1)}$ of a curve:

$$\frac{dx^r}{ds} = \lambda_{(0)}{}^r, \quad \frac{\delta \lambda_{(0)}{}^r}{\delta s} = \kappa_{(1)} \lambda_{(1)}{}^r, \quad \epsilon_{(1)} \lambda_{(1)n} \lambda_{(1)}{}^n = 1.$$

EXERCISES II

1. For cylindrical coordinates in Euclidean 3-space, write down the metric form by inspection of a diagram showing a general infinitesimal displacement, and calculate all the Christoffel symbols of both kinds.

2. If a_{rs} and b_{rs} are covariant tensors, show that the roots of the determinantal equation

$$|Xa_{rs} - b_{rs}| = 0$$

are invariants.

3. Is the form

$$dx^2 + 3dx\,dy + 4dy^2 + dz^2$$

positive-definite?

4. If X^r, Y^r are unit vectors inclined at an angle θ, prove that

$$\sin^2\theta = (a_{rm}a_{sn} - a_{rs}a_{mn})X^r Y^s X^m Y^n.$$

5. Show that, if θ is the angle between the normals to the surfaces $x^1 =$ const., $x^2 =$ const., then

$$\cos\theta = \frac{a^{12}}{\sqrt{a^{11}a^{22}}}.$$

6. Let x^1, x^2, x^3 be rectangular Cartesian coordinates in Euclidean 3-space, and let x^1, x^2 be taken as coordinates on a surface $x^3 = f(x^1, x^2)$. Show that the Christoffel symbols of the second kind for the surface are

$$\left\{\begin{matrix} r \\ mn \end{matrix}\right\} = \frac{f_r f_{mn}}{1 + f_p f_p},$$

the suffixes taking the values 1, 2, and the subscripts indicating partial derivatives.

7. Write down the differential equations of the geodesics on a sphere, using colatitude θ and azimuth ϕ as coordinates. Integrate the differential equations and obtain a finite equation

$$A\sin\theta\cos\phi + B\sin\theta\sin\phi + C\cos\theta = 0,$$

where A, B, C are arbitrary constants.

8. Find in integrated form the geodesic null lines in a V_3 for which the metric form is

$$(dx^1)^2 - R^2[(dx^2)^2 + (dx^3)^2],$$

R being a function of x^1 only.

9. Show that, for a normal coordinate system, the Christoffel symbols

$$[\rho N, \sigma],\ [\rho\sigma, N],\ [\rho N, N],\ [NN, N]$$
$$\left\{\begin{matrix}\rho\\ N\sigma\end{matrix}\right\}, \left\{\begin{matrix}N\\ \rho\sigma\end{matrix}\right\}, \left\{\begin{matrix}\rho\\ NN\end{matrix}\right\}, \left\{\begin{matrix}N\\ N\rho\end{matrix}\right\}, \left\{\begin{matrix}N\\ NN\end{matrix}\right\},$$

have tensor character with respect to transformations of the coordinates $x^1, \ldots, x^{N-1}$.

10. If θ, ϕ are colatitude and azimuth on a sphere, and we take

$$x^1 = \theta \cos \phi,\ x^2 = \theta \sin \phi,$$

calculate all the Christoffel symbols for the coordinate system x^1, x^2 and show that they vanish at the point $\theta = 0$.

11. If vectors T^r and S_r undergo parallel propagation along a curve, show that $T^n S_n$ is constant along that curve.

12. Deduce from 2.201 that the determinant $a = |a_{mn}|$ transforms according to

$$a' = aJ^2,\ J = \left|\frac{\partial x^r}{\partial x'^s}\right|.$$

13. Using local Cartesians and applying the result of the previous exercise (No. 12), prove that, if the metric form is positive-definite, then the determinant $a = |a_{mn}|$ is always positive.

14. In a plane, let x^1, x^2 be the distances of a general point from the points with rectangular coordinates (1, 0), (-1, 0), respectively. (These are bipolar coordinates.) Find the line element for these coordinates, and find the conjugate tensor a^{mn}.

15. Given $\Phi = a_{mn}dx^m dx^n$, with $a_{11} = a_{22} = 0$ but $a_{12} \neq 0$, show that Φ may be written in the form

$$\Phi = \epsilon\Psi_1^2 - \epsilon\Psi_2^2 + \Phi_2,$$

where Φ_2 is a homogeneous quadratic form in $dx^3, dx^4, \ldots, dx^N$, where $\epsilon = \pm 1$ such that $\epsilon a_{12} > 0$, and where

$$\Psi_1 = (2\epsilon a_{12})^{-\frac{1}{2}} [a_{12}(dx^1 + dx^2) + (a_{13} + a_{23})\, dx^3 + \ldots + (a_{1N} + a_{2N})dx^N],$$

$$\Psi_2 = (2\epsilon a_{12})^{-\frac{1}{2}} [a_{12}(-dx^1 + dx^2) + (a_{13} - a_{23})\, dx^3 + \ldots + (a_{1N} - a_{2N})\, dx^N].$$

16. Find the null geodesics of a 4-space with line element

$$ds^2 = \gamma(dx^2 + dy^2 + dz^2 - dt^2),$$

where γ is an arbitrary function of x, y, z, t.

17. In a space V_N the metric tensor is a_{mn}. Show that the null geodesics are unchanged if the metric tensor is changed to b_{mn}, where $b_{mn} = \gamma a_{mn}$, γ being a function of the coordinates.

18. Are the relations

$$T_{|rs} = T_{|sr},$$
$$T_{r|sk} = T_{r|ks},$$

true (*a*) in curvilinear coordinates in Euclidean space, (*b*) in a general Riemannian space?

19. Consider a V_N with indefinite metric form. For all points P lying on the cone of geodesic null lines drawn from O, the definition 2.611 for Riemannian coordinates apparently breaks down. Revise the definition of Riemannian coordinates so as to include such points.

CHAPTER III

CURVATURE OF SPACE

3.1. The curvature tensor. The idea of curvature is simple and familiar in Euclidean geometry. A line is curved if it deviates from a straight line, and a surface is curved if it deviates from a plane. However, it is usually possible to discover whether or not a surface is curved by purely intrinsic operations on the surface. Let us think of the simplest of curved surfaces—a sphere. Imagine a two-dimensional being who moves in the spherical surface, and cannot perceive anything outside that surface. The operations he performs consist solely of measurements of distances (and hence angles) in the surface. From their geodesic (shortest-distance) property he can construct the great circles on his sphere. If he measures the angles of a spherical triangle formed of great circles, he finds that the sum of the angles is greater than two right angles. This result tells him that his two-dimensional region is not a plane.

This simple example illustrates our point of view in discussing the curvature of Riemannian space. Curvature is regarded as something intrinsic to the space, and not as something to be measured by comparison of the space with another space. Nor do we think of the Riemannian space as necessarily embedded in a Euclidean space, as we are tempted to do when we discuss the intrinsic geometry of a sphere.

In creating a geometrical theory of Riemannian N-space, we have to generalize our familiar concepts in two ways. First, we have to pass from three to N dimensions; secondly, we have to consider the possibility of an intrinsic curvature, such as is found in a surface drawn in Euclidean 3-space.

We define flatness as follows: *A space is said to be flat if it is possible to choose coordinates for which the metric form is**

3.101. $$\Phi = \epsilon_1(dx^1)^2 + \epsilon_2(dx^2)^2 + \ldots + \epsilon_N(dx^N)^2,$$
each ϵ *being* $+1$ *or* -1.

A space which is not flat is called *curved.*

Exercise. Explain why the surfaces of an ordinary cylinder and an ordinary cone are to be regarded as "flat" in the sense of our definition.

Our next task is to develop a test to tell us whether a given space is flat or curved. By "given space" we mean a space in which the metric form $\Phi = a_{mn}dx^m dx^n$ is given. The question is: Can we, or can we not, by transformation of coordinates reduce Φ to the form 3.101?

We attack this question by means of the formulae for covariant derivatives given in 2.5. Let T_r be an arbitrary covariant vector field. Its covariant derivative is given by 2.521:

3.102. $$T_{r|m} = \frac{\partial T_r}{\partial x^m} - \left\{ {p \atop r\,m} \right\} T_p.$$

This is a covariant tensor of the second order, and we can obtain its covariant derivative by means of the formula 2.523. In writing a second-order covariant derivative, we omit all vertical strokes except the first: thus

3.103. $$T_{r|mn} = \frac{\partial T_{r|m}}{\partial x^n} - \left\{ {q \atop r\,n} \right\} T_{q|m} - \left\{ {q \atop m\,n} \right\} T_{r|q}.$$

Interchanging m and n and subtracting, we get (since $\left\{ {q \atop m\,n} \right\} = \left\{ {q \atop n\,m} \right\}$),

*It should be clearly understood that a space V_N is flat only if it is possible to reduce Φ to the form 3.101 throughout V_N. Every space is *elementarily flat* in the sense that it is always possible to reduce Φ to the form 3.101 at a single assigned point. This was done in 2.606.

3.104. $T_{r|mn} - T_{r|nm}$

$$= \frac{\partial}{\partial x^n}\left(\frac{\partial T_r}{\partial x^m} - \left\{ {p \atop r\,m} \right\} T_p\right) - \left\{ {q \atop r\,n} \right\}\left(\frac{\partial T_q}{\partial x^m} - \left\{ {p \atop q\,m} \right\} T_p\right)$$

$$- \frac{\partial}{\partial x^m}\left(\frac{\partial T_r}{\partial x^n} - \left\{ {p \atop r\,n} \right\} T_p\right) + \left\{ {q \atop r\,m} \right\}\left(\frac{\partial T_q}{\partial x^n} - \left\{ {p \atop q\,n} \right\} T_p\right).$$

The partial derivatives of the components of T_r cancel out, and we get

3.105. $$T_{r|mn} - T_{r|nm} = R^s_{.rmn} T_s,$$

where

3.106. $$R^s_{.rmn} = \frac{\partial}{\partial x^m}\left\{ {s \atop r\,n} \right\} - \frac{\partial}{\partial x^n}\left\{ {s \atop r\,m} \right\}$$

$$+ \left\{ {p \atop r\,n} \right\}\left\{ {s \atop p\,m} \right\} - \left\{ {p \atop r\,m} \right\}\left\{ {s \atop p\,n} \right\}.$$

Now T_s is an arbitrary covariant vector and the left-hand side of 3.105 is a covariant tensor of the third order. It follows from the tests of 1.6 that $R^s_{.rmn}$ has the tensor character shown. It is called the *mixed curvature tensor*.

Exercise. What are the values of $R^s_{.rmn}$ in an Euclidean plane, the coordinates being rectangular Cartesians? Deduce the values of the components of this tensor for polar coordinates from its tensor character, or else by direct calculation.

Let us now suppose that a vector field $T^r(u, v)$ is given over a 2-space V_2 with equations $x^r = x^r(u, v)$, immersed in a Riemannian V_N. By taking absolute derivatives along the parametric lines of u and v respectively, we obtain the vector fields

$$\frac{\delta T^r}{\delta u}, \quad \frac{\delta T^r}{\delta v}.$$

These fields may in turn be differentiated absolutely, yielding fields

$$\frac{\delta^2 T^r}{\delta u^2}, \quad \frac{\delta^2 T^r}{\delta u \delta v}, \quad \frac{\delta^2 T^r}{\delta v \delta u}, \quad \frac{\delta^2 T^r}{\delta v^2}.$$

If we were dealing with the usual partial differentiation operators $(\partial/\partial u, \partial/\partial v)$, the order in which the operators are applied would be a matter of indifference, i.e.

$$\frac{\partial^2 T^r}{\partial u \partial v} = \frac{\partial^2 T^r}{\partial v \partial u}.$$

(We say that these operators *commute.*) However, the result 3.105 might lead us to suspect that the operators $(\delta/\delta u, \delta/\delta v)$ do not commute in general. A straightforward calculation, based on 2.511, leads to

3.107.
$$\frac{\delta^2 T^r}{\delta u \delta v} - \frac{\delta^2 T^r}{\delta v \delta u} = R^r_{.pmn} T^p \frac{\partial x^m}{\partial u} \frac{\partial x^n}{\partial v}.$$

Thus the operators $(\delta/\delta u, \delta/\delta v)$ do not commute in general.

The formulae 3.105 and 3.107 are important.

We are now in a position to give a partial answer to the question raised above regarding the flatness of a space. If the space is flat, then there exists a coordinate system such that 3.101 is true. But then the Christoffel symbols vanish, and hence by 3.106 all the components of the curvature tensor vanish. Hence *the conditions*

3.108.
$$R^s_{.rmn} = 0$$

are necessary for flatness. These conditions are also sufficient, but we shall postpone the proof of this to 3.5.

A word with regard to the structure of the formula 3.106 will help if we wish to remember it. The curvature tensor involves derivatives of the metric tensor up to the *second* order and the expression 3.106 starts with differentiation with respect to the *second* subscript. Noting the relative heights of the suffixes, we have then no difficulty in remembering the first term. To get the second term, interchange m and n. The first Christoffel symbol in the third term is the same as that in the first term, except that s is replaced by the dummy p; the rest of the term then fits in uniquely so as to preserve the heights of unrepeated suffixes and to show the repeated suffix once as a superscript and once as a subscript. To get the last term, interchange m and n. It is, of course, unwise to attempt to remember too many formulae, but the curvature tensor plays such an important part in tensor calculus that it is worth while paying special attention to it.

The mixed curvature tensor has some symmetry properties. First, it is skew-symmetric in the last two subscripts:

3.109.
$$R^{s}_{.rmn} = -R^{s}_{.rnm}.$$

Secondly, it has a cyclic symmetry in its subscripts:

3.110.
$$R^{s}_{.rmn} + R^{s}_{.mnr} + R^{s}_{.nrm} = 0.$$

These results are easy to show from 3.106. The work may be shortened by using Riemannian coordinates with origin at the point in V_N under consideration. Then the Christoffel symbols vanish, and

3.111.
$$R^{s}_{.rmn} = \frac{\partial}{\partial x^m}\begin{Bmatrix} s \\ r\ n \end{Bmatrix} - \frac{\partial}{\partial x^n}\begin{Bmatrix} s \\ r\ m \end{Bmatrix}.$$

This formula holds only at the origin of the Riemannian coordinates.

We may lower the superscript of the mixed curvature tensor in the usual way, and get the *covariant curvature tensor*, or *Riemann tensor*.

3.112.
$$R_{prmn} = a_{ps} R^{s}_{.rmn}.$$

A little manipulation with Christoffel symbols leads to the formulae

3.113.
$$R_{rsmn} = \frac{\partial}{\partial x^m}[sn, r] - \frac{\partial}{\partial x^n}[sm, r] + \begin{Bmatrix} p \\ s\ m \end{Bmatrix}[rn, p] - \begin{Bmatrix} p \\ s\ n \end{Bmatrix}[rm, p],$$

3.114.
$$R_{rsmn} = \tfrac{1}{2}\left(\frac{\partial^2 a_{rn}}{\partial x^s \partial x^m} + \frac{\partial^2 a_{sm}}{\partial x^r \partial x^n} - \frac{\partial^2 a_{rm}}{\partial x^s \partial x^n} - \frac{\partial^2 a_{sn}}{\partial x^r \partial x^m}\right) + a^{pq}\left([rn, p]\,[sm, q] - [rm, p]\,[sn, q]\right).$$

The covariant form of the curvature tensor has the symmetries already given for the mixed form, but it has others. It is skew-symmetric with respect to its first two subscripts, and symmetric with respect to its two pairs of subscripts. All the symmetry properties are listed as follows:

3.115. $$R_{rsmn} = -R_{srmn}, \quad R_{rsmn} = -R_{rsnm}, \quad R_{rsmn} = R_{mnrs},$$

3.116. $$R_{rsmn} + R_{rmns} + R_{rnsm} = 0.$$

Exercise. Show that in a V_2 all the components of the covariant curvature tensor either vanish or are expressible in terms of R_{1212}.

How many independent components has the tensor R_{rsmn}? We cannot answer this question by simply counting the equations 3.115 and 3.116, because they overlap. We reason as follows. By the first two of 3.115 a component vanishes unless $r \neq s$ and $m \neq n$. Denote by (rs) a combination r, s, with r and s distinct and with no consideration of order. The number of (rs) is

$$M = \tfrac{1}{2}N(N-1),$$

N being the number of dimensions of the space. There is of course the same number of combinations (mn). If the first two of 3.115 were the only conditions on R_{rsmn}, we would have M^2 independent components. But the last of 3.115 cuts down this number by the number of combinations of M things taken 2 at a time, viz. $\frac{1}{2}M(M-1)$. Thus if the three relations of 3.115 were the only identities, we would have for the number of independent components

3.117. $$M^2 - \tfrac{1}{2}M(M-1) = \tfrac{1}{2}M(M+1) = \tfrac{1}{8}N(N-1)(N^2-N+2).$$

Turning to 3.116, it is easy to see that unless r, s, m, n are all distinct the identity is included in 3.115. Further, we get only one identity from a given combination. Thus 3.116 give a number of new identities equal to the number of combinations of N things taken 4 at a time, i.e.

3.118. $$\frac{1}{24}N(N-1)(N-2)(N-3).$$

Subtracting this number from 3.117, we get this result: *The number of independent components of the covariant curvature tensor in a space of N dimensions is*

3.119.
$$\frac{1}{12}N^2(N^2-1).$$

The following table shows how rapidly the number of independent components increases with N:

Number of dimensions of space	2	3	4	5
Number of independent components of R_{rsmn}	1	6	20	50 .

In addition to the relations 3.115, 3.116 satisfied by the covariant curvature tensor, there are certain identities satisfied by its covariant derivative. Consider the covariant derivative of 3.114, calculated at the origin of Riemannian coordinates; it is the same as the partial derivative, i.e.

3.120.
$$R_{rsmn|t}=\tfrac{1}{2}\left(\frac{\partial^3 a_{rn}}{\partial x^t\partial x^s\partial x^m}+\frac{\partial^3 a_{sm}}{\partial x^t\partial x^r\partial x^n}-\frac{\partial^3 a_{rm}}{\partial x^t\partial x^s\partial x^n}-\frac{\partial^3 a_{sn}}{\partial x^t\partial x^r\partial x^m}\right).$$

Permuting the last three subscripts cyclically, and adding, we obtain

3.121.
$$R_{rsmn|t}+R_{rsnt|m}+R_{rstm|n}=0.$$

This is a tensor equation, and so it is true in general, since it is true for one particular coordinate system. The subscript r may be raised in the usual way, giving

3.122.
$$R^r_{.\,smn|t}+R^r_{.\,snt|m}+R^r_{.\,stm|n}=0.$$

This is known as the *Bianchi identity*. It may also be derived from 3.106, using Riemannian coordinates.

Exercise. Using the fact that the absolute derivative of the fundamental tensor vanishes, prove that 3.107 may be written

3.123. $$\frac{\delta^2 T_r}{\delta u \delta v} - \frac{\delta^2 T_r}{\delta v \delta u} = R_{rpmn} T^p \frac{\partial x^m}{\partial u} \frac{\partial x^n}{\partial v},$$

where T_r is any covariant vector and $T^p = a^{pq} T_q$.

3.2. The Ricci tensor, the curvature invariant, and the Einstein tensor. We may contract the mixed curvature tensor 3.106 and so obtain a covariant tensor of the second order

3.201. $$R_{rm} = R^n_{.rmn},$$

or equivalently

3.202. $$R_{rm} = a^{sn} R_{srmn}.$$

It follows from the identities 3.115 that R_{rm} is a symmetric tensor. It is called the *Ricci tensor.*

By 3.106 we have

3.203.
$$R_{rm} = \frac{\partial}{\partial x^m}\left\{\begin{matrix} n \\ r\ n \end{matrix}\right\} - \frac{\partial}{\partial x^n}\left\{\begin{matrix} n \\ r\ m \end{matrix}\right\} + \left\{\begin{matrix} p \\ r\ n \end{matrix}\right\}\left\{\begin{matrix} n \\ p\ m \end{matrix}\right\} - \left\{\begin{matrix} p \\ r\ m \end{matrix}\right\}\left\{\begin{matrix} n \\ p\ n \end{matrix}\right\}.$$

Now, by 2.541,

3.204. $$\left\{\begin{matrix} n \\ r\ n \end{matrix}\right\} = \tfrac{1}{2}\frac{\partial}{\partial x^r} \ln a,$$

assuming a to be positive, and so

3.205.
$$R_{rm} = \tfrac{1}{2}\frac{\partial^2}{\partial x^r \partial x^m}\ln a - \tfrac{1}{2}\left\{\begin{matrix} p \\ rm \end{matrix}\right\}\frac{\partial}{\partial x^p}\ln a - \frac{\partial}{\partial x^n}\left\{\begin{matrix} n \\ r\ m \end{matrix}\right\} + \left\{\begin{matrix} p \\ r\ n \end{matrix}\right\}\left\{\begin{matrix} n \\ p\ m \end{matrix}\right\}.$$

If a is negative, 3.205 holds with a replaced by $-a$. This form makes the symmetry of the Ricci tensor obvious.

Taking its symmetry into consideration, the number of independent components of the Ricci tensor in a space of N dimensions is $\frac{1}{2}N(N+1)$. Thus there are 3 components if $N = 2$, 6 components if $N = 3$, and 10 components if $N = 4$. In the case $N = 2$ (e.g. in the intrinsic geometry of a surface in Euclidean 3-space) we have

3.206. $R_{11} = a^{22}R_{2112},\ R_{12} = a^{12}R_{2121},\ R_{22} = a^{11}R_{1221}.$

But, for $N = 2$,

3.207. $$a^{11} = \frac{a_{22}}{a},\quad a^{12} = -\frac{a_{21}}{a},\quad a^{22} = \frac{a_{11}}{a},$$

and so

3.208. $$\frac{R_{11}}{a_{11}} = \frac{R_{12}}{a_{12}} = \frac{R_{22}}{a_{22}} = -\frac{R_{1212}}{a}.$$

Thus, in a 2-space, the components of the Ricci tensor are proportional to the components of the metric tensor. This is not true in general in spaces of higher dimensionality.

The *curvature invariant* R is defined as

3.209. $$R = a^{mn}R_{mn} = R^{n}_{.n}.$$

It follows from 3.208 that for $N = 2$ we have

3.210. $$R = -\frac{2}{a}R_{1212}.$$

This result, like 3.208, is restricted to two-dimensional space. This shows that familiarity with the properties of curved surfaces in Euclidean 3-space is not of much assistance in understanding the properties of curved spaces of higher dimensionality—there are too many simplifications in the case $N = 2$.

There is an identity of considerable importance in the theory of relativity, which is obtained from the Bianchi identity in the form 3.121. We multiply that equation by $a^{rn}a^{sm}$; this gives

3.211. $$a^{sm}R_{sm|t} - a^{sm}R_{st|m} - a^{rn}R_{rt|n} = 0.$$

(Use has been made of the skew-symmetry of R_{rsmn} with respect to the last two subscripts.) This may be written

3.212. $$R_{|t} - 2R^{n}_{.t|n} = 0,$$

or

3.213. $$(R^{n}_{.t} - \tfrac{1}{2}\,\delta^{n}_{t}R)_{|n} = 0.$$

The *Einstein tensor* $G^{n}_{.t}$ is defined by

3.214. $$G^{n}_{.t} = R^{n}_{.t} - \tfrac{1}{2}\,\delta^{r}_{t}R.$$

Thus 3.213 may be written

3.215. $$G^{n}_{.t|n} = 0;$$

this may be expressed in words by saying that *the divergence of the Einstein tensor vanishes.*

3.3. Geodesic deviation. Imagine a two-dimensional observer living in a two-dimensional space. He wants to explore the properties of his space by measuring distances. From the property of stationary length he is able to construct geodesics. What can he find out by following one geodesic? Not very much. We are so used to thinking of 2-spaces as surfaces immersed in a Euclidean 3-space that we are inclined to think of the geodesics of a 2-space as curved lines (e.g. the great circles on a sphere). But when we consider a 2-space in itself there is no master curve with which to compare a geodesic of the 2-space; the idea of the curvature of a geodesic disappears—indeed we regard the geodesic as the "straightest" curve we can draw in the 2-space.

Thus, as our two-dimensional observer travels along a geodesic, he has nothing interesting to report, except for one possibility: the geodesic may meet itself, as a great circle on a sphere meets itself. This is a property of the space "in the large," i.e. it does not belong to the domain of differential geometry.

Much more interesting results are obtained if the observer considers not one geodesic, but two. Suppose that these two geodesics start from a common point and make a small angle with one another. Consider two points, one on each geodesic, equidistant from the common point; we shall call such points "corresponding points." How does the small distance between corresponding points vary as we move along the geodesics? This is the problem of *geodesic deviation*, and if we can solve it we get a good insight into the nature of the space.

Think, for example, of the application of the method of geodesic deviation to a plane and to a sphere. In a plane, the distance between corresponding points on the two geodesics (straight lines) increases steadily, being in fact proportional to the distance from the common point of the two geodesics. In a sphere, the distance between corresponding points on the two geodesics (great circles) increases at first, but after a while it has a maximum value, and then decreases to zero. Thus the study of geodesic deviation enables us to distinguish very simply between a plane and a sphere.

Exercise. Would the study of geodesic deviation enable us to distinguish between a plane and a right circular cylinder?

To introduce the idea of geodesic deviation as simply as possible, we have talked about a 2-space. The same idea applies in V_N, and we shall now develop basic formulae for this general case.

Consider a singly infinite family of geodesics in V_N, forming a V_2. Let u be a parameter varying along each geodesic of the family, and let v be a parameter constant along each geodesic of the family, but varying as we pass from one geodesic to another. The equations of V_2 may be written

3.301. $$x^r = x^r(u, v);$$

the geodesics are the parametric lines of u in V_2.

Let us make the parameter u more precise in the following way. Draw a curve AA' in V_2 (Fig. 5) cutting the geodesics

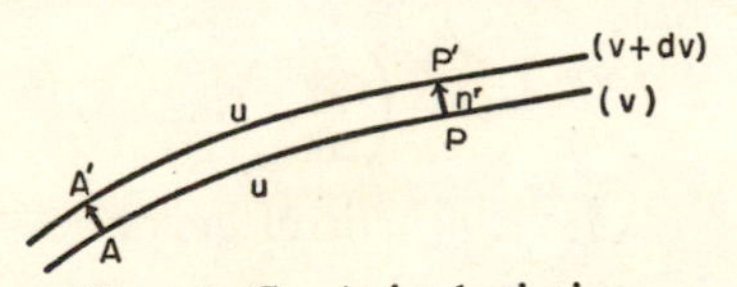

FIG. 5. Geodesic deviation.

orthogonally, and let u be arc-length measured along each geodesic from AA'. Then, since the curves v = const. are geodesics in V_N, we have all over V_2, by 2.513,

3.302. $$\frac{\delta p^r}{\delta u} = 0,$$

where $p^r = \partial x^r/\partial u$; p^r is the field of unit tangent vectors to the geodesics, satisfying the equation

3.303. $$a_{mn}p^m p^n = \pm 1.$$

Consider two adjacent geodesics of the family, C and C', with parameters v and $v + dv$. We shall say that points P on C and P' on C' are corresponding points if they have equal values of u, i.e. are equidistant from the cross-curve AA'. Let η^r be the infinitesimal vector PP', so that

3.304. $$\eta^r = \frac{\partial x^r}{\partial v} dv.$$

In setting up the geodesic normal coordinate system in 2.6, we had occasion to compare two adjacent geodesics. With a slight change in notation to suit the present case, the equation 2.653 may be written

3.305. $$a_{mn}p^m \eta^n = 0;$$

the equation holds at the point P. This equation tells us that the deviation η^r is perpendicular to C. Equation 3.305 may also be written

3.306. $$a_{mn}\frac{\partial x^m}{\partial u}\frac{\partial x^n}{\partial v} = 0.$$

This equation holds all over V_2, since P may be any point on V_2.

Let us now see how η^r changes as we pass along C. We have

3.307. $$\frac{\delta \eta^r}{\delta u} = \frac{\delta}{\delta u}\left(\frac{\partial x^r}{\partial v}\right) dv,$$

since dv is a constant for the pair of geodesics, C and C'. Now

3.308. $$\frac{\delta}{\delta u}\frac{\partial x^r}{\partial v} = \frac{\partial^2 x^r}{\partial u \partial v} + \left\{ {r \atop m\,n} \right\} \frac{\partial x^m}{\partial v}\frac{\partial x^n}{\partial u} = \frac{\delta}{\delta v}\frac{\partial x^r}{\partial u},$$

and so 3.307 may be written

3.309. $$\frac{\delta \eta^r}{\delta u} = \frac{\delta p^r}{\delta v} dv.$$

We take the absolute derivative of this equation with respect to u, and use 3.107. Thus

3.310.
$$\frac{\delta^2 \eta^r}{\delta u^2} = \frac{\delta}{\delta u}\frac{\delta p^r}{\delta v}\, dv$$
$$= \left(\frac{\delta}{\delta v}\frac{\delta p^r}{\delta u} + R^r_{.\,smn} p^s p^m \frac{\partial x^n}{\partial v}\right) dv.$$

The first term on the right-hand side vanishes by 3.302. Rearranging the terms, and replacing u by s (arc length along the geodesic) we obtain the *equation of geodesic deviation*

3.311.
$$\frac{\delta^2 \eta^r}{\delta s^2} + R^r_{.\,smn} p^s \eta^m p^n = 0.$$

We recall that η^r is the infinitesimal normal displacement to the neighbouring geodesic, and p^r is the unit tangent vector to the geodesic.

Equation 3.311 contains N ordinary differential equations of the second order for the N quantities η^r. These equations determine η^r as functions of s if the initial values of η^r and $\delta\eta^r/\delta s$ (or $d\eta^r/ds$) are given.

Exercise. For rectangular Cartesians in Euclidean 3-space, show that the general solution of 3.311 is $\eta^r = A^r s + B^r$, where A^r, B^r are constants. Verify this result by elementary geometry.

3.4. Riemannian curvature. The great charm of classical geometry lies in the interplay of visual intuition and precise analytical arguments. In passing to a Riemannian N-space, much of the intuition must be left behind. But it is worth an effort to seek to build up in this general geometry at least a shadow of that type of thought which has proved so powerful in the differential geometry of ordinary surfaces. To this end the curvature tensor itself is of little use. The concept of Riemannian curvature is much more helpful. Since we wish our definition to be applicable to a space of indefinite metric form, we must first clear out of the way some preliminaries connected with indicators.

Let X^r and Y^r be two vectors, given at a point. Write

$$Z^r = AX^r + BY^r, \tag{3.401}$$

and let A, B be invariants which take arbitrary values. We say that the vector Z^r so defined is *coplanar* with X^r and Y^r. The totality of infinitesimal displacements in the directions defined by these vectors Z^r determine an *elementary* 2-*space* at the point.

Suppose now that X^r and Y^r are orthogonal unit vectors with indicators $\epsilon(X)$ and $\epsilon(Y)$ respectively. Let Z^r and W^r be two more unit vectors, orthogonal to one another and coplanar with X^r and Y^r; let their indicators be $\epsilon(Z)$ and $\epsilon(W)$ respectively. We shall prove* that

$$\epsilon(X)\epsilon(Y) = \epsilon(Z)\epsilon(W). \tag{3.402}$$

We have

$$Z^r = AX^r + BY^r,\ W^r = A'X^r + B'Y^r, \tag{3.403}$$

and hence

$$\begin{aligned} \epsilon(Z) &= a_{mn}Z^mZ^n = A^2\epsilon(X) + B^2\epsilon(Y), \\ \epsilon(W) &= a_{mn}W^mW^n = A'^2\epsilon(X) + B'^2\epsilon(Y). \end{aligned} \tag{3.404}$$

We have also

$$0 = a_{mn}Z^mW^n = \epsilon(X)AA' + \epsilon(Y)BB'. \tag{3.405}$$

Consequently

$$\begin{aligned} \epsilon(Z)\epsilon(W) &= [\epsilon(X)A^2 + \epsilon(Y)B^2][\epsilon(X)A'^2 + \epsilon(Y)B'^2] \\ &\qquad - [\epsilon(X)AA' + \epsilon(Y)BB']^2 \\ &= \epsilon(X)\epsilon(Y)(AB' - A'B)^2. \end{aligned} \tag{3.406}$$

Since the indicators are each ± 1, it follows that 3.402 is true, and also

$$AB' - A'B = \pm 1. \tag{3.407}$$

We now define *the Riemannian curvature associated with an elementary* 2-*space* V_2 *to be the invariant*

*This may also be easily proved by Sylvester's theorem (see Appendix A).

3.408. $$K = \epsilon(X)\epsilon(Y)R_{rsmn}X^rY^sX^mY^n,$$

where X^r, Y^r *are orthogonal unit vectors in* V_2 *and* $\epsilon(X)$, $\epsilon(Y)$ *their indicators.*

Let us show that K does not depend on the particular pair of orthogonal unit vectors chosen in V_2. If Z^r, W^r are two other orthogonal unit vectors, the corresponding Riemannian curvature given by 3.408 is

3.409. $$K' = \epsilon(Z)\epsilon(W)R_{rsmn}Z^rW^sZ^mW^n.$$

To compare this with 3.408, we use 3.402 and 3.403. Using the symmetry equations 3.115, we easily obtain

3.410. $$R_{rsmn}Z^rW^sZ^mW^n = (AB' - A'B)^2R_{rsmn}X^rY^sX^mY^n,$$

and by virtue of 3.407 we have $K' = K$, as required.

It is sometimes advisable to express K in terms of any two vectors in the elementary 2-space, rather than a pair of orthogonal unit vectors. If ξ^r, η^r are any two vectors in the elementary 2-space, then

3.411. $$K = \frac{R_{rmsn}\xi^r\eta^s\xi^m\eta^n}{(a_{pu}a_{qv} - a_{pv}a_{qu})\xi^p\eta^q\xi^u\eta^v}.$$

To prove this we write

3.412. $$\xi^r = AX^r + BY^r, \quad \eta^r = A'X^r + B'Y^r$$

where X^r, Y^r are orthogonal unit vectors in the elementary 2-space. The numerator in 3.411 becomes

$$(AB' - A'B)^2R_{rsmn}X^rY^sX^mY^n,$$

and the denominator becomes

$$\epsilon(X)\epsilon(Y)(AB' - A'B)^2.$$

The truth of 3.411 is then evident.

When the given manifold is of two dimensions ($N = 2$), the elementary 2-space at a point is unique. Thus, with any point of a V_2 there is associated a unique Riemannian *curvature* K. If this V_2 has a positive definite metric, we can set up here a connection between the formalism of tensor calculus and a familiar concept of differential geometry. Let us recall that a surface (V_2) in Euclidean 3-space has two principal radii of curvature,* say R_1 and R_2. The Gaussian curvature of this V_2 is

3.413. $$G = \frac{1}{R_1 R_2}.$$

The great contribution of Gauss to differential geometry was his proof that G can actually be expressed in terms of intrinsic properties of the surface, without reference to the 3-space in which it lies. He established, in fact, a result which may be stated as follows:

3.414. $$G = \lim_{S \to 0} E/S.$$

Here S is the area of a small geodesic quadrilateral, and E the excess of the sum of its angles over four right angles.

Now suppose that we take a vector in V_2 and propagate it parallelly (cf. equation 2.512) round the sides of a small geodesic quadrilateral. On completion of the circuit, the vector will have undergone a small change which, as will be seen in 3.5, is given by equation 3.519. From this formula it is not hard to calculate the small angle through which the vector has been turned: it turns out to be KS, where K is the Riemannian curvature and S the area of the quadrilateral.† But under parallel propagation along a geodesic, a vector makes a constant angle with the geodesic; following the vector round the small quadrilateral, it is easy to see that the angle through which the vector has turned on completion of the circuit is E, the excess of the angle-sum over four right angles. Hence

3.415. $$KS = E.$$

Comparison with 3.414 gives $K = G$: *for a* V_2 *with positive definite metric, the Riemannian curvature is equal to the Gaussian curvature, defined by* 3.414.

The above geometrical interpretation of the Riemannian curvature K holds only in a space of two dimensions with a

*Cf. C. E. Weatherburn, *Differential Geometry*, I (Cambridge University Press, 1939), Sect. 29.

†See equation 3.533.

positive definite metric. To get an interpretation in V_N, we refer to the equation of geodesic deviation 3.311. Let μ^r be the unit vector in the direction of the deviation η^r, and let η be the magnitude of η^r, so that

3.416. $$\eta^r = \eta\mu^r, \quad a_{mn}\mu^m\mu^n = \epsilon(\mu).$$

Let us substitute this in 3.311; we get

3.417. $$\frac{d^2\eta}{ds^2}\mu^r + 2\frac{d\eta}{ds}\frac{\delta\mu^r}{\delta s} + \eta\frac{\delta^2\mu^r}{\delta s^2} + \eta R^r_{.smn}p^s\mu^m p^n = 0,$$

where p^r is the unit tangent vector to the geodesic from which the deviation η^r is measured. Now from the second of 3.416 we have

3.418. $$\mu_m\mu^m = \epsilon(\mu), \quad \mu_m\frac{\delta\mu^m}{\delta s} = 0, \quad \frac{\delta\mu_m}{\delta s}\frac{\delta\mu^m}{\delta s} + \mu_m\frac{\delta^2\mu^m}{\delta s^2} = 0.$$

Hence, if we multiply 3.417 across by μ_r, we get

3.419. $$\epsilon(\mu)\frac{d^2\eta}{ds^2} + \eta\mu_r\frac{\delta^2\mu^r}{\delta s^2} + \eta\epsilon(p)\epsilon(\mu)K = 0,$$

where K is the Riemannian curvature for the 2-element defined by η^r and p^r. Let us suppose that the two neighbouring geodesics involved in the geodesic deviation start from a common point O, and let s be measured from O. Then $\eta \to 0$ as $s \to 0$, and so by 3.419 we have

3.420. $$\lim_{s\to 0}\frac{d^2\eta}{ds^2} = 0.$$

Now return to 3.417 and let s tend to zero. On account of 3.420 and the fact that $d\eta/ds$ does *not* tend to zero at O (if it did, the geodesics would not separate at all), we deduce

3.421. $$\lim_{s\to 0}\frac{\delta\mu^r}{\delta s} = 0.$$

Hence, by 3.418,

3.422. $$\lim_{s\to 0}\mu_r\frac{\delta^2\mu^r}{\delta s^2} = 0.$$

We return to 3.419, divide by η and let s tend to zero. This gives

3.423.
$$\lim_{s\to 0}\frac{1}{\eta}\frac{d^2\eta}{ds^2} = -\epsilon K,$$

where ϵ is the indicator of either geodesic (the same for both since they are adjacent) and K is the Riemannian curvature corresponding to the 2-element defined by the two geodesics.

Suppose now that we expand η in a power series in s, the distance from the common point of the two geodesics. If we define θ by

3.424.
$$\theta = \lim_{s\to 0}\frac{d\eta}{ds},$$

it is easy to see that the expansion takes the form

3.425.
$$\eta = \theta(s - \tfrac{1}{6}\epsilon K s^3 + \ldots).$$

This formula gives us an insight into the geometrical meaning of the sign of K. For simplicity, take a space with positive

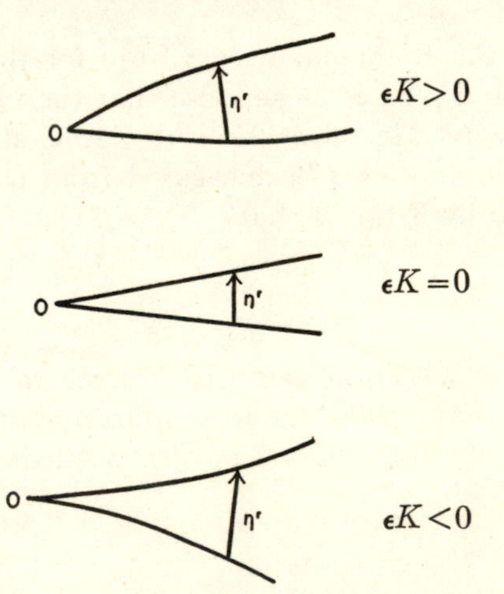

FIG. 6. Behaviour of diverging geodesics, relative to the sign of ϵK.

definite metric, so that $\epsilon = 1$. The quantity θ, defined by 3.424, is the small angle between the two geodesics at their common point—this follows at once from the formula 2.309 which defines angle. Thus the term θs on the right of 3.425 represents the deviation with which we are familiar in Euclidean space. When we bring the next term into consideration, we see that a *positive* Riemannian curvature implies a *convergence* of the geodesics, and a *negative* Riemannian curvature a *divergence* (convergence and divergence being interpreted relative to the behaviour of straight lines in Euclidean space).

The behaviour of the geodesics relative to the sign of ϵK is shown in Fig. 6.

3.5. Parallel propagation. We shall now discuss some questions connected with parallel propagation. By definition (cf. equation 2.512) a vector with contravariant components T^r is propagated parallelly along a curve $x^r = x^r(u)$ provided the following equation is satisfied:

3.501. $$\frac{\delta T^r}{\delta u} = 0.$$

If $T_r = a_{rm}T^m$, then 3.501 is equivalent to

3.502. $$\frac{\delta T_r}{\delta u} = 0.$$

Reference may be made to 2.511 and 2.515 for the definition of the operator $\delta/\delta u$.

Let us now investigate how the result of parallel propa-

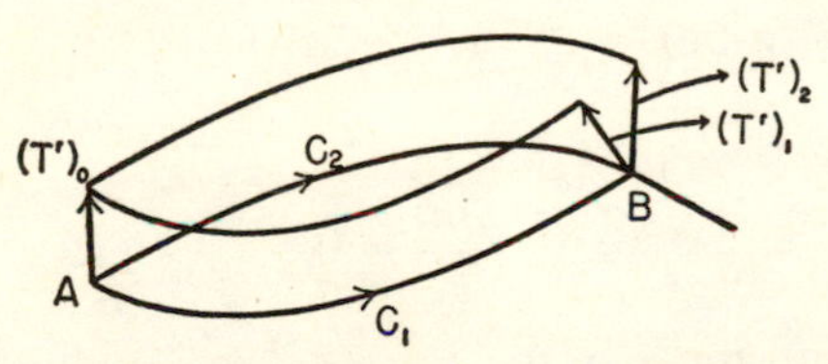

FIG. 7. Parallel propagation of vector from A to B along different paths, C_1 and C_2.

gation of a vector depends on the curve along which the propagation is carried out. Let A, B (Fig. 7) be two points in V_N, and C_1, C_2 two curves joining them. On each of these curves let us take a parameter u running between the same limits

u_1, u_2 for both curves. We may then write as the equations of the two curves

3.503. $$C_1:\ x^r = f^r(u),\quad C_2:\ x^r = g^r(u),\quad u_1 \leqslant u \leqslant u_2,$$

with $u = u_1$ at A and $u = u_2$ at B. Let us take any vector $(T_r)_0$ at A and propagate it parallelly to B, using first the curve C_1 and secondly the curve C_2 as path of propagation. Let the resulting vectors at B be $(T^r)_1$ and $(T^r)_2$ respectively. We seek to evaluate the difference

3.504. $$(\Delta T^r)_B = (T^r)_2 - (T^r)_1.$$

Let us construct a continuous family of curves joining A and B, the curves C_1 and C_2 being members of the family.* Let the parameter u vary between the same limits u_1, u_2 for all these curves; and let v be a parameter which is fixed on each curve, and varies continuously from $v = v_1$ on C_1 to $v = v_2$ on C_2. Then the equations of the family may be written;

3.505. $$\begin{aligned} x^r &= h^r(u, v), \\ u_1 \leqslant\ &u \leqslant u_2, \\ v_1 \leqslant\ &v \leqslant v_2. \end{aligned}$$

We may also regard 3.505 as the equations of a V_2 on which u and v are coordinates. Note that

3.506. $$\begin{aligned} h^r(u, v_1) &= f^r(u), \quad h^r(u, v_2) = g^r(u), \\ h^r(u_1, v) &= f^r(u_1) = g^r(u_1), \\ h^r(u_2, v) &= f^r(u_2) = g^r(u_2), \end{aligned}$$

and hence

3.507. $$\frac{\partial x^r}{\partial v} = 0$$

for $u = u_1$ and for $u = u_2$.

*This can always be done if V_N is simply connected (like an infinite plane or the surface of a sphere). But if V_N is multiply connected (like the surface of an infinite cylinder or a torus), it is possible to join A and B by curves which cannot be continuously transformed into one another. Curves which can be so transformed are called *reconcilable*. We shall understand that the curves C_1, C_2 are reconcilable.

Starting with the vector $(T^r)_0$ at A, let us propagate it parallelly along all the curves v = constant. This gives us a vector field $T^r(u, v)$ over V_2, single-valued except possibly at B. This field satisfies

3.508. $$\frac{\delta T^r}{\delta u} = 0,$$

and consequently

3.509. $$\frac{\delta^2 T^r}{\delta v \delta u} = 0.$$

Furthermore, we have

3.510. $$\frac{\delta T^r}{\delta v} = 0 \text{ for } u = u_1,$$

since T^r is independent of v at A, and 3.507 holds. We now take a second vector Y_r, choosing it arbitrarily at B, and propagating it parallelly back along the curves v = constant to A. This defines a vector field $Y_r(u, v)$, satisfying

3.511. $$\frac{\delta Y_r}{\delta u} = 0$$

over V_2.

Consider the invariant $T^r Y_r$. Since Y_r is single-valued at B, we have

3.512. $$(\Delta T^r)_B (Y_r)_B = \int_{v_1}^{v_2} \left(\frac{\partial}{\partial v}(T^r Y_r) \right)_{u=u_2} dv = \int_{v_1}^{v_2} \left(\frac{\delta T^r}{\delta v} Y_r \right)_{u=u_2} dv.$$

This last integrand may be written

3.513. $$\left(\frac{\delta T^r}{\delta v} Y_r \right)_{u=u_2} = \left(\frac{\delta T^r}{\delta v} Y_r \right)_{u=u_1} + \int_{u_1}^{u_2} \frac{\partial}{\partial u} \left(\frac{\delta T^r}{\delta v} Y_r \right) du.$$

The first term on the right vanishes, by 3.510. The second term reduces as follows, use being made of 3.511 and 3.107:

3.514.

$$\int_{u_1}^{u_2} \frac{\partial}{\partial u}\left(\frac{\delta T^r}{\delta v} Y_r\right) du = \int_{u_1}^{u_2} \frac{\delta^2 T^r}{\delta u \delta v} Y_r du$$

$$= \int_{u_1}^{u_2}\left(\frac{\delta^2 T^r}{\delta v \delta u} + R^r_{.pmn} T^p \frac{\partial x^m}{\partial u}\frac{\partial x^n}{\partial v}\right) Y_r du.$$

The first part of the integrand vanishes by 3.509. When we substitute back through 3.513 in 3.512, we get

3.515. $$(\Delta T^r)_B (Y_r)_B = \iint Y_r R^r_{.pmn} T^p \frac{\partial x^m}{\partial u}\frac{\partial x^n}{\partial v} du dv,$$

the double integral being taken over the 2-space V_2 between the curves C_1 and C_2. *This formula gives us the effect of path on the result of parallel propagation*, since $(Y_r)_B$ is arbitrary.

We shall use this result to investigate the effect of propagating a vector parallelly around a closed circuit. Still using the same notation and Fig. 7, we have at B the three vectors

$$(T^r)_1, \ (T^r)_2, \ Y_r.$$

We recall that the first two of these vectors are the results of parallel propagation of $(T^r)_0$ at A along C_1 and C_2 respectively. Let us propagate these three vectors *back* along C_2 to A. Now it is obvious from the form of the equations of parallel propagation that if we propagate a vector parallelly along a curve from A to B, and then propagate it parallelly back along the *same* curve from B to A, we arrive back at A with precisely the same vector as we started with. Let us denote by $(T^r)_0 + (\Delta T^r)_A$ the vector obtained by propagating the vector $(T^r)_0$ around the circuit formed of C_1 and C_2 reversed. Then, under parallel propagation back along C_2, the three vectors mentioned above come to A with the values

$$(T^r)_0 + (\Delta T^r)_A, \ (T^r)_0, \ (Y_r)_{A,2}$$

the final notation being used to indicate that Y_r has been propagated along C_2 to A. Since parallel propagation does not alter the invariant $T^r Y_r$, it follows that

3.516. $$(\Delta T^r)_A (Y_r)_{A,2} = -(\Delta T^r)_B (Y_r)_B.$$

For the right-hand side of this equation we have the expression

3.515, and so, in a sense, 3.516 gives an evaluation of the effect $(\Delta T^r)_A$ of parallel propagation around a closed circuit. However, we can improve the statement by noting that we can choose $(Y_r)_{A,2}$ arbitrarily, provided we then determine Y_r at B by parallel propagation along C_2. With this in mind, let us restate our result as follows:

Let A be any point in V_N and V_2 a 2-space passing through A (Fig. 8). *Let C be a closed curve on V_2, passing through A, with*

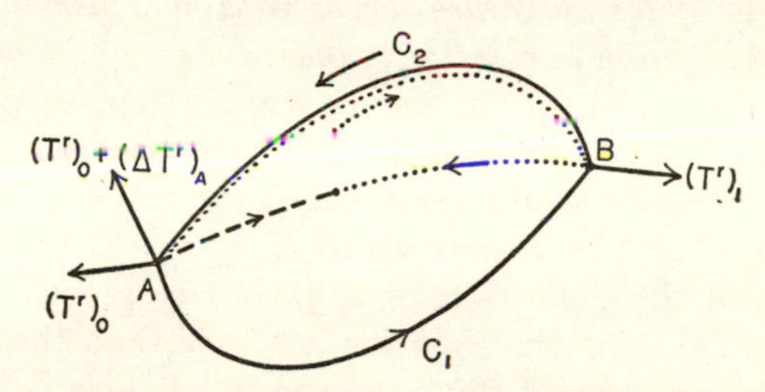

FIG. 8. Effect of parallel propagation round closed circuit $C_1 + C_2$, illustrating equation (3.517).
Path of propagation of T^r shown - - - - - -
Path of propagation of Y_r shown . . . . . .

a certain sense indicated on it. Let B be any other point on C, joined to A by a set of curves lying in V_2. Let $(Y_r)_0$ be chosen arbitrarily at A, propagated parallelly to B along C, but against the assigned sense of C, and then propagated parallelly from B back towards A along the given set of curves. Then, as a result of parallel propagation around C in the assigned sense, any vector T^r assigned at A receives an increment ΔT^r satisfying

$$\textbf{3.517.} \qquad \Delta T^r.(Y_r)_0 = -\iint Y_r R^r_{.pmn} T^p \frac{\partial x^m}{\partial u}\frac{\partial x^n}{\partial v}\,du dv,$$

the double integral being taken over the part of V_2 bounded by C.

If the circuit C shrinks toward zero size, the principal part of the preceding integral may be written

$$\textbf{3.518.} \qquad -(Y_r)_0 \iint R^r_{.pmn} T^p \frac{\partial x^m}{\partial u}\frac{\partial x^n}{\partial v}\,du dv.$$

Since $(Y_r)_0$ is arbitrary, we see that for an infinitesimal circuit in the form of an infinitesimal parallelogram with adjacent edges $d_{(1)}x^r$, $d_{(2)}x^r$, the infinitesimal increment in T^r may be written

$$\textbf{3.519.} \qquad \Delta T^r = -R^r_{.pmn} T^p d_{(1)}x^m d_{(2)}x^n.$$

We draw an important conclusion from 3.515. If $R^r_{.pmn} = 0$, the integral vanishes. But $(Y_r)_B$ is arbitrary; therefore $(\Delta T^r)_B = 0$. In words, *if the curvature tensor vanishes, then the result of parallel propagation of a vector is independent of the path of propagation*. Equivalently, if the curvature tensor vanishes, a vector is unchanged by parallel propagation around a closed circuit. It is easily seen that these statements apply to both contravariant and covariant vectors.

We recall the definition of a flat space given in 3.1. (cf. equation 3.101). In that section we saw that the condition 3.108—the vanishing of the curvature tensor—is necessary for flatness. We shall now show that that condition is also sufficient, i.e. if the curvature tensor vanishes, then there exists a coordinate system such that the metric form is as in 3.101. But first we need to prove the following theorem:

If $R^r_{.pmn} = 0$ in V_N, a vector $(T_r)_0$ assigned at a point A defines a vector field T_r throughout V_N by parallel propagation no matter what paths of propagation are used. Then $\int_A^B T_n dx^n$, where B is any second point in V_N, is independent of the path of integration joining A to B.

The truth of the first sentence in this theorem has already been established. To prove the second statement, we take two curves C_1 and C_2 joining A and B, and fill in a family of curves as in 3.505 and Fig. 7. For any one of these curves (v = constant) we write

$$\textbf{3.520.} \qquad I(v) = \int_{u_1}^{u_2} T_n dx^n = \int_{u_1}^{u_2} T_n \frac{\partial x^n}{\partial u} du.$$

Then

3.521.
$$\frac{dI}{dv} = \int_{u_1}^{u_2} \frac{\partial}{\partial v}\left(T_n \frac{\partial x^n}{\partial u} \right) du$$
$$= \int_{u_1}^{u_2} \frac{\delta T_n}{\delta v}\frac{\partial x^n}{\partial u} du + \int_{u_1}^{u_2} T_n \frac{\delta}{\delta v}\frac{\partial x^n}{\partial u} du.$$

Now $\delta T_r/\delta v = 0$, since T_r is propagated parallelly along *all* curves in V_N. Also, using $\delta T_r/\delta u = 0$, we have

3.522.
$$\int_{u_1}^{u_2} T_n \frac{\delta}{\delta v}\frac{\partial x^n}{\partial u} du = \int_{u_1}^{u_2} T_n \frac{\delta}{\delta u}\frac{\partial x^n}{\partial v} du$$
$$= \int_{u_1}^{u_2} \frac{\partial}{\partial u}\left(T_n \frac{\partial x^n}{\partial v} \right) du = \left[T_n \frac{\partial x^n}{\partial v} \right]_{u_1}^{u_2} = 0,$$

since $\partial x^r/\partial v = 0$ at A and at B. Therefore $dI/dv = 0$, and so $I(v)$ is independent of v. Consequently the integral has the same value for C_1 and C_2; the theorem is proved.

In a V_N with $R^r_{.pmn} = 0$ we shall now set up a system of coordinates for which the fundamental form is as in 3.101. We take a point A, and at A we take a set of N orthogonal unit vectors $X_{(m)r}$. At A these satisfy

3.523.
$$a^{pq} X_{(m)p} X_{(n)q} = E_{mn},$$

where $E_{mn} = 0$ for $m \neq n$, and $E_{mn} = \epsilon_n$ for $m = n$, ϵ_n being the indicator of the vector $X^r_{(n)}$. We define throughout V_N N vector fields by parallel propagation of $X_{(m)r}$. Then the relations 3.523 are satisfied throughout V_N. We define at any point P

3.524.
$$y_r = \int_A^P X_{(r)n} dx^n.$$

As we have seen above, the value of y_r at P is independent of the choice of the curve of integration AP. Therefore, for an arbitrary infinitesimal displacement of P we have

3.525.
$$dy_r = X_{(r)n} dx^n,$$

and so

3.526.
$$\frac{\partial y_r}{\partial x^n} = X_{(r)n}.$$

Hence by 5.523,

3.527.
$$a^{pq}\frac{\partial y_m}{\partial x^p}\frac{\partial y_n}{\partial x^q} = E_{mn}.$$

We may use y_r as a system of coordinates in V_N. Let b_{mn} be the corresponding fundamental tensor, and b^{mn} its conjugate tensor. But the latter is a contravariant tensor, and so, by the law of tensor transformation,

3.528.
$$b^{mn} = a^{pq}\frac{\partial y_m}{\partial x^p}\frac{\partial y_n}{\partial x^q} = E_{mn}.$$

From this it easily follows that $b_{mn} = E_{mn}$, and so the metric form is

3.529.
$$\Phi = \epsilon_1(dy_1)^2 + \epsilon_2(dy_2)^2 + \ldots + \epsilon_N(dy_N)^2.$$

But this is of the form 3.101. Thus we have proved that *the conditions* $R^r_{.pmn} = 0$ *are sufficient for flatness.*

It remains to clear up a point in connection with equation 3.415. We wish to show that, if $N = 2$ and the metric form is positive definite, a vector is turned through an angle KS after parallel propagation around a small rectangle of area S. Let P be any point, and λ^r, μ^r two perpendicular unit vectors at P. Let us construct a small rectangle, two sides of which emanate from P in the directions of λ^r and μ^r. If the lengths of these sides are $ds_{(1)}$, $ds_{(2)}$, respectively, we may put, in 3.519, $d_{(1)}x^m = \lambda^m ds_{(1)}$, $d_{(2)}x^n = \mu^n ds_{(2)}$. Then $ds_{(1)}ds_{(2)} = S$, the area of the rectangle (or, to be precise, its principal part), and 3.519 may be written

3.530.
$$\Delta T^r = -R^r_{.pmn}T^p\lambda^m\mu^n S.$$

Since the angle between two vectors is unaltered by parallel propagation, we can compute the angle through which T^r is turned by taking $T^r = \lambda^r$. Then, multiplication of 3.530 by μ_r, gives, on account of 3.408,

3.531.
$$\mu_r\Delta T^r = KS.$$

Hence, since $\mu_r T^r = \mu_r\lambda^r = 0$, we have

3.532. $$\mu_r(T^r + \Delta T^r) = KS.$$

The magnitude of a vector is unchanged by parallel propagation; hence the magnitude of $T^r + \Delta T^r$ is the same as that of λ^r, viz. unity. If θ is the angle through which T^r is turned, counted positive in the sense of passage from λ^r to μ^r, then $\mu_r(T^r + \Delta T^r) = \cos\left(\frac{1}{2}\pi - \theta\right) = \sin\theta$, and so, replacing $\sin\theta$ by θ, since we are interested only in the principal part, we have

3.533. $$\theta = KS,$$

the required result, from which 3.415 follows.

SUMMARY III

Non-commutative property of covariant differentiation:

$$T_{r|mn} - T_{r|nm} = R^s_{.rmn} T_s.$$

Non-commutative property of absolute differentiation:

$$\frac{\delta^2 T^r}{\delta u \delta v} - \frac{\delta^2 T^r}{\delta v \delta u} = R^r_{.pmn} T^p \frac{\partial x^m}{\partial u} \frac{\partial x^n}{\partial v}.$$

Parallel propagation round a small circuit:

$$\Delta T^r = -R^r_{.smn} T^s d_{(1)} x^m d_{(2)} x^n.$$

Mixed curvature tensor:

$$R^s_{.rmn} = \frac{\partial}{\partial x^m}\begin{Bmatrix} s \\ r\,n \end{Bmatrix} - \frac{\partial}{\partial x^n}\begin{Bmatrix} s \\ r\,m \end{Bmatrix} + \begin{Bmatrix} p \\ r\,n \end{Bmatrix}\begin{Bmatrix} s \\ p\,m \end{Bmatrix} - \begin{Bmatrix} p \\ r\,m \end{Bmatrix}\begin{Bmatrix} s \\ p\,n \end{Bmatrix}.$$

Covariant curvature tensor:

$$R_{rsmn} = \frac{1}{2}\left(\frac{\partial^2 a_{rn}}{\partial x^s \partial x^m} + \frac{\partial^2 a_{sm}}{\partial x^r \partial x^n} - \frac{\partial^2 a_{rm}}{\partial x^s \partial x^n} - \frac{\partial^2 a_{sn}}{\partial x^r \partial x^m}\right) + a^{pq}([rn, p][sm, q] - [rm, p][sn, q]).$$

Identities:

$$R_{rsmn} = -R_{srmn}, \quad R_{rsmn} = -R_{rsnm}, \quad R_{rsmn} = R_{mnrs},$$

$$R_{rsmn} + R_{rmns} + R_{rnsm} = 0.$$

Number of independent components in V_N: $\frac{1}{12}N^2(N^2 - 1)$.

Conditions for flatness: $R_{rsmn} = 0$.

Bianchi identity:

$$R^{r}_{.\,smn|t} + R^{r}_{.\,snt|m} + R^{r}_{.\,stm|n} = 0.$$

Ricci tensor:

$$\begin{aligned} R_{rm} = R_{mr} = R^{n}_{.\,rmn} \\ &= \tfrac{1}{2}\frac{\partial^2}{\partial x^r \partial x^m}\ln a - \tfrac{1}{2}\left\{ {p \atop r\,m} \right\}\frac{\partial}{\partial x^p}\ln a \\ &\quad - \frac{\partial}{\partial x^n}\left\{ {n \atop r\,m} \right\} + \left\{ {p \atop r\,n} \right\}\left\{ {n \atop m\,p} \right\}. \end{aligned}$$

Einstein tensor:

$$G^{n}_{.\,t} = R^{n}_{.\,t} - \tfrac{1}{2}\,\delta^{n}_{t}R, \quad G^{n}_{.\,t|n} = 0.$$

Geodesic deviation:

$$\frac{\delta^2\eta^r}{\delta s^2} + R^{r}_{.\,msn}p^m\eta^s p^n = 0.$$

Riemannian curvature:

$$\begin{aligned} K &= \epsilon(X)\epsilon(Y)R_{rsmn}X^rY^sX^mY^n \\ &= \frac{R_{rsmn}\xi^r\eta^s\xi^m\eta^n}{(a_{pu}a_{qv} - a_{pv}a_{qu})\xi^p\eta^q\xi^u\eta^v}. \end{aligned}$$

Spherical excess: $E = KS$.

EXERCISES III

1. Taking polar coordinates on a sphere of radius a, calculate the curvature tensor, the Ricci tensor, and the curvature invariant.

2. Take as manifold V_2 the surface of an ordinary right circular cone, and consider one of the circular sections. A vector in V_2 is propagated parallelly round this circle. Show that its direction is changed on completion of the circuit. Can you reconcile this result with the fact that V_2 is flat?

3. Consider the equations

$$(R_{mn} - \theta a_{mn})X^n = 0,$$

where R_{mn} is the Ricci tensor in a $V_N (N > 2)$, a_{mn} the metric tensor, θ an invariant, and X^n a vector. Show that, if these equations are to be consistent, θ must have one of a certain set of N values, and that the vectors X^r corresponding to different values of θ are perpendicular to one another. (The directions of these vectors are called the *Ricci principal directions.*)

4. What becomes of the Ricci principal directions (see above) if $N = 2$?

5. Suppose that two spaces V_N, V'_N have metric tensors a_{mn}, a'_{mn} such that $a'_{mn} = ka_{mn}$, where k is a constant. Write down the relations between the curvature tensors, the Ricci tensors, and the curvature invariants of the two spaces.

6. For an orthogonal coordinate system in a V_2 we have

$$ds^2 = a_{11}(dx^1)^2 + a_{22}(dx^2)^2.$$

Show that

$$\frac{1}{a} R_{1212} = -\tfrac{1}{2} \frac{1}{\sqrt{a}} \left[\frac{\partial}{\partial x^1}\left(\frac{1}{\sqrt{a}} \frac{\partial a_{22}}{\partial x^1} \right) + \frac{\partial}{\partial x^2}\left(\frac{1}{\sqrt{a}} \frac{\partial a_{11}}{\partial x^2} \right) \right].$$

7. Suppose that in a V_3 the metric is

$$ds^2 = (h_1 dx^1)^2 + (h_2 dx^2)^2 + (h_3 dx^3)^2,$$

where h_1, h_2, h_3 are functions of the three coordinates. Calculate the curvature tensor in terms of the h's and their derivatives. Check your result by noting that the curvature tensor will vanish if h_1 is a function of x^1 only, h_2 a function of x^2 only, and h_3 a function of x^3 only.

8. In relativity we encounter the metric form

$$\Phi = e^{\alpha}(dx^1)^2 + e^{x^1}[(dx^2)^2 + \sin^2 x^2 (dx^3)^2] - e^{\gamma}(dx^4)^2,$$

where α and γ are functions of x^1 and x^4 only.

Show that the complete set of non-zero components of the Einstein tensor (see equation (3.214)) for the form given above are as follows:

$$
\begin{aligned}
G_1^1 &= e^{-\alpha}\left(-\tfrac{1}{4}-\tfrac{1}{2}\gamma_1\right)+e^{-x^1},\\
G_2^2 &= e^{-\alpha}\left(-\tfrac{1}{4}-\tfrac{1}{2}\gamma_{11}-\tfrac{1}{4}\gamma_1{}^2-\tfrac{1}{4}\gamma_1+\tfrac{1}{4}\alpha_1+\tfrac{1}{4}\alpha_1\gamma_1\right)\\
&\qquad + e^{-\gamma}\left(\tfrac{1}{2}\alpha_{44}+\tfrac{1}{4}\alpha_4{}^2-\tfrac{1}{4}\alpha_4\gamma_4\right),\\
G_3^3 &= G_2^2,\\
G_4^4 &= e^{-\alpha}\left(-\tfrac{3}{4}+\tfrac{1}{2}\alpha_1\right)+e^{-x^1},\\
e^{\alpha}G_4^1 &= -e^{\gamma}G_1^4 = -\tfrac{1}{2}\alpha_4.
\end{aligned}
$$

The subscripts on α and γ indicate partial derivatives with respect to x^1 and x^4.

9. If we change the metric tensor from a_{mn} to $a_{mn}+b_{mn}$, where b_{mn} are small, calculate the principal parts of the increments in the components of the curvature tensor.

10. If we use normal coordinates in a Riemannian V_N, the metric form is as in equation 2.630. For this coordinate system, express the curvature tensor, the Ricci tensor, and the curvature invariant in terms of the corresponding quantities for the $(N-1)$-space $x^N=$ const. and certain additional terms. Check these additional terms by noting that they must have tensor character with respect to transformations of the coordinates $x^1, x^2, \ldots, x^{N-1}$.

11. Prove that

$$T^{mn}{}_{|mn} = T^{mn}{}_{|nm},$$

where T^{mn} is not necessarily symmetric.

CHAPTER IV

SPECIAL TYPES OF SPACE

4.1. Space of constant curvature. A general Riemannian space of N dimensions is bound to remain somewhat elusive as far as geometrical intuition is concerned. It is only when we specialize the type of space that simple and interesting properties emerge. We shall devote this section to spaces of constant curvature, passing on in the next section to the more specialized concept of flat spaces.

We lead up to the idea of a space of constant curvature by defining an *isotropic* point in a Riemannian space. It is a point at which the Riemannian curvature is the same for all 2-elements. This means that the quantity K in 3.411 is independent of the choice of the vectors ξ^r, η^r. So, if we define the tensor T_{rsmn} by the equation

4.101. $$T_{rsmn} = R_{rsmn} - K(a_{rm}\, a_{sn} - a_{rn}\, a_{sm}),$$

we have

4.102. $$T_{rsmn}\xi^r\eta^s\xi^m\eta^n = 0.$$

This is an identity in ξ^r, η^r, and so

4.103. $$T_{rsmn} + T_{msrn} + T_{rnms} + T_{mnrs} = 0.$$

On examining 4.101, we see that the tensor T_{rsmn} satisfies the same identical relations 3.115, 3.116 as R_{rsmn}. Thus

4.104. $$T_{rsmn} = -T_{srmn} = -T_{rsnm} = T_{mnrs},$$

4.105. $$T_{rsmn} + T_{rmns} + T_{rnsm} = 0.$$

From the equations 4.103, 4.104, and 4.105 it is possible to show that T_{rsmn} vanishes. This is done in the following steps.

First, by 4.104, we may change 4.103 to

4.106. $$T_{rsmn} + T_{rnms} = 0,$$

and, by 4.105, this may be changed to

4.107. $$T_{rsmn} - T_{rmsn} - T_{rsnm} = 0,$$

or

4.108. $$2T_{rsmn} - T_{rmsn} = 0.$$

Interchanging m and s, and adding the resulting equation, we get

4.109. $$T_{rsmn} + T_{rmsn} = 0.$$

Adding this to 4.108, we get $T_{rsmn} = 0$. Thus, it follows from 4.101 that *at an isotropic point the Riemannian curvature satisfies*

4.110. $$K(a_{rm}\, a_{sn} - a_{rn}\, a_{sm}) = R_{rsmn}.$$

Since a V_2 is isotropic (although in a rather trivial sense), the equation 4.110 holds throughout any V_2.

Exercise. Deduce from 4.110 that the Gaussian curvature of a V_2 with positive-definite metric is given by

$$G = \frac{R_{1212}}{a_{11}\, a_{22} - a_{12}^2}.$$

Now we come to a remarkable theorem, due to Schur: *If a Riemannian space $V_N (N > 2)$ is isotropic at each point in a region, then the Riemannian curvature is constant throughout that region.* In brief, isotropy implies homogeneity.

The proof of this theorem is based on 4.110, valid at each isotropic point, and Bianchi's identity in the form 3.121. We wish to deduce from 4.110 that K is a constant. Covariant differentiation of 4.110 with respect to x^t gives

4.111. $$K_{|t}(a_{rm}\, a_{sn} - a_{rn}\, a_{sm}) = R_{rsmn|t},$$

since the covariant derivative of the metric tensor vanishes. Permuting the subscripts m, n, t cyclically, and adding, the right-hand side disappears by 3.121 and we have

4.112.
$$\begin{aligned} &K_{|t}(a_{rm}\,a_{sn} - a_{rn}\,a_{sm}) \\ + &K_{|m}(a_{rn}\,a_{st} - a_{rt}\,a_{sn}) \\ + &K_{|n}(a_{rt}\,a_{sm} - a_{rm}\,a_{st}) = 0. \end{aligned}$$

Multiplication by $a^{rm}\,a^{sn}$ yields

4.113.
$$(N-1)(N-2)K_{|t} = 0.$$

But $N > 2$, and so $K_{|t} = 0$; but this is simply $\partial K/\partial x^t = 0$ and so K is a constant. Schur's theorem is proved.

A space for which K is constant (with respect to choice of 2-element at a point, and therefore by Schur's theorem with respect to choice of point) is called a *space of constant curvature.* The basic relation for a space of constant curvature is, as in 4.110,

4.114.
$$R_{rsmn} = K(a_{rm}\,a_{sn} - a_{rn}\,a_{sm}),$$

where K is the constant curvature.

It will be observed that in the proof of Schur's theorem we found it necessary to assume $N > 2$, and were led to 4.114 with K a constant. As already seen, 4.114 holds everywhere in any V_2, K being in general not a constant, but a function of position.

Exercise. Prove that, in a space V_N of constant curvature K,

4.115.
$$R_{mn} = -(N-1)Ka_{mn}, \quad R = -N(N-1)K.$$

The study of geodesic deviation in a space of constant curvature is interesting. We go back to the general equation of geodesic deviation 3.311 and substitute, by 4.114,

4.116.
$$R^{r}_{.\,smn} = K(\delta^{r}_{m}a_{sn} - \delta^{r}_{n}a_{sm}).$$

Noting that the unit tangent vector p^r is perpendicular to the deviation vector η^r, so that $a_{sm}p^s\eta^m = 0$, we see that the equation of geodesic deviation reduces to

4.117.
$$\frac{\delta^2\eta^r}{\delta s^2} + \epsilon K\eta^r = 0,$$

where ϵ is the indicator of the tangent to the geodesic.

This is not as simple an equation as might appear at first sight. We recall that $\delta^2\eta^r/\delta s^2$ is a complicated thing, involving Christoffel symbols and their derivatives. So any hope of treating 4.117 as a linear differential equation of the second order with constant coefficients seems over-optimistic. However, a simple device enables us to do this very thing. We introduce a unit vector field X_r, propagated parallelly along the geodesic, so that $\delta X_r/\delta s = 0$. If we multiply 4.117 by X_r, with of course the usual summation convention, the resulting equation may be written

4.118. $$\frac{d^2}{ds^2}(X_r\eta^r) + \epsilon K(X_r\eta^r) = 0.$$

The general solution of this is

4.119a. $X_r\eta^r = A \sin s\sqrt{\epsilon K} + B \cos s\sqrt{\epsilon K}$, if $\epsilon K > 0$,

4.119b. $X_r\eta^r = As + B$, if $K = 0$,

4.119c. $X_r\eta^r = A \sinh s\sqrt{-\epsilon K} + \mathrm{B} \cosh s\sqrt{-\epsilon K}$, if $\epsilon K < 0$,

where A and B are infinitesimal constants.

Consider now two adjacent geodesics drawn from a common point, from which we shall measure s. For $s = 0$, we have $\eta^r = 0$, and so the above equations become

4.120a. $X_r\eta^r = A \sin s\sqrt{\epsilon K}$, if $\epsilon K > 0$,

4.120b. $X_r\eta^r = As$, if $K = 0$,

4.120c. $X_r\eta^r = A \sinh s\sqrt{-\epsilon K}$, if $\epsilon K < 0$.

There is a remarkable difference between 4.120a and the other two equations. For 4.120a we find $X_r\eta^r = 0$ when $s = \pi/\sqrt{\epsilon K}$, but for the other equations $X_r\eta^r$ vanishes only for $s = 0$. Since X_r may be chosen arbitrarily at any one point of the geodesic, it follows that $\eta^r = 0$ in the case 4.120a when $s = \pi/\sqrt{\epsilon K}$. Thus, *if* $\epsilon K > 0$, *two adjacent geodesics issuing from a point intersect again at a distance*

4.121.
$$s = \frac{\pi}{\sqrt{\epsilon K}}$$

from the initial point.

Exercise. By taking an orthogonal set of N unit vectors propagated parallelly along the geodesic, deduce from 4.120a that the magnitude η of the vector η^r is given by

4.122.
$$\eta = C \mid \sin s\sqrt{\epsilon K} \mid ,$$

where C is a constant.

Let us investigate the consequences of the result expressed in 4.121, assuming that the metric form of the space of constant curvature is positive-definite, so that $\epsilon = 1$, and that the space is of *positive* constant curvature ($K > 0$). Consider the family of all geodesics drawn out from a point O. Any two adjacent geodesics of this family intersect at a distance

4.123.
$$s = \frac{\pi}{\sqrt{K}} \cdot$$

It follows that *all* the geodesics drawn out from O meet at a common point (say O'). From 4.122 it follows that the small angle at which two adjacent geodesics separate at O is

4.124.
$$\chi = \left(\frac{d\eta}{ds}\right)_{s=0} = C\sqrt{K},$$

and the angle at which they meet at O' is

4.125.
$$\chi' = -\left(\frac{d\eta}{ds}\right)_{s=\pi/\sqrt{K}} = C\sqrt{K}.$$

Thus $\chi = \chi'$. Hence it follows that, since all geodesics issuing from O fill all possible directions at O, all the geodesics coming in at O' fill all possible directions at O'. If we continue on any one of these geodesics, we shall arrive back at O after travelling a further distance $\pi/\sqrt{K}$.

The situation described above is familiar in the case of a sphere. All geodesics (great circles) drawn from a pole O meet again at the opposite pole O'. The geodesic distance OO' is $\pi/\sqrt{K} = \pi R$, if R is the radius of the sphere. But in a space of positive constant curvature a remarkable thing may occur of which we are not warned by our familiarity with the geometry of the sphere. *The point O' may be the point O itself.* To distinguish the two cases, the following terms are used:

Point O' different from O: the space is *antipodal* (or *spherical*).

Point O' coincident with O: the space is *polar* (or *elliptic*).

A mere knowledge of the line element or metric form does not tell us whether a space is antipodal or polar. It is a property "in the large." Given the metric form, we can find by direct calculation whether this form is positive-definite and whether the space is of positive constant curvature. If it is, then the space is either antipodal or polar, but we cannot tell which without further information.

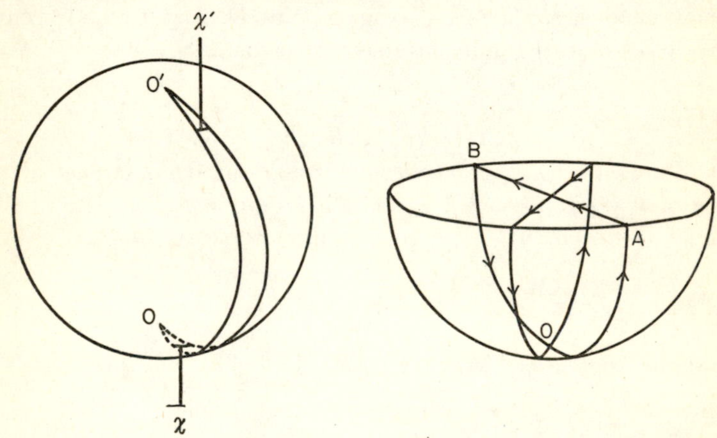

FIG. 9a. Antipodal, or spherical, 2-space.

FIG. 9b. Polar, or elliptic, 2-space.

To make the concept of a polar space more real, let us consider a model of such a space. The simplest model of an antipodal space is a sphere constructed in ordinary Euclidean space of three dimensions (Fig. 9a). Let us slice this sphere along the equator, throwing away the northern hemisphere and retaining the southern hemisphere. Let us agree that points A, B (Fig. 9b) at the ends of an equatorial diameter shall be regarded as identical. Then a geodesic (great circle) coming up from the south pole O to A jumps across to B and continues from B to O. All the geodesics (great circles) drawn out from O return to O, and do not intersect until they get back there.

We shall now use the idea of geodesic deviation to find an expression for the fundamental form in a 3-space of positive curvature K. We shall assume the metric form to be positive-definite.

Since any such Riemannian space is infinitesimally Euclidean, we can fix the initial direction of a geodesic OP issuing from a point O by means of the usual polar angles relative to an orthogonal triad at O (Fig. 10). Further, we can appeal to Euclidean geometry to get an expression for the small angle χ between two adjacent geodesics OP, OQ with directions (θ, ϕ),

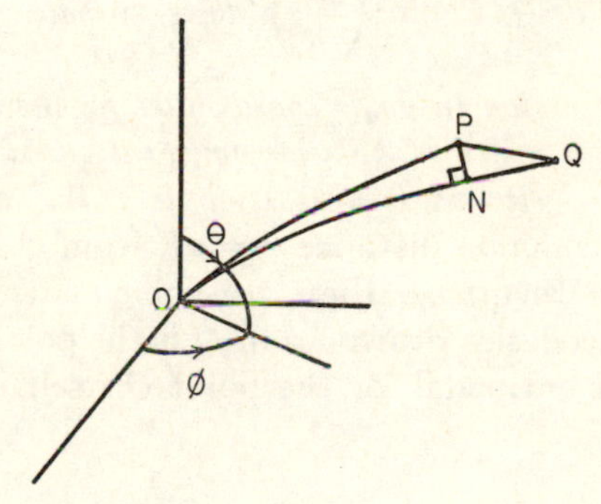

FIG. 10. Line element PQ, in a 3-space of constant curvature.

$(\theta + d\theta, \phi + d\phi)$ respectively. The value is

4.126. $$\chi = (d\theta^2 + \sin^2\theta d\phi^2)^{\frac{1}{2}}.$$

This follows from the fact that this expression represents the distance between adjacent points on a unit sphere in Euclidean 3-space.

Let us attach to P the coordinates r, θ, ϕ, where r is the geodesic distance OP, and to the adjacent point Q the coordinates $r + dr$, $\theta + d\theta$, $\phi + d\phi$. Let PN be the perpendicular dropped from P on the geodesic OQ. Then

4.127. $$ON = OP = r, \; NQ = dr, \; NP = \eta,$$

where η is the magnitude of the geodesic deviation. By 4.122 we have

4.128. $$\eta = C \sin (r/R), \quad R = 1/\sqrt{K},$$

R being the "radius of curvature" of the space. As in 4.124, the angle χ is C/R. Comparison with 4.126 gives

4.129. $$C = R(d\theta^2 + \sin^2\theta d\phi^2)^{\frac{1}{2}}.$$

From the infinitesimal right-angled triangle PNQ we have

$$PQ^2 = NQ^2 + NP^2.$$

Hence, denoting by ds the infinitesimal distance PQ, we have

4.130. $$ds^2 = dr^2 + R^2 \sin^2 \left(\frac{r}{R} \right) (d\theta^2 + \sin^2\theta d\phi^2).$$

This is the expression in polar coordinates for the line-element or metric form of a space of positive constant curvature $1/R^2$. We note that if $r = \pi R$ and $dr = 0$, then $ds = 0$. This means that points at the common distance $r = \pi R$ from O have zero distance between them, i.e. they are coincident. This meeting point of the geodesics drawn from O is the pole opposite to O if the space is antipodal, or the point O itself if the space is polar.

Exercise. Examine the limit of the form 4.130 as R tends to infinity, and interpret the result.

4.2. Flat space. In 3.1 we defined a flat space, and saw that the vanishing of the curvature tensor is a necessary condition for flatness. In 3.5 we saw that this condition was also sufficient for flatness. This means that if the curvature tensor vanishes, it is possible to choose coordinates y_r so that the metric form may be written, as in 3.529,

4.201. $$\Phi = \epsilon_1(dy_1)^2 + \epsilon_2(dy_2)^2 + \ldots + \epsilon_N(dy_N)^2,$$

where the ϵ's are $+1$ or -1. The presence of these ϵ's is a nuisance notationally, so we resort to the subterfuge of introducing coordinates which are imaginary when the corresponding ϵ's are negative. We write

4.202. $$z_1 = \sqrt{\epsilon_1} y_1, \quad z_2 = \sqrt{\epsilon_2} y_2, \ldots, \quad z_N = \sqrt{\epsilon_N} y_N,$$

and obtain

4.203. $$\Phi = dz_n dz_n.$$

Any system of coordinates for which Φ takes this form will be called *homogeneous*. In the Euclidean plane or 3-space rectangular Cartesian coordinates are homogeneous coordinates; but no imaginary coordinates occur in those cases because the metric form is positive-definite. In the space-time manifold of relativity one of the four homogeneous coordinates is imaginary, as shown in 4.230 below.

If we look back to the argument by which 3.529 was obtained, we see that there was considerable arbitrariness in the way the coordinates y_r were defined. It is obvious, in fact, that infinitely many systems of homogeneous coordinates exist in a given flat space. We shall now show that a linear transformation

4.204. $$z'_m = A_{mn} z_n + A_m,$$

transforms homogeneous system z_r into another homogeneous system z'_r provided the coefficients of the transformation satisfy certain conditions.

Exercise. Show that a transformation of a homogeneous coordinate system into another homogeneous system is necessarily linear. (Use the transformation equation 2.507 for Christoffel symbols, noting that all Christoffel symbols vanish when the coordinates are homogeneous).

The coordinates z'_r are homogeneous if and only if

4.205. $$\Phi = dz'_m dz'_m.$$

But from 4.204 we have

4.206. $$dz'_m dz'_m = A_{mp} A_{mq} dz_p dz_q,$$

and Φ is given by 4.203 since z_r are homogeneous. We deduce that z'_r are homogeneous if and only if

4.207. $$A_{mp} A_{mq} dz_p dz_q = dz_m dz_m,$$

identically in the dz's. We can write

4.208. $$dz_m dz_m = \delta_{pq} dz_p dz_q,$$

where δ_{pq} is the Kronecker delta defined in 1.207, with both suffixes written as subscripts for reasons which will appear later. Since the quadratic form on the left of 4.207 is to be identical with the quadratic form on the right of 4.208, and since the coefficients are symmetric in p and q, we obtain, as necessary and sufficient conditions for the homogeneity of z'_r,

4.209. $$A_{mp} A_{mq} = \delta_{pq}.$$

A linear transformation 4.204 whose coefficients satisfy 4.209 is said to be *orthogonal*. We may state this result: *Given one homogeneous coordinate system z_r, all other coordinate systems z'_r obtained from z_r by an orthogonal transformation are also homogeneous.* Conversely, *if a linear transformation carries one homogeneous coordinate system into another homogeneous coordinate system, the transformation is necessarily orthogonal.*

Exercise. If z_r, z'_r are two systems of rectangular Cartesian coordinates in Euclidean 3-space, what is the geometrical interpretation of the constants in 4.204 and of the orthogonality conditions 4.209?

We shall now prove a useful theorem: *If A_{mn} satisfy* 4.209, *they also satisfy*

4.210. $$A_{pm} A_{qm} = \delta_{pq}.$$

To prove this, we multiply 4.204 by A_{mp} and use 4.209. This gives

4.211. $$A_{mp} z'_m = z_p + A_{mp} A_m,$$

which may be written

4.212. $$z_p = A_{mp}z'_m + A'_p, \quad A'_p = -A_{mp}A_m.$$

Hence

4.213. $$dz_p dz_p = A_{mp}A_{np}dz'_m dz'_n.$$

But since 4.209 is satisfied, by assumption, the transformation is orthogonal and so

4.214. $$dz_p dz_p = dz'_p dz'_p = \delta_{mn}dz'_m dz'_n$$

identically, and hence, by comparison of 4.213 and 4.214, we obtain $A_{mp}A_{np} = \delta_{mn}$, which is 4.210 in different notation. The result is established.

The Jacobian of the transformation 4.204 is the determinant $|A_{mn}|$. (Refer to 1.202 for the definition of the Jacobian.) By the rule for the multiplication of determinants we have

$$|A_{mn}|^2 = |A_{mn}| \cdot |A_{rs}| = |A_{mn}A_{ms}| = |\delta_{ns}| = 1.$$

Hence $|A_{mn}|$ is either $+1$ or -1. We shall call an orthogonal transformation *positive** or *negative* according as the Jacobian is $+1$ or -1. There is a notable difference between the two types of transformation. A positive orthogonal transformation may be regarded as the result of the application of an infinite number of infinitesimal transformations, each positive orthogonal, starting from the identical transformation $z'_m = z_m$. At each stage, the Jacobian of the resultant transformation is $+1$. Such a procedure is impossible in the case of a negative orthogonal transformation, because the Jacobian of the identical transformation is $+1$, and it cannot change to -1 in a continuous process. The positive orthogonal transformation corresponds to a translation and rotation of axes in the Euclidean plane or 3-space. A negative orthogonal transformation corresponds to a translation and rotation of axes, followed by the reversal in direction of an odd number of axes, i.e., a reflection.

Let us now consider *geodesics* in a flat space. Using homogeneous coordinates, we see that the fundamental tensor is

*"Positive" transformations are also called ***proper.***

4.215. $$a_{mn} = \delta_{mn}.$$

Since these values are constants, all the Christoffel symbols vanish, and the differential equations 2.424 of a geodesic reduce to

4.216. $$\frac{d^2 z_r}{ds^2} = 0.$$

Integration gives

4.217. $$\frac{dz_r}{ds} = \alpha_r, \quad z_r = \alpha_r s + \beta_r.$$

The constants α_r are not independent of one another. We have $dz_m dz_m = \epsilon ds^2$, where ϵ is the indicator of the geodesic, and hence

4.218. $$\alpha_m \alpha_m = \epsilon.$$

In a general curved space, the determination of the geodesic distance between two points requires the integration of the differential equations of the geodesic. In general this cannot be done explicitly, and so the concept of the finite distance between two points plays a very minor role in the geometry of curved spaces. In flat space, on the other hand, the differential equations of a geodesic are already integrated in 4.217. An explicit expression for finite geodesic distance follows easily. If we pass along a geodesic, giving to the arc length s a finite increment Δs, we have, from the last of 4.217 and from 4.218,

4.219. $$\Delta z_r = \alpha_r \Delta s, \quad \Delta z_n \Delta z_n = \epsilon \Delta s^2.$$

Thus *the finite geodesic distance Δs between two points in flat space is equal to the square root of the absolute value of the sum of the squares of the finite coordinate differences Δz_r.*

The fact that the differential equations of a geodesic can be integrated as in 4.217 makes the geometry of a flat space very much simpler (and richer in interesting results) than the geometry of a curved space. Many terms familiar in ordinary Euclidean geometry in three dimensions can be given significant and simple definitions in a flat space of N dimensions. We shall now make some of these definitions.

A *straight line* is defined as a curve with parametric equations of the form

4.220. $$z_r = A_r u + B_r,$$

where the A's and B's are constants and u is a parameter. For such a curve we have

4.221. $$dz_r = A_r du, \quad \epsilon ds^2 = dz_n dz_n = A_n A_n du^2,$$
$$ds = \sqrt{\epsilon A_n A_n}\, du.$$

Unless $A_n A_n = 0$, it follows that dz_r/ds is constant; hence 4.216 is satisfied, and so the *straight line is a geodesic.* If $A_n A_n = 0$, it follows from 2.445 and 2.446 that the straight line is a geodesic null line. Thus the totality of straight lines includes all geodesics and all geodesic null lines.

A *plane* is defined by the equation

4.222. $$A_n z_n + B = 0,$$

where A_n and B are constants. It is easy to see that this surface (itself a space of $N - 1$ dimensions) has the familiar property of the plane in Euclidean 3-space, viz., *the straight line joining any two points in a plane lies entirely in the plane.*

A plane may also be called an $(N - 1)$-*flat.* This name suggests the definition of an $(N - 2)$-flat, an $(N - 3)$-flat, and so on. An $(N - 2)$-flat is defined as the totality of points whose coordinates satisfy the *two* linear equations

4.223. $$A_n z_n + B = 0, \quad C_n z_n + D = 0.$$

An $(N - 3)$-flat has *three* equations, and so on.

Exercise. Show that a one-flat is a straight line.

A sphere is defined by the equation

4.224. $$z_n z_n = \pm R^2,$$

R being a real constant, called the *radius* of the sphere. If the metric is positive definite, the $+$ sign must be chosen in 4.224. If the metric is indefinite either sign may be chosen, and there exist two families of spheres; these spheres have the unexpected property of extending to infinity.

In developing the geometry of a flat space of N dimensions, one naturally tries to carry over the familiar theory of three-dimensional Euclidean geometry as far as possible. A warning is not out of place regarding some possible properties of a flat space of N dimensions:

(1) The metric form may not be positive definite. In such cases we find points which are at zero distance from one another, but which are not coincident.

(2) The number of dimensions may be greater than three. In such cases a 2-flat, although in some ways analogous to the Euclidean plane, does not divide the space into two parts.

(3) The space may be topologically different from Euclidean space. For example, we can have a flat 2-space which has the topology of a torus; such a 2-space has a finite area.

In any serious discussion of flat spaces, the possibilities (1) and (2) must be considered. The situation (3) is extremely interesting, but we shall not discuss it further. In what follows we shall suppose that the space has Euclidean topology. This means that the coordinates y_r, for which the fundamental form is as in 4.201, take all values from $-\infty$ to $+\infty$, and that to each set of values of y_r there corresponds one distinct point.

Let us briefly consider a flat space of three dimensions with an indefinite metric form reducible to

4.225. $$\Phi = dy_1^2 + dy_2^2 - dy_3^2.$$

The corresponding homogeneous coordinates are

4.226. $$z_1 = y_1,\ z_2 = y_2,\ z_3 = iy_3.$$

Turning to the general equations 2.445 and 2.446 for geodesic null lines, we see at once that the geodesic null lines drawn from the origin have the equations

4.227. $$z_r = A_r u, \quad A_n A_n = 0.$$

Hence all points on this family of geodesic null lines satisfy an equation which may be written in either of the following forms:

4.228. $$z_n z_n = 0, \quad y_1^2 + y_2^2 - y_3^2 = 0.$$

Comparing the former equation with 4.224, we recognize that this surface is a sphere of zero radius. However, on account of the indefinite character of the metric form, it does not consist of a single point, like a sphere of zero radius in Euclidean space. The second equation of 4.228 shows that this surface is a cone; it is called a *null cone.* It extends to infinity, since the equation is satisfied by $y_1 = 0$ and any equal values of y_2 and y_3.

There is of course a null cone with vertex at any point; we took the vertex at the origin for simplicity in obtaining 4.228.

It is possible to make a model of the null cone 4.228 in Euclidean 3-space. Let y_r be rectangular Cartesian coordinates in Euclidean 3-space. The null cone then appears as in Fig. 11.

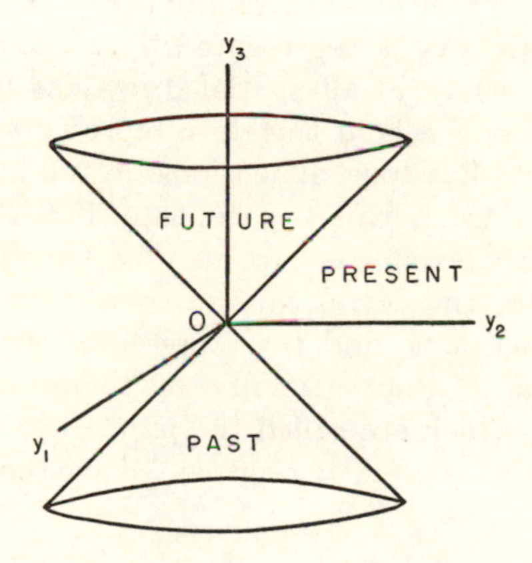

FIG. 11. The null cone in space-time.

It is a right circular cone, with a semi-vertical angle of 45°. In using this model we must bear in mind that the Euclidean line element $dy_1^2 + dy_2^2 + dy_3^2$ of the Euclidean space in which we make the model has no significance in the geometry for which the metric form is 4.225.

In the special theory of relativity, space-time is a four-dimensional space with an indefinite metric form, reducible to

4.229. $$dy_1^2 + dy_2^2 + dy_3^2 - dy_4^2$$

by suitable choice of coordinates. Of the four real coordinates y_r, the first three are rectangular Cartesian coordinates in the physical space of the observer, and the fourth y_4 is ct, where c is the constant velocity of light and t is the time as measured by the observer. The homogeneous coordinates are

4.230. $$z_1 = y_1,\ z_2 = y_2,\ z_3 = y_3,\ z_4 = iy_4 = ict.$$

The imaginary character of z_4 explains the introduction of the concept of "imaginary time" into the theory of relativity. It is merely a notational convenience, in order that we may use homogeneous coordinates.

A point in the space-time manifold is called an *event*. The history of a free particle is represented by a geodesic, and the history of a light ray is represented by a geodesic null line. Rays of light issuing in all spatial directions from an event give us a null cone, which therefore represents a light wave. Since space-time has four dimensions, a null cone cannot be represented in a Euclidean 3-space, as in Fig. 11. However, if we consider only events occurring in a single plane in the observer's space, the corresponding space-time manifold has only three dimensions, and the representation is possible. It is clear from Fig. 11 that this null cone divides space-time into three portions, which are called the *past*, the *present*, and the *future*, relative to the event represented by the vertex of the null cone.

Exercise. Show that the null cone with vertex at the origin in space-time has the equation

4.231. $$y_1^2 + y_2^2 + y_3^2 - y_4^2 = 0.$$

Prove that this null cone divides space-time into three regions such that

a. Any two points (events) both lying in one region can be joined by a continuous curve which does not cut the null cone.

b. All continuous curves joining any two given points (events) which lie in different regions, cut the null cone.

Show further that the three regions may be further classified into past, present, and future as follows: If A and B are any two points in the past, then the straight segment AB lies entirely in the past. If A and B are any two points in the future, then the straight segment AB lies entirely in the future. If A is any point in the present, there exists at least one point B in the present such that the straight segment AB cuts the null cone.

4.3. Cartesian tensors. Contravariant tensors were defined in 1.3 and covariant and mixed tensors in 1.4. The definitions were based on the formulae of transformation of the components corresponding to a general transformation of the coordinates $x'^r = f^r(x^1, x^2, \ldots, x^N)$, as in 1.201. These definitions did not involve the metrical geometry of the space in which x'^r and x^r are coordinates. However, the geometry appeared quickly in Chapter II, and the tensor concept proved of great service in the study of Riemannian space. Although special coordinate systems were introduced in 2.6, they were used only for special purposes.

But when we came to the study of flat spaces in 4.2, we found it advantageous to use homogeneous coordinates because in terms of them the metric form was particularly simple. As long as we keep to homogeneous coordinates, the transformations involved are not general transformations, but orthogonal transformations, as expressed in 4.204 and 4.209. For the discussion of flat spaces it is therefore wise to review our definitions of tensors. To avoid confusion, we keep the word "tensor" (unqualified) to denote quantities transforming according to the laws set down in Chapter I when the coordinates undergo a general transformation, and define as *Cartesian tensors* quantities which transform according to the same laws when the coordinates undergo an orthogonal transformation, i.e. when we pass from one set of homogeneous coordinates to another. In accordance with the notation used in the present chapter, homogeneous coordinates will be denoted by z_r, and the orthogonal transformation, with the relations between the coefficients, will be written

4.301. $$z'_m = A_{mn}z_n + A_m, \quad A_{mp}A_{mq} = \delta_{pq}.$$

In using the transformation laws of Chapter I for Cartesian tensors we are of course to substitute z for x in the partial derivatives, and depress the superscripts on the x's so that they become subscripts on the z's.

We use the word "Cartesian" because homogeneous coordinates in a flat space are analogous to rectangular Cartesian coordinates in a Euclidean plane or 3-space, and indeed reduce to rectangular Cartesians when the space reduces to the Euclidean plane or 3-space.

In order to qualify as a *tensor*, a set of quantities has to satisfy certain laws of transformation when the coordinates undergo a general transformation. This is a much more stringent condition than that which we impose on a set of quantities in order that it may qualify as a *Cartesian tensor*; in the latter case only *orthogonal* transformations are involved. Consequently, *every tensor is a Cartesian tensor, but the converse is not true.* (This is an abuse of language comparable to the following: "All horses are black horses, but the converse is not true." To avoid it, we might call the tensors of Chapter I "complete tensors." A simpler plan is to regard the expression "Cartesian tensor" as a single noun, not divisible into "Cartesian" and "tensor.")

Let us now investigate Cartesian tensors more closely. As in 4.212, the transformation 4.301 may be written

4.302. $$z_n = A_{mn}z'_m + A'_n.$$

From 4.301 and 4.302 we obtain

4.303. $$\frac{\partial z'_m}{\partial z_n} = A_{mn} = \frac{\partial z_n}{\partial z'_m}.$$

This relation leads to a remarkable simplification in the theory of Cartesian tensors. *For Cartesian tensors there is no distinction between contravariant and covariant components.* To establish this, consider the transformation formula 1.305 for a contravariant tensor of the second order. With the change from x to z noted above, this reads

4.304. $$T'^{rs} = T^{mn} \frac{\partial z'_r}{\partial z_m} \frac{\partial z'_s}{\partial z_n}.$$

By 4.303, this may be written

4.305. $$T'^{rs} = T^{mn} \frac{\partial z_m}{\partial z'_r} \frac{\partial z_n}{\partial z'_s}.$$

But this is the formula 1.408 for the transformation of a *covariant* tensor of the second order, viz.

4.306. $$T'_{rs} = T_{mn} \frac{\partial z_m}{\partial z'_r} \frac{\partial z_n}{\partial z'_s}.$$

Similar results hold for mixed tensors and tensors of other orders. *In every case the law of transformation remains unchanged when a subscript of the tensor is raised or a superscript lowered.* Consequently there is no point in using both superscripts and subscripts when dealing with Cartesian tensors. We shall use *subscripts* exclusively, and the words "contravariant" and "covariant" disappear. We have already anticipated this plan in writing the Kronecker delta in the form δ_{mn}, and also in writing the coordinates z_r with a subscript. If we restrict the orthogonal transformation 4.301 by deleting the constant A_m, the coordinates z_r are themselves the components of a Cartesian tensor of the first order, i.e. a vector. If we restore the constant A_m, this is no longer true, but the finite differences Δz_r of the coordinates of two points are the components of a Cartesian tensor.

Since the metric form in a flat space is

4.307. $$\Phi = dz_m dz_m,$$

the metric tensor is

4.308. $$a_{mn} = \delta_{mn}.$$

The conjugate tensor, previously written a^{mn}, is seen without difficulty to be precisely a_{mn} itself, i.e. the Kronecker delta.

Thus the artifice of lowering and raising suffixes, introduced in 2.2, ceases to have any significance, because the artifice now involves merely multiplication by the Kronecker delta, which does not alter the tensor at all.

Since a_{mn} are constants, all Christoffel symbols vanish, and the elaborate formulae for the differentiation of tensors (see 2.5) become extremely simple. The situation may be summed up by saying that *ordinary or partial derivatives of Cartesian tensors are themselves Cartesian tensors.* There is therefore no point in retaining the notations

$$\frac{\delta T_r}{\delta u}, \quad T_{rs|m};$$

we shall write instead

$$\frac{dT_r}{du}, \quad T_{rs,\, m},$$

using a subscript m, preceded by a comma, to denote partial differentiation with respect to the coordinate z_m.

A word of warning may be given here. It may be desirable, in dealing with a flat space, to use coordinates other than homogeneous coordinates, just as we frequently use polar coordinates in a plane. Once we abandon homogeneous coordinates, we must revert to general tensors. In fact, the technique of Cartesian tensors is available only when both the following conditions are fulfilled:

(i) The space is flat.

(ii) The coordinates are homogeneous.

In 4.2 we saw that there are two types of orthogonal transformations—positive transformations with $|A_{mn}| = +1$, and negative transformations with $|A_{mn}| = -1$. We are about to introduce tensors which are Cartesian tensors in a restricted sense, i.e. they obey the tensor laws of transformation with respect to a *positive* orthogonal transformation, but not with respect to a negative one. We shall call such quantities *oriented Cartesian tensors*.

Let us start with the simplest case—a flat space of two dimensions. Consider the *permutation symbols** ϵ_{mn} defined by

4.309.
$$\epsilon_{11} = 0, \qquad \epsilon_{12} = 1,$$
$$\epsilon_{21} = -1, \qquad \epsilon_{22} = 0.$$

These quantities do not involve any coordinate system, any more than the Kronecker delta does. Nevertheless, it is profitable to associate them with a coordinate system z_r. The permutation symbols associated with another coordinate system z'_r will be marked with a prime (ϵ'_{mn}), but will be defined by the same formulae 4.309. Let us test ϵ_{mn} for tensor character by seeing whether the equation

4.310.
$$\epsilon'_{rs} = \epsilon_{mn} \frac{\partial z_m}{\partial z'_r} \frac{\partial z_n}{\partial z'_s}$$

is true.

By 4.302, the right-hand side of 4.310 may be written

$$\epsilon_{mn} A_{rm} A_{sn}$$

or, by 4.309,

$$A_{r1} A_{s2} - A_{r2} A_{s1}.$$

Explicitly, the values of these quantities are as follows:

$$r = 1, s = 1;\ A_{11} A_{12} - A_{12} A_{11} = 0;$$
$$r = 1, s = 2;\ A_{11} A_{22} - A_{12} A_{21} = |A_{pq}|\ ;$$
$$r = 2, s = 1;\ A_{21} A_{12} - A_{22} A_{11} = -\ |A_{pq}|\ ;$$
$$r = 2, s = 2;\ A_{21} A_{22} - A_{22} A_{21} = 0.$$

If the orthogonal transformation is positive (as we shall now suppose), we have $|A_{pq}| = 1$. Then the four components of the right-hand side of 4.310 are equal respectively to the four components of the left-hand side of that equation, as given by 4.309. Therefore, *in a space of two dimensions* ϵ_{mn} *are the components of an oriented Cartesian tensor.*

In a space of three dimensions permutation symbols ϵ_{mnr} are defined by the following conditions:

(i) $\epsilon_{mnr} = 0$ if two of the suffixes are equal.

*For a general discussion of the permutation symbols, compare 7.1.

(ii) $\epsilon_{mnr} = 1$ if the sequence of numbers *mnr* is the sequence 123, or an even permutation of the sequence.

(iii) $\epsilon_{mnr} = -1$ if the sequence of numbers *mnr* is an odd permutation of the sequence 123.

This definition tells us that the non-zero components of ϵ_{mnr} are as follows:

4.311. $$\epsilon_{123} = \epsilon_{231} = \epsilon_{312} = 1, \quad \epsilon_{132} = \epsilon_{213} = \epsilon_{321} = -1.$$

We shall now show that ϵ_{mnr} are the components of an oriented Cartesian tensor, that is,

4.312. $$\epsilon'_{stu} = \epsilon_{mnr} \frac{\partial z_m}{\partial z'_s} \frac{\partial z_n}{\partial z'_t} \frac{\partial z_r}{\partial z'_u}.$$

This can be proved directly as we proved 4.310 in two dimensions but the work is tedious to write out. We therefore adopt a more general method. Consider the equation

4.313. $$|A_{pq}| = \epsilon_{mnr} A_{1m} A_{2n} A_{3r};$$

it is true, because the right-hand side consists of products of elements of the determinant $|A_{pq}|$, one element from each row, with the proper sign prefixed according to the algebraic rule for the expansion of a determinant. Now if we permute the suffixes 1 and 2 in the right-hand side of 4.313, forming the expression

4.314. $$\epsilon_{mnr} A_{2m} A_{1n} A_{3r},$$

we have the expansion of the determinant formed from $|A_{pq}|$ by interchanging the first and second rows. Therefore the expression 4.314 is equal to $-|A_{pq}|$. If we make further permutations of the suffixes 123, the resulting expression will be equal to $|A_{pq}|$ if the total number of permutations of the suffixes 123 is even, and it will be equal to $-|A_{pq}|$ if the total number of permutations is odd. Therefore the expression

4.315. $$\epsilon_{mnr} A_{sm} A_{tn} A_{ur}$$

is equal to $|A_{pq}|$ if *stu* is an even permutation of 123, and equal to $-|A_{pq}|$ if *stu* is an odd permutation of 123. Moreover, the expression 4.315 vanishes if two of the numbers *stu* are equal to one another, for then we have the expansion of a determinant with two identical rows. These statements establish the identity of the expression 4.315 with the expression $\epsilon_{stu}|A_{pq}|$, and so we have

4.316. $$\epsilon_{stu}|A_{pq}| = \epsilon_{mnr}A_{sm}A_{tn}A_{ur}.$$

This is a purely algebraic result for any set of quantities A_{mn}. Let us suppose that these quantities are the coefficients of a positive orthogonal transformation, so that $|A_{pq}| = 1$. Since the permutation symbols are defined by the same rules for all coordinate systems, we may write $\epsilon_{stu} = \epsilon'_{stu}$, and so

4.317. $$\epsilon'_{stu} = \epsilon_{mnr}A_{sm}A_{tn}A_{ur},$$

which is the same as 4.312. Therefore *in a space of three dimensions the permutation symbols* ϵ_{mnr} *are the components of an oriented Cartesian tensor.*

The definition of the permutation symbols is immediately extended to cover the case of a space of N dimensions. In V_N the permutation symbol has N suffixes,

$$\epsilon_{m_1 m_2 \ldots m_N},$$

and vanishes unless they are all different from one another. It is equal to $+1$ or -1 according as the number of permutations required to transform $m_1 m_2 \ldots m_N$ into $12 \ldots N$ is even or odd. There is no difficulty in extending the above proof to establish that *the permutation symbols in* V_N *are the components of an oriented Cartesian tensor.* It is obvious that the product of two permutation symbols is a Cartesian tensor which is not oriented. Any permutation symbol is, of course, skew-symmetric in any pair of suffixes. For example, in four dimensions

$$\epsilon_{mnrs} = -\epsilon_{mrns} = \epsilon_{msnr}.$$

Exercise. In a space of two dimensions prove the relation

4.318. $$\epsilon_{mp}\epsilon_{mq} = \delta_{pq}.$$

On account of the depression of all suffixes to the subscript position for Cartesian tensors, contraction is carried out at the subscript level. Thus, if X_n is a vector, X_nX_n is an invariant. So also is $X_{n,n}$, the divergence of the vector X_n. If X_n and Y_n are two vectors, then X_nY_n is an invariant, the scalar product of the vectors. Further, if T_{mn} and S_{mn} are two tensors, then

$$U_{ns} = T_{mn}S_{ms}$$

is a tensor. Similar remarks apply to oriented Cartesian tensors. For example, $\epsilon_{mnr}\epsilon_{mnr}$ is an invariant; its value is 6.

Quantities familiar in ordinary vector theory can be reconstructed with great ease in the present notation, which has five advantages over the ordinary notation. These advantages in most cases outweigh the slightly greater brevity of the ordinary notation. The advantages may be listed as follows:

1. The notation is explicit, i.e., it shows each component instead of using a single symbol to denote a vector.

2. The vector or invariant character of the expressions is immediately obvious to the eye.

3. The notation covers tensors as well as vectors.

4. There is no restriction to a positive definite metric form or line element.

5. Extensions to spaces of more than three dimensions are easy.

The treatment of oriented Cartesian tensors involving permutation symbols is particularly interesting. The vector product P_m of two vectors X_m and Y_m in three dimensions is defined by

4.319. $$P_m = \epsilon_{mnr}X_nY_r.$$

The analogous formula for two dimensions is

4.320. $$P = \epsilon_{mn}X_mY_n;$$

this tells us immediately that $X_1Y_2 - X_2Y_1$ is an oriented Cartesian invariant. Equation 4.320 suggests a formula in three dimensions

4.321. $$P = \epsilon_{mnr}X_mY_nZ_r.$$

This is the mixed triple product of three vectors; the formula tells us at once that it is an oriented Cartesian invariant, since it is formed by contraction from oriented Cartesian tensors.

It is interesting to see what happens to the idea of the vector product when we pass to a space of four dimensions. The following formulae suggest themselves:

4.322. $$P = \epsilon_{mnrs}X_mY_nZ_rW_s,$$

4.323. $$P_m = \epsilon_{mnrs}X_nY_rZ_s,$$

4.324. $$P_{mn} = \epsilon_{mnrs}X_rY_s,$$

4.325. $$P_{mnr} = \epsilon_{mnrs}X_s.$$

Equation 4.322 gives us the four-dimensional analogue of the mixed triple product; it tells us that the determinant formed from four vectors is an oriented Cartesian invariant. Equations 4.323 and 4.324 give us two extensions to four dimensions of the familiar vector product 4.319. In 4.323 a vector is formed from three given vectors; in 4.324 a skew-symmetric tensor is formed.

Exercise. Write out the six independent non-zero components of P_{mn} as given by 4.324.

The operator *curl* is important in ordinary vector analysis. In a space of three dimensions the curl of a vector field is defined by the formula

4.326. $$Y_m = \epsilon_{mnr}X_{r,n}.$$

In two dimensions the analogous formula is

4.327. $$Y = \epsilon_{mn}X_{n,m},$$

yielding an invariant. In four dimensions the analogous operation applied to a vector field yields a skew-symmetric tensor of the second order:

4.328. $$Y_{mn} = \epsilon_{mnrs}X_{s,r}.$$

The following relation is very useful for the treatment of continued vector multiplication in three dimensions:

4.329. $$\epsilon_{mrs}\epsilon_{mpq} = \delta_{rp}\delta_{sq} - \delta_{rq}\delta_{sp}.$$

This may be verified by writing out all components numerically, but the following procedure is a little shorter: The left-hand side is zero unless the numbers r, s are distinct and (p, q) is a permutation of (r, s), viz. $p = r$, $q = s$ or $p = s$, $q = r$. In the former case the left-hand side is 1, and in the latter case -1. Precisely the same remarks apply to the right-hand side, and thus 4.329 is established.

Exercise. Translate the well-known vector relations

$$\mathbf{A} \times (\mathbf{B} \times \mathbf{C}) = \mathbf{B}\,(\mathbf{A} \cdot \mathbf{C}) - \mathbf{C}\,(\mathbf{A} \cdot \mathbf{B}),$$
$$\nabla \times (\nabla \times \mathbf{V}) = \nabla\,(\nabla \cdot \mathbf{V}) - \nabla^2\mathbf{V},$$

into Cartesian tensor form, and prove them by use of 4.329.

4.4. A space of constant curvature regarded as a sphere in a flat space. An ordinary sphere can be regarded in two different ways. First, as a surface in Euclidean space of three dimensions, its curvature being put in evidence by the fact that a tangent line at any point deviates from the surface. Secondly, as a 2-space, without consideration of any points other than the points in the 2-space itself. In this case the curvature is put in evidence by measurements made entirely in the surface, these measurements showing that the angle-sum of a geodesic triangle exceeds two right angles.

We think of an N-space of constant curvature primarily as a manifold in itself. We do not think of going out of the manifold, as we do when we step off an ordinary sphere into the surrounding Euclidean space. However, we now inquire whether it is not possible to consider a space of constant curvature as a surface embedded in a flat space of one more dimension, just as an ordinary sphere is embedded in Euclidean 3-space.

Let us take a flat space V_N with metric form which reduces to

4.401. $$\Phi = dz_n dz_n,$$

for homogeneous coordinates z_r. The equation

4.402.
$$z_n z_n = C$$
defines a sphere V_{N-1} in V_N, C being a constant. This equation may be solved for $z^2{}_N$:

4.403.
$$z^2{}_N = C - z_\mu z_\mu.$$
(Greek suffixes have the range of values 1, 2, . . ., $N - 1$.) For any infinitesimal displacement in V_{N-1} we have

4.404.
$$z_N dz_N = - z_\mu dz_\mu,$$
and so the metric form of V_N for a displacement in V_{N-1} is

4.405.
$$\begin{aligned} \Phi &= dz_\mu dz_\mu + (dz_N)^2 \\ &= dz_\mu dz_\mu + z_\mu dz_\mu \cdot z_\nu dz_\nu / z^2{}_N \\ &= a_{\mu\nu} dz_\mu dz_\nu, \end{aligned}$$
where

4.406.
$$a_{\mu\nu} = \delta_{\mu\nu} + \frac{z_\mu z_\nu}{C - z_\rho z_\rho}.$$

The coordinates z_ρ are $N - 1$ coordinates in the manifold V_{N-1}; $a_{\mu\nu}$ is the metric tensor of that manifold. It is not hard to calculate the Christoffel symbols of V_{N-1} for this system of coordinates. Direct computation from 2.421 with Greek suffixes gives

4.407.
$$[\mu\nu, \rho] = \frac{\delta_{\mu\nu} z_\rho}{C - z_\sigma z_\sigma} + \frac{z_\mu z_\nu z_\rho}{(C - z_\sigma z_\sigma)^2}.$$

Consider the point P on the sphere with coordinates $z_\rho = 0$, $z_N = C^{\frac{1}{2}}$. Denoting values there with the subscript 0, we have
$$(a_{\mu\nu})_0 = \delta_{\mu\nu},$$

4.408.
$$[\mu\nu, \rho]_0 = 0,$$
$$\left(\frac{\partial}{\partial z_\sigma}[\mu\nu, \rho]\right)_0 = C^{-1}\delta_{\mu\nu}\delta_{\rho\sigma}.$$

Hence, by (3.113) with Greek suffixes, we have for the Riemann tensor of V_{N-1} at P

4.409.
$$(R_{\rho\sigma\mu\nu})_0 = C^{-1}(\delta_{\nu\sigma}\delta_{\mu\rho} - \delta_{\mu\sigma}\delta_{\nu\rho})$$
$$= C^{-1}(a_{\nu\sigma}a_{\mu\rho} - a_{\mu\sigma}a_{\nu\rho})_0.$$

Comparison of this result with 4.110 shows that P is an isotropic point of V_{N-1} with Riemannian curvature C^{-1} at that point. This has been proved only for a special point P with coordinates $z_\rho = 0$, $z_N = C^{\frac{1}{2}}$. However, by means of an orthogonal transformation

4.410.
$$z'_m = A_{mn}z_n, \quad A_{mp}A_{mq} = \delta_{pq},$$

the coordinates of any point on V_{N-1} may be given these values. (The situation is essentially the same as if we were dealing with an ordinary sphere in Euclidean 3-space. We could rotate the axes of rectangular Cartesian coordinates so that two of the coordinates of any assigned point P on the sphere become zero.) It follows that every point of V_{N-1} is an isotropic point with Riemannian curvature C^{-1}. Since the radius R of the sphere is defined by $R^2 = C$, we have this result: *A sphere of radius R in a flat V_N is itself a space of $N - 1$ dimensions with constant Riemannian curvature* $1/R^2$.

SUMMARY IV

Space of constant curvature K:

$$R_{rsmn} = K(a_{rm}a_{sn} - a_{rn}a_{sm}),$$

$$\frac{\delta^2\eta^r}{\delta s^2} + \epsilon K\eta^r = 0,$$

$$ds^2 = dr^2 + R^2 \sin^2\left(\frac{r}{R}\right)(d\theta^2 + \sin^2\theta\, d\phi^2), \quad K = 1/R^2.$$

Flat space with homogeneous coordinates:

$$\Phi = dz_n dz_n.$$

Orthogonal transformation:

$$z'_m = A_{mn}z_n + A_m, \quad A_{mp}A_{mq} = \delta_{pq};$$
$$z_m = A_{nm}z'_n + A'_m, \quad A_{pm}A_{qm} = \delta_{pq}.$$

(Positive transformation if $|A_{mn}| = 1$.)

Cartesian tensors:

$$T'_{rs} = T_{mn}\frac{\partial z_m}{\partial z'_r}\frac{\partial z_n}{\partial z'_s} = T_{mn}A_{rm}A_{sn}.$$

Scalar product: X_nY_n. Divergence: $X_{n,n}$.

In three dimensions:

$$\epsilon_{mrs}\epsilon_{mpq} = \delta_{rp}\delta_{sq} - \delta_{rq}\delta_{sp}.$$

Vector product: $\epsilon_{mnr}X_nY_r$. Curl: $\epsilon_{mnr}X_{r,n}$.

EXERCISES IV

1. Show that, in a 3-space of constant negative curvature $-1/R^2$ and positive-definite metric form, the line element in polar coordinates is

$$ds^2 = dr^2 + R^2\sinh^2\left(\frac{r}{R}\right)(d\theta^2 + \sin^2\theta d\phi^2).$$

2. Show that the volume of an antipodal 3-space of positive definite metric form and positive constant curvature $1/R^2$ is $2\pi^2R^3$. (Use the equation 4.130 to find the area of a sphere r = constant in polar coordinates. Multiply by dr and integrate for $0 \leqslant r \leqslant \pi R$ to get the volume.) What is the volume if the space is polar?

3. By direct calculation of the tensor R_{rsmn} verify that 4.130 is the metric form of a space of constant curvature.

4. Show that if V_N has a positive-definite metric form and constant positive curvature K, then coordinates y^r exist so that

$$ds^2 = \frac{dy^m dy^m}{(1 + \frac{1}{4}Ky^ny^n)^2}.$$

(Starting with a general coordinate system x^r, take at any point P the coordinates

$$y^r = p^r\frac{2}{\sqrt{K}}\tan\left(\tfrac{1}{2}r\sqrt{K}\right),$$

where p^r are the components of the unit tangent vector (dx^r/ds) at O to the geodesic OP and r is the geodesic distance OP.)

5. Show that in a flat V_N the straight line joining any two points of a P-flat $(P < N)$ lies entirely in the P-flat.

6. Show that in four dimensions the transformation

$$\begin{aligned} z_1' &= z_1 \cosh \phi + i z_4 \sinh \phi, \\ z_2' &= z_2, \quad z_3' = z_3, \\ z_4' &= -i z_1 \sinh \phi + z_4 \cosh \phi, \end{aligned}$$

is orthogonal, ϕ being any constant. Putting $z_1 = x$, $z_2 = y$, $z_3 = z$, $z_4 = ict$, $\tanh \phi = v/c$, obtain the transformation connecting (x', y', z', t') and (x, y, z, t). This is the *Lorentz transformation* of the special theory of relativity.

7. Prove that in a flat space a plane, defined by 4.222, is itself a flat space of $N - 1$ dimensions.

8. Show that in a flat space of positive-definite metric form a sphere of zero radius consists of a single point, but that if the metric form is indefinite a sphere of zero radius extends to infinity.

9. Prove that in two dimensions

$$\epsilon_{mn}\epsilon_{pq} = \delta_{mp}\delta_{nq} - \delta_{mq}\delta_{np}.$$

10. If, in a space of four dimensions, F_{mn} is a skew-symmetric Cartesian tensor, and $\hat{F}_{mn}$ is defined by

$$\hat{F}_{mn} = \tfrac{1}{2}\,\epsilon_{rsmn}F_{rs},$$

prove that the differential equations

$$F_{mn,r} + F_{nr,m} + F_{rm,n} = 0$$

may be written

$$\hat{F}_{mn,n} = 0.$$

11. Write out explicitly and simplify the expressions

$$F_{mn}F_{mn}, \quad \epsilon_{mnrs}F_{mn}F_{rs},$$

where F_{mn} is a skew-symmetric oriented Cartesian tensor. What is the tensor character of these expressions?

12. Show that in a flat space with positive-definite metric form all spheres have positive constant curvature. Show that if the metric form is indefinite then some spheres have positive constant curvature and some have negative constant curvature. Discuss the Riemannian curvature of the null cone.

13. Show that in any space of three dimensions the permutation symbols transform according to

$$\epsilon'_{mnr} = \epsilon_{stu}\, J' \frac{\partial x^s}{\partial x'^m} \frac{\partial x^t}{\partial x'^n} \frac{\partial x^u}{\partial x'^r}, \quad J' = \left| \frac{\partial x'^p}{\partial x^q} \right|,$$

or

$$\epsilon'_{mnr} = \epsilon_{stu}\, J \frac{\partial x'^m}{\partial x^s} \frac{\partial x'^n}{\partial x^t} \frac{\partial x'^r}{\partial x^u}, \quad J = \left| \frac{\partial x^p}{\partial x'^q} \right|.$$

Using the result of Exercises II, 12, deduce that in a Riemannian 3-space the quantities η_{mnr} and η^{mnr} defined by

$$\eta_{mnr} = \epsilon_{mnr}\sqrt{a}, \quad \eta^{mnr} = \epsilon_{mnr}/\sqrt{a}, \quad a = |a_{pq}|,$$

are components of covariant and contravariant oriented tensors respectively.

14. Translate into Cartesian tensor form and thus verify the following well known vector relations:

$$\begin{aligned}
\nabla \cdot (\phi \mathbf{V}) &= \phi \nabla \cdot \mathbf{V} + \mathbf{V} \cdot \nabla \phi, \\
\nabla \times (\phi \mathbf{V}) &= \phi \nabla \times \mathbf{V} - \mathbf{V} \times \nabla \phi, \\
\nabla \cdot (\mathbf{U} \times \mathbf{V}) &= \mathbf{V} \cdot (\nabla \times \mathbf{U}) - \mathbf{U} \cdot (\nabla \times \mathbf{V}), \\
\nabla \times (\mathbf{U} \times \mathbf{V}) &= \mathbf{V} \cdot \nabla \mathbf{U} - \mathbf{U} \cdot \nabla \mathbf{V} + \mathbf{U} \nabla \cdot \mathbf{V} - \mathbf{V} \nabla \cdot \mathbf{U}, \\
\nabla \; (\mathbf{U} \cdot \mathbf{V}) &= \mathbf{U} \cdot \nabla \mathbf{V} + \mathbf{V} \cdot \nabla \mathbf{U} + \mathbf{U} \times (\nabla \times \mathbf{V}) + \mathbf{V} \times (\nabla \times \mathbf{U}), \\
\nabla \times (\nabla \phi) &= 0, \\
\nabla \cdot (\nabla \times \mathbf{V}) &= 0, \\
\nabla \times (\nabla \times \mathbf{V}) &= \nabla \nabla \cdot \mathbf{V} - \nabla^2 \mathbf{V}, \\
\nabla \cdot \mathbf{r} &= 3, \\
\nabla \times \mathbf{r} &= 0, \\
\mathbf{V} \cdot \nabla \mathbf{r} &= \mathbf{V},
\end{aligned}$$

where **r** is the vector with components equal to the Cartesian coordinates z_1, z_2, z_3.

15. Prove that

$$\epsilon_{amn}\epsilon_{ars} + \epsilon_{ams}\epsilon_{anr} = \epsilon_{amr}\epsilon_{ans}.$$

CHAPTER V

APPLICATIONS TO CLASSICAL DYNAMICS

5.1. Physical components of tensors. Tensor calculus came into prominence with the development of the general theory of relativity by Einstein in 1916. It provides the only suitable mathematical language for general discussions in that theory. But actually the tensor calculus is older than that. It was invented by the Italian mathematicians Ricci and Levi-Civita, who published in 1900 a paper showing its applications in geometry and classical mathematical physics. It can also be used in the special theory of relativity, which is that simpler form of the theory of relativity covering physical phenomena which do not involve gravitation. Thus tensor calculus comes near to being a universal language in mathematical physics. Not only does it enable us to express general equations very compactly, but it also guides us in the selection of physical laws, by indicating automatically invariance with respect to the transformation of coordinates.

The present chapter is devoted to the use of tensor calculus in classical mechanics. The "space" of classical mechanics is a Euclidean space of three dimensions. In choosing a system of coordinates, as a general rule it is best to use rectangular Cartesians. If we restrict ourselves to these, the only transformations we have to consider are orthogonal transformations. If we further restrict ourselves to axes with one orientation (say right-handed axes), the Jacobian of the transformation is $+1$. The tensors which present themselves are then oriented Cartesian tensors, and the equations of mechanics are tensor equations in that sense.

On the other hand, if we are dealing with problems in which certain surfaces play an important part, it may be advisable to abandon Cartesian coordinates in favour of other systems in which the equations of the surfaces in question take simple forms. Thus, if a sphere is involved, spherical polar coordinates are indicated; if a cylinder is involved, cylindrical coordinates. In order to take into consideration all possible systems of curvilinear coordinates, it is wise to forget that we are dealing with a flat space and consider general transformations of coordinates as in 1.2. With respect to such transformations the entities of mechanics (velocity, momentum, and so on) behave like tensors in the general sense, and the equations of mechanics are tensor equations in the general sense. To help in distinguishing rectangular Cartesians from curvilinear coordinates, we shall write z_r (with a subscript) for rectangular Cartesians, and x^r (with a superscript) for curvilinear coordinates. The Latin suffixes have the range of values 1, 2, 3, since we are dealing with a 3-space.

Consider any familiar vector, such as the velocity of a particle. If we use rectangular Cartesians z_r, we may denote its components by Z_r. We recall that as long as we restrict ourselves to rectangular Cartesians, the contravariant components of a vector are the same as the covariant components. Thus we are entitled, if we like, to write $Z^r = Z_r$. The quantities Z_r are called the *physical components of the vector along the coordinate axes.*

If we now introduce curvilinear coordinates x^r, we may define a contravariant vector X^r and a covariant vector X_r by the transformation formulae (cf. 1.302 and 1.402)

5.101. $$X^r = Z^s \frac{\partial x^r}{\partial z_s}, \quad X_r = Z_s \frac{\partial z_s}{\partial x_r}.$$

These quantities are called *the contravariant and covariant components* of the vector in question for the coordinate system x^r. We do not use the word "physical" in connection with these components, because in general they have no direct physical meaning; indeed they may have physical dimensions different from those of the physical components Z_r.

So far we have considered the physical components of a vector along the coordinate axes, and also its contravariant and covariant components for a curvilinear coordinate system. We shall now introduce a third set of components, namely, the physical components in assigned directions, and, in particular, the physical components along the parametric lines of orthogonal curvilinear coordinates.

Let x^r be a curvilinear coordinate system with metric tensor a_{mn}. Let X^r be the vector whose components are under discussion, and let λ^r be an assigned unit vector. We define *the physical component of the vector* X^r *in the direction of* λ^r to be the invariant $a_{mn}X^m\lambda^n$. By the usual rules for raising and lowering suffixes, we have

5.102. $$a_{mn}X^m\lambda^n = X^n\lambda_n = X_n\lambda^n.$$

Let us now introduce rectangular Cartesians z_r and see what this definition means. Let Z_r and ζ_r be the components of X^r and λ^r respectively in this coordinate system. Then, since the expressions in 5.102 are invariants, we have

5.103. $$X^n\lambda_n = Z^n\zeta_n = Z_n\zeta_n.$$

But $Z_n\zeta_n$ is the scalar product of the vectors Z_r and ζ_r. It is, in fact, the projection of Z_r on the direction of ζ_r, and so is the component of Z_r in that direction, the word "component" being used in the sense commonly understood in discussing the resolution of forces, velocities, and other vectors in mechanics. Thus the invariant $a_{mn}X^m\lambda^n$ represents the physical component of X^r in the direction of λ^r in the usual sense of orthogonal projection.

Now suppose that the curvilinear coordinates x^r are orthogonal coordinates, so that as in 2.658 the line element is

5.104. $$ds^2 = (h_1dx^1)^2 + (h_2dx^2)^2 + (h_3dx^3)^2.$$

Let us take the unit vector λ^r in the direction of the parametric line of x^1, so that

5.105. $$\lambda^1 = dx^1/ds,\quad \lambda^2 = 0,\quad \lambda^3 = 0.$$

Since it is a *unit* vector, we have

5.106. $$1 = a_{mn}\lambda^m\lambda^n = h_1^2(\lambda^1)^2,$$

and so

5.107. $$\lambda^1 = 1/h_1.$$

Lowering superscripts by means of the metric tensor of 5.104, we have also

5.108. $$\lambda_1 = h_1,\quad \lambda_2 = 0,\quad \lambda_3 = 0.$$

Hence, by 5.102, the physical component of the vector X^r along the parametric line of x^1 is X_1/h_1 or h_1X^1. Collecting similar results for the other parametric lines, we have the following result: *For orthogonal curvilinear coordinates with line element* 5.104, *the physical components of a vector X^r are*

5.109. $$X_1/h_1,\quad X_2/h_2,\quad X_3/h_3,$$

or equivalently

5.110. $$h_1X^1,\quad h_2X^2,\quad h_3X^3.$$

Let us now consider the components of a tensor of the second order. We start with rectangular Cartesians z_r, then pass to general curvilinear coordinates x^r, and finally specialize these to orthogonal coordinates. Usually in mechanics the components Z_{rs} for Cartesians z_r present themselves first. We call them *the physical components along the coordinate axes.* Then for curvilinear coordinates x^r we define *contravariant and covariant components* by the transformation equations

5.111. $$X^{mn} = Z^{rs}\frac{\partial x^m}{\partial z_r}\frac{\partial x^n}{\partial z_s} = Z_{rs}\frac{\partial x^m}{\partial z_r}\frac{\partial x^n}{\partial z_s},$$
$$X_{mn} = Z_{rs}\frac{\partial z_r}{\partial x^m}\frac{\partial z_s}{\partial x^n}.$$

We can define mixed components also, using the fact that for a Cartesian tensor it is permissible to push a suffix up or down:

5.112. $$X^m_{\cdot n} = Z^r_{\cdot s}\frac{\partial x^m}{\partial z_r}\frac{\partial z_s}{\partial x^n} = Z_{rs}\frac{\partial x^m}{\partial z_r}\frac{\partial z_s}{\partial x^n}.$$

Next, we introduce two unit vectors (which may coincide), with contravariant components λ^r, μ^r in the curvilinear coordinates, and define *the physical component of the tensor along these directions* to be the invariant

$$X_{mn}\lambda^m\mu^n. \tag{5.113}$$

Finally, taking x^r to be orthogonal curvilinear coordinates, we select the unit vector λ^r along a parametric line, and the unit vector μ^r along the same parametric line or along another. We obtain from 5.113, by the same type of argument as we use for a vector, nine *physical components along the parametric lines*. These physical components are expressed in terms of covariant components as follows:

$$\begin{matrix} X_{11}/h_1^2, & X_{12}/h_1h_2, & X_{13}/h_1h_3, \\ X_{21}/h_2h_1, & X_{22}/h_2^2, & X_{23}/h_2h_3, \\ X_{31}/h_3h_1, & X_{32}/h_3h_2, & X_{33}/h_3^2. \end{matrix} \tag{5.114}$$

In terms of contravariant components they are as follows:

$$\begin{matrix} h_1^2X^{11}, & h_1h_2X^{12}, & h_1h_3X^{13}, \\ h_2h_1X^{21}, & h_2^2X^{22}, & h_2h_3X^{23}, \\ h_3h_1X^{31}, & h_3h_2X^{32}, & h_3^2X^{33}. \end{matrix} \tag{5.115}$$

The procedure set out above is logical. It enables us to pass from the physical components along coordinate axes to contravariant and covariant components in curvilinear coordinates, and finally to physical components along parametric lines of orthogonal coordinates. However, the procedure may be considerably shortened by making proper use of the tensor idea. To illustrate this, we shall now consider a simple fundamental problem.

Problem: To obtain, for spherical polar coordinates (r, θ, ϕ), contravariant and covariant components of velocity, and also the physical components along the parametric lines.

Taking z_r to be rectangular Cartesians, and putting $x^1 = r$, $x^2 = \theta$, $x^3 = \phi$, we have the transformation

5.116.
$$z_1 = x^1 \sin x^2 \cos x^3,$$
$$z_2 = x^1 \sin x^2 \sin x^3,$$
$$z_3 = x^1 \cos x^2.$$

The physical components of velocity along the coordinate axes are

5.117.
$$V_r = dz_r/dt.$$

Let v^r and v_r be the contravariant and covariant components of velocity. The formulae 5.101 tell us that

5.118.
$$v^r = V_s \frac{\partial x^r}{\partial z_s}, \quad v_r = V_s \frac{\partial z_s}{\partial x^r}.$$

To carry out these computations we need the 18 partial derivatives $\partial z_s/\partial x^r$ and $\partial x^r/\partial z_s$, to be calculated from 5.116. We sidestep this complicated calculation in the following way. We look at the formula 5.117, and ask ourselves: Is there any simple contravariant or covariant vector which reduces to dz_r/dt when the coordinates are reduced to rectangular Cartesians? The answer is immediate: dx^r/dt is such a contravariant vector. So we boldly write

5.119.
$$v^r = dx^r/dt,$$

and justify this statement as follows. Equation 5.119 is a tensor equation, and so it is true for all coordinate systems if it is true for one. But it is true for rectangular Cartesians by 5.117, since V_r and v^r are different representations of the same vector, and so also are dz_r/dt and dx^r/dt. Therefore 5.119 is true, in the sense that it gives the values of v^r demanded by 5.118.

To get the covariant components v_r, we introduce the metric tensor a_{mn}. The line element in spherical polars is

5.120.
$$ds^2 = dr^2 + r^2 d\theta^2 + r^2 \sin^2\theta \, d\phi^2$$
$$= (dx^1)^2 + (x^1 dx^2)^2 + (x^1 \sin x^2 \, dx^3)^2,$$

and so

5.121.
$$a_{11} = 1, \; a_{22} = (x^1)^2, \; a_{33} = (x^1 \sin x^2)^2,$$
$$a_{mn} = 0 \text{ for } m \neq n.$$

Lowering the superscript in the usual way, we get

5.122. $$v_r = a_{rm}v^m,$$

or explicitly, from 5.119,

5.123. $$v_1 = dx^1/dt,\ v_2 = (x^1)^2dx^2/dt,\ v_3 = (x^1 \sin x^2)^2dx^3/dt.$$

Once we have seen our way through this reasoning, we may shorten the work by omitting the symbols z_r and x^r and working entirely with r, θ, ϕ. We reason as follows: The contravariant components of velocity are the time-derivatives of the coordinates as in 5.119, and so

5.124. $$v^1 = dr/dt,\ v^2 = d\theta/dt,\ v^3 = d\phi/dt.$$

(Note that only the first component has the dimensions of velocity—length divided by time; the other two components have the dimensions of angular velocity.) The line element is as in 5.120, so that

5.125. $$a_{11} = 1,\ a_{22} = r^2,\ a_{33} = r^2 \sin^2 \theta,$$
$$a_{mn} = 0 \text{ for } m \neq n.$$

On lowering the superscripts in 5.124, we get the covariant components

5.126. $$v_1 = dr/dt,\ v_2 = r^2d\theta/dt,\ v_3 = r^2 \sin^2 \theta\, d\phi/dt.$$

Comparison of 5.104 and 5.120 gives

5.127. $$h_1 = 1,\ h_2 = r,\ h_3 = r \sin \theta.$$

Hence, by 5.110, the physical components of velocity along the parametric lines of spherical polar coordinates are

5.128. $$v_r = dr/dt,\ v_\theta = rd\theta/dt,\ v_\phi = r \sin \theta\, d\phi/dt.$$

These components are easily checked by considering the components of a general infinitesimal displacement along the parametric lines. Note that v_r, v_θ, v_ϕ all have the dimensions of velocity.

The preceding problem has been discussed at some length to show how tensor ideas are actually used. Tensor theory sets up a logical but clumsy procedure. Then, by a little judicious guessing, usually very easy, we short-circuit tedious computations, and arrive quickly at the required answer. We appeal to tensor theory to justify the guess and hence the answer.

5.2. Dynamics of a particle. Our purpose is to discuss the dynamics of a particle using general curvilinear coordinates x^r in ordinary Euclidean space. The plan is to write down tensorial expressions which reduce to known quantities when the curvilinear coordinates reduce to rectangular Cartesians.

As we have already seen, the contravariant components of velocity are

5.201. $$v^r = dx^r/dt.$$

Acceleration is commonly defined as the time-derivative of velocity. However, dv^r/dt are not the components of a tensor, and so cannot represent acceleration in curvilinear coordinates. Instead, we write tentatively for the contravariant components f^r of acceleration

5.202. $$f^r = \delta v^r/\delta t,$$

where δ indicates the absolute derivative as in 2.511. We verify that this is correct, because the expression on the right is a vector which reduces to dv^r/dt when the coordinates are rectangular Cartesians.

Let us now carry out an important resolution of acceleration into components along the tangent and first normal of the trajectory of the particle, using curvilinear coordinates. If ds is an element of arc of the trajectory, then the unit tangent vector is

5.203. $$\lambda^r = dx^r/ds.$$

Let ν^r be the contravariant components of the unit first normal and κ the curvature of the trajectory (reciprocal of the radius of curvature). Let us recall the first Frenet formula 2.705,

5.204. $$\delta\lambda^r/\delta s = \kappa\nu^r.$$

Let a_{mn} be the metric tensor for the curvilinear coordinate system x^r, and let v be the magnitude of the velocity, so that

5.205. $$v^2 = a_{mn}\, v^m v^n = v_n v^n.$$

Then

5.206. $$v^2 = a_{mn} \frac{dx^m}{dt} \frac{dx^n}{dt} = \left(\frac{ds}{dt}\right)^2,$$

and so, by 5.204,

5.207. $$f^r = \frac{\delta v^r}{\delta t} = \frac{\delta}{\delta t}\left(\frac{dx^r}{ds}\frac{ds}{dt}\right) = \frac{\delta}{\delta t}(v\lambda^r)$$
$$= \frac{\delta}{\delta s}(v\lambda^r)\frac{ds}{dt} = v\frac{dv}{ds}\lambda^r + \kappa v^2 \nu^r.$$

Thus the acceleration is resolved into a component $v\, dv/ds$ along the tangent and a component κv^2 along the first normal.

The Newtonian law of motion tells us that the product of the mass and the acceleration is equal to the applied force. Treating the mass as an invariant, we write tentatively

5.208. $$mf^r = F^r$$

as the equation of motion of a particle, F^r being the contravariant component of force. This is correct, because it is a tensor equation which reduces for Cartesian coordinates to the law stated above in words.

The question now arises: How are we to express 5.208 in a form suitable for integration, in order that the motion under a given force may be determined? In 5.1 we saw how to circumvent tedious transformations in computing contravariant and covariant components of velocity. The left-hand side of 5.208 presents no difficulty, since it can be computed from 5.202. In fact, the explicit form of 5.208 is

5.209. $$m\left(\frac{d^2x^r}{dt^2} + \left\{ {r \atop s\ n} \right\} \frac{dx^s}{dt}\frac{dx^n}{dt}\right) = F^r,$$

in contravariant form. In covariant form, 5.208 reads

5.210. $$mf_r = F_r,$$

or, explicitly,

5.211. $$m\left(a_{rs}\frac{d^2x^s}{dt^2}+[sn,\,r]\frac{dx^s}{dt}\frac{dx^n}{dt}\right)=F_r.$$

Referring to 2.431 and 2.438, we get, after a change in notation, an alternative (and very important) form of 5.210, namely,

5.212. $$mf_r\equiv\frac{d}{dt}\frac{\partial T}{\partial\dot{x}^r}-\frac{\partial T}{\partial x^r}=F_r,$$

where

5.213. $$T=\tfrac{1}{2}ma_{pq}\dot{x}^p\dot{x}^q,\quad \dot{x}^p=dx^p/dt.$$

We note that $T=\frac{1}{2}mv^2$, the kinetic energy of the particle. The equation 5.212 is called the *Lagrangian equation of motion.* It was obtained by Lagrange long before the invention of the tensor calculus.

However, the question still remains whether F^r or F_r must be computed from the physical components of force along the coordinate axes of Cartesians by transformations of the type 5.101. Happily, a short-cut can be found here also. Consider the differential expression

5.214. $$dW=F_n dx^n,$$

where F_r is the covariant force and dx^r an arbitrary infinitesimal displacement. This expression is clearly an invariant. If the curvilinear coordinates become rectangular Cartesians, dW is the work done in the displacement corresponding to dx^r. But since the expression is invariant, its value is the same for all coordinate systems. Hence the device is to calculate the work done in an arbitrary infinitesimal displacement corresponding to increments dx^r in the curvilinear coordinates, and read off the coefficients in the linear differential form representing the work. These coefficients are the covariant components F_r. The contravariant components may then be found by raising the subscript by means of the conjugate tensor a^{mn}:

5.215. $$F^r = a^{rn}F_n.$$

If the particle has a potential energy V, a function of its position (as it has in many physical problems), then the procedure is simpler still. For, in that case,

5.216. $$F_r = -\frac{\partial V}{\partial x^r}.$$

This is justified by the tensor character of this equation, which states (if the coordinates are rectangular Cartesians) that the force is the gradient of the potential energy, with sign reversed—the well-known relation by which in fact the potential energy may be defined.

Let us now consider the dynamics of a particle which is constrained to move on a smooth surface S. We choose in space a system of curvilinear coordinates x^r such that the equation of S is $x^3 = C$, a constant. Then x^α (Greek suffixes having the range 1, 2) form a system of curvilinear coordinates on S. The motion of the particle must satisfy the general equations of motion, for which we have the three equivalent forms 5.209, 5.211, and 5.212, F^r or F_r being the total force acting on the particle, i.e. the resultant of the applied force and the reaction which keeps the particle on the surface. Let us concentrate our attention on the formula 5.212. Since $x^3 = C$ throughout the whole motion, we are to put

5.217. $$x^3 = C,\ \dot{x}^3 = 0,\ \ddot{x}^3 = 0,$$

the dots indicating differentiation with respect to the time. It is easy to see that, for the first two components of 5.212, it is a matter of indifference whether we make the substitutions 5.217 *before* or *after* carrying out the differential operations required in 5.212. With the substitution 5.217, we get

5.218. $$T = \tfrac{1}{2}ma_{\alpha\beta}\dot{x}^\alpha\dot{x}^\beta,$$

and the equation of motion is

5.219. $$\frac{d}{dt}\frac{\partial T}{\partial \dot{x}^\alpha} - \frac{\partial T}{\partial x^\alpha} = F_\alpha.$$

Here F_α is such that $F_\alpha\, dx^\alpha$ is the work done in an infinitesimal displacement on the surface. If the surface is smooth, the reaction is normal to the surface, and does no work in such a displacement. Consequently F_α may be computed from the work done by the applied force alone. The quantities F_α are called the covariant components of *intrinsic force.*

The surface S is itself a Riemannian space of two dimensions, with metric form

5.220. $$ds^2 = a_{\alpha\beta}dx^\alpha dx^\beta.$$

The motion of the particle on S defines the vectors

5.221. $$v^\alpha = \frac{dx^\alpha}{dt}, \qquad f^\alpha = \frac{\delta v^\alpha}{\delta t},$$

where the δ-operation is performed with respect to the metric tensor $a_{\alpha\beta}$. We call v^α the *intrinsic velocity* (contravariant components) and f^α the *intrinsic acceleration* (contravariant components). If we want covariant components, we can get them by lowering the superscripts in the usual way with $a_{\alpha\beta}$.

The Lagrangian form 5.219 of the equations of motion may be transformed at once into either of the following forms:

5.222. $$mf^\alpha = F^\alpha, \quad mf_\alpha = F_\alpha.$$

The following important conclusion may be drawn: Given the metric form 5.220 for a smooth surface, and the vector F_α (or F^α) of intrinsic force, the study of the motion of the particle on the surface may be carried out intrinsically, i.e., without reference to the Euclidean 3-space in which the surface is embedded.

Exercise. If μ^α are the contravariant components of a unit vector in a surface S, show that $\mu^\alpha f_\alpha$ is the physical component of acceleration in the direction tangent to S defined by μ^α.

Let us pursue further the dynamics of a particle moving on a smooth surface S, treating the question from the intrinsic standpoint. Considered as a curve in S, the trajectory has two Frenet formulae (cf. 2.712):

5.223. $$\frac{\delta\lambda^\alpha}{\delta s} = \bar{\kappa}\nu^\alpha, \qquad \frac{\delta\nu^\alpha}{\delta s} = -\bar{\kappa}\lambda^\alpha.$$

Here the Greek suffixes have the range 1, 2; λ^α is the unit tangent vector in S; ν^α is the unit first normal; and $\bar{\kappa}$ is the first curvature. These formulae must not be confused with the Frenet formulae of the trajectory in space, of which there are three. To avoid confusion, we have written $\bar{\kappa}$ for the curvature of the trajectory considered as a curve in S, to distinguish it from κ, the curvature of the trajectory in space which appeared in 5.204. The curvature $\bar{\kappa}$ is a measure of the deviation of the curve from a tangent geodesic in S; it is usually called the *geodesic curvature*.

If we carry through in S the transformation made in 5.207 for space, we get

5.224. $$f^\alpha = v\frac{dv}{ds}\lambda^\alpha + \bar{\kappa}v^2\nu^\alpha.$$

Thus the equations of motion 5.219 may be written

5.225. $$m\left(v\frac{dv}{ds}\lambda^\alpha + \bar{\kappa}v^2\nu^\alpha\right) = F^\alpha.$$

If, in particular, there is no applied force, we have $F^\alpha = 0$. Multiplying 5.225 first by λ_α and secondly by ν_α, we obtain

5.226. $$v\frac{dv}{ds} = 0, \qquad \bar{\kappa}v^2 = 0.$$

Assuming that the particle is not at rest, we have $v \neq 0$, and therefore $\bar{\kappa} = 0$. Since this implies that the curve is a geodesic, we deduce the following result: *Under no applied force, a particle on a smooth surface moves along a geodesic with constant speed.*

As another special case, suppose that the vector F^α in the surface is perpendicular to the trajectory and of constant magnitude C. Then we may write

5.227. $$F^\alpha = C\nu^\alpha.$$

Substituting in 5.225, we easily deduce

5.228. $$v\frac{dv}{ds} = 0, \qquad m\bar{\kappa}v^2 = C.$$

The first equation tells us that the speed is constant; the second equation then tells us that $\bar{\kappa}$ is a constant. This means that the trajectory is a *geodesic circle* on the surface. (On a sphere, for example, a geodesic circle is a circle in the ordinary sense, but not in general a great circle. On a general surface, a geodesic circle is not necessarily a closed curve, nor is it the locus of points equidistant from a fixed point.)

Emphasis is again laid on the essential similarity between the dynamics of a particle moving on a smooth surface and the dynamics of a particle moving in space. It is true that there is one dimension less for the surface, and it is a curved space, not a flat one. However, these differences are not noticeable in the tensor formulation of the equations of motion. This formulation shows us how easily we might set up a dynamical theory for a particle moving in a curved space of three or more dimensions. However, the theory of relativity has introduced a fundamental difference in our concept of time, which makes t unsuitable as an independent variable in equations of motion.

In relativity we deal with a Riemannian 4-space (space-time). The equations of motion of a particle are

5.229. $$m\frac{\delta\lambda^r}{\delta s} = X^r,$$

where m is an invariant (the proper mass of the particle), λ^r is the unit tangent vector to the trajectory in space-time, ds is an element of arc as defined by the metric of space-time, and X^r is a vector in space-time corresponding to force in Newtonian mechanics; it is called the force 4-vector.

Exercise. Show that in relativity the force 4-vector X^r lies along the first normal of the trajectory in space-time. Express the first curvature in terms of the proper mass m of the particle and the magnitude X of X^r.

Let us return to the dynamics of a particle in Euclidean 3-space. Let z_r be rectangular Cartesians, v_r the components of velocity of a particle along the coordinate axes, and F_r the components of force along the coordinate axes. We define the *vector moment* of the force about the origin to be the vector

5.230. $$M_r = \epsilon_{rmn} z_m F_n.$$

(For the permutation symbol, see 4.311.) With this is associated the skew-symmetric *tensor moment*

5.231. $$M_{rs} = \epsilon_{rsn} M_n = z_r F_s - z_s F_r.$$

In the same way the *vector of angular momentum* h_r and the *tensor of angular momentum* h_{rs} are defined by

5.232. $$h_r = m \epsilon_{rmn} z_m v_n,$$

5.233. $$h_{rs} = \epsilon_{rsn} h_n = m(z_r v_s - z_s v_r),$$

where m is the mass of the particle. If we differentiate these equations, remember that the acceleration is $f_r = dv_r/dt$, and use the equation of motion 5.208, we obtain the following equivalent forms for the *equation of angular momentum* for a particle:

5.234. $$\frac{dh_r}{dt} = M_r \quad \text{or} \quad \frac{dh_{rs}}{dt} = M_{rs}.$$

It is to be noted that now we are tied down to rectangular Cartesians. The extension of 5.234 to curvilinear coordinates involves complications into which we shall not go. This arises from the fact that actually *two* points are involved—the particle itself and the origin about which the moments are taken. It is only relations involving a single point and its immediate neighbourhood that can be translated easily into curvilinear coordinates.

5.3. Dynamics of a rigid body. It is usual, in discussing the dynamics of a rigid body, to use a good deal of physical intuition derived from our experience with motions in space. Such methods, especially those which employ vector notation, are successful in leading us quickly and easily to the differential equations of motion. However, the use of spatial intuition obscures the deeper mathematical structure of the argument, and so fails to open up new trains of thought. The situation is parallel to that in geometry. The study of Euclidean geometry of three dimensions by methods based on our spatial intuition does not lead on to new ideas; but once we introduce the analytic method, the possibility of discussing space of higher dimensionality is opened up.

Let us then study the dynamics of a rigid body in N-dimensional flat space; we have only to put $N = 3$ in order to recover the familiar theory. We shall use homogeneous coordinates z_r.

Rigidity of a body implies the constancy of the distances between the particles which compose the body. Consider two points $P^{(1)}$ and $P^{(2)}$ with coordinates $z_r^{(1)}$ and $z_r^{(2)}$, respectively. The square of the distance between them is

5.301. $$(P^{(1)}P^{(2)})^2 = (z_m^{(1)} - z_m^{(2)})(z_m^{(1)} - z_m^{(2)}).$$

If the body undergoes an infinitesimal displacement in time dt, the coordinates of a point in the body change from z_r to $z_r + dz_r$, where $dz_r = (dz_r/dt)dt$. In view of the rigidity of the body, differentiation of 5.301 gives

5.302. $$(z_m^{(1)} - z_m^{(2)})(dz_m^{(1)} - dz_m^{(2)}) = 0.$$

Obviously, if this condition is satisfied for every pair of points, the displacement is a possible one for a rigid body.

The simplest infinitesimal displacement of a rigid body is a *translation*, which is described mathematically by writing

5.303. $$dz_r = a_r dt,$$

where a_r are the same for all points of the body. When we substitute

5.304. $$dz_r^{(1)} = a_r dt, \quad dz_r^{(2)} = a_r dt$$

in 5.302, the equation is satisfied. This verifies that the infinitesimal translation satisfies the condition of rigidity.

Consider now an infinitesimal displacement which leaves unmoved the particle of the body which is situated at the origin ($z_r = 0$). If this is a rigid body displacement, the distance of any point from the origin remains unchanged, and so it follows from 5.302 that

5.305. $$z_m^{(1)} dz_m^{(1)} = 0, \quad z_m^{(2)} dz_m^{(2)} = 0.$$

When these are used in 5.302, we get

5.306. $$z_m^{(1)} dz_m^{(2)} + z_m^{(2)} dz_m^{(1)} = 0.$$

In a continuous motion of a rigid body, with the origin fixed, the displacement dz_r of a point with coordinates z_r is

5.307. $$dz_r = v_r(z_1, z_2, \ldots; t)dt,$$

where v_r is the velocity at time t of the particle at z_r. Then 5.306 may be written

5.308. $$z_m^{(1)} v_m^{(2)} + z_m^{(2)} v_m^{(1)} = 0.$$

Noting that $v_r^{(2)}$ (the velocity at $z_r^{(2)}$) is independent of the point $z_r^{(1)}$, we obtain by differentiation

5.309. $$z_m^{(2)} \frac{\partial^2 v_m^{(1)}}{\partial z_r^{(1)} \partial z_s^{(1)}} = 0.$$

Since $z_r^{(2)}$ may be chosen arbitrarily in the body, without changing the choice of $z_r^{(1)}$, and the partial derivative is independent of $z_r^{(2)}$, it follows that the partial derivative vanishes. Therefore $v_r^{(1)}$ is a linear function of $z_r^{(1)}$. But it vanishes at the origin. Therefore it is a homogeneous linear function, and, dropping the superscript (1), we may write

5.310. $$v_r = -\omega_{rm} z_m.$$

The coefficients ω_{rm} are independent of the coordinates, i.e. they are functions of t only. But as in 5.305, we have

5.311. $$z_r v_r = 0,$$

and so

5.312. $$\omega_{rm} z_r z_m = 0.$$

From this it follows that ω_{rs} is skew-symmetric:

5.313. $$\omega_{rs} = -\omega_{sr}.$$

From 5.310 and the vector character of v_r and z_r (for transformations which do not change the origin), it follows that ω_{rs}

is a Cartesian tensor of the second order. It is called the *angular velocity tensor*. It is clear that if v_r is given by 5.310 with ω_{rs} arbitrary but skew-symmetric, then the corresponding motion satisfies the condition of rigidity. Let us sum up as follows: *If a rigid body rotates about the origin as fixed point, the velocity of any point z_r is given by equation* 5.310, *in which the coefficients ω_{rs} are skew-symmetric functions of the time.*

Exercise. Show that if a rigid body rotates about the point $z_r = b_r$ as fixed point, the velocity of a general point of the body is given by

5.314. $$v_r = -\omega_{rm}(z_m - b_m).$$

Let us now put $N = 3$ and introduce the *angular velocity vector*

5.315. $$\omega_r = \tfrac{1}{2}\epsilon_{rmn}\omega_{mn}.$$

Obviously this is an oriented Cartesian tensor. It is easily seen that the angular velocity tensor is expressed in terms of the angular velocity vector by

5.316. $$\omega_{rs} = \epsilon_{rsn}\omega_n.$$

Explicitly, the relations are

5.317. $$\omega_1 = \omega_{23}, \quad \omega_2 = \omega_{31}, \quad \omega_3 = \omega_{12}.$$

In terms of the angular velocity vector, 5.310 becomes

5.318. $$v_r = \epsilon_{rmn}\omega_m z_n,$$

which is the vector product of the vectors ω_r and z_r. If we put $z_r = \theta\omega_r$ in 5.318, it is clear that $v_r = 0$, no matter what value θ may have. Thus, in addition to the origin, a line of points in the body is instantaneously at rest, given by the equation

5.319. $$z_r = \theta\omega_r.$$

This line is called the *instantaneous axis*.

So far we have considered only the *kinematics* of a rigid body. As a basis for *dynamics* let us take D'Alembert's principle in the form

5.320. $$\Sigma m f_n dz_n = \Sigma F_n dz_n.$$

Let us now return to N-space. The repetition of the suffix implies summation over the range $1, 2, \ldots, N$. The sign Σ indicates a different summation, viz., a summation over all the particles of the rigid body. In 5.320, m is the mass of a typical particle, f_r its acceleration, F_r the external force acting on it, and dz_r an infinitesimal displacement (called a *virtual displacement*) which is arbitrary except for the condition that the displacements must not violate the conditions of rigidity of the body. The internal reactions of the body do not appear, since they do no work in such a system of displacements. The virtual displacements are not necessarily those which actually occur in the motion of the body in time dt.

First, take dz_r to be the displacements corresponding to an infinitesimal translation. This means that $dz_r = da_r$, where da_r are the same for all the particles, but otherwise arbitrary. Then 5.320 may be written

5.321. $$(\Sigma m f_n - \Sigma F_n) da_n = 0,$$

and it follows that

5.322. $$\Sigma m f_r = \Sigma F_r.$$

These are the *equations of motion for translation.* They may also be written

5.323. $$\frac{d}{dt}[\Sigma m v_r] = \Sigma F_r,$$

or, in words, *the rate of change of the total linear momentum is equal to the total external force.*

Next, let us take dz_r to correspond to an infinitesimal rotation about the origin. This may be done by writing, as in 5.310,

5.324. $$dz_n = -\eta \Omega_{np} z_p,$$

where η is an arbitrary infinitesimal invariant, and Ω_{rm} an arbitrary skew-symmetric tensor. Substitution in 5.320 gives

5.325. $$\Omega_{np}\Sigma(mf_n z_p) = \Omega_{np}\Sigma F_n z_p,$$

and hence, since Ω_{np} is arbitrary,

5.326. $$\Sigma m(f_n z_p - f_p z_n) = \Sigma(F_n z_p - F_p z_n).$$

We now introduce the tensor moment of forces M_{np}, and the tensor of angular momentum h_{np}, defined by (cf. 5.231, 5.233)

5.327. $$M_{np} = \Sigma(z_n F_p - z_p F_n), \quad h_{np} = \Sigma m(z_n v_p - z_p v_n).$$

Since $v_r = dz_r/dt$, $f_r = dv_r/dt$, equation 5.326 may be written

5.328. $$\frac{d}{dt} h_{np} = M_{np}.$$

Thus for a rigid body in N dimensions, turning about a fixed point *the rate of change of angular momentum is equal to the moment of external forces about the origin.*

To compute the angular momentum, we use 5.310, where ω_{rm} is the angular velocity tensor of the body. Then

5.329. $$\begin{aligned} h_{np} &= \Sigma m(\omega_{nq} z_q z_p - \omega_{pq} z_q z_n) \\ &= J_{nprq}\omega_{rq}, \end{aligned}$$

where

5.330. $$J_{nprq} = \Sigma m(\delta_{nr} z_p z_q - \delta_{pr} z_n z_q).$$

This may be called the *fourth-order moment of inertia tensor.* The equations of motion 5.328 now may be written

5.331. $$\frac{d}{dt}(J_{nprq}\omega_{rq}) = M_{np}.$$

Let us see what becomes of this formula when $N = 3$. If we use 5.316, define the moment vector, as in 5.231, by $M_s = \frac{1}{2}\epsilon_{snp}M_{np}$, and multiply 5.331 by $\frac{1}{2}\epsilon_{snp}$, we get

5.332. $$\frac{d}{dt}(I_{st}\omega_t) = M_s,$$

where

5.333. $$I_{st} = \tfrac{1}{2} J_{nprq}\epsilon_{rqt}\epsilon_{snp}.$$

On substituting from 5.330, this becomes

5.334. $$I_{st} = \Sigma m \epsilon_{ptq} \epsilon_{psn} z_q z_n,$$

or, by 4.329,

5.335. $$I_{st} = \delta_{st} \Sigma m z_q z_q - \Sigma m z_s z_t.$$

Explicitly, we have

5.336. $$\begin{gathered} I_{11} = \Sigma m(z_2^2 + z_3^2),\ I_{22} = \Sigma m(z_3^2 + z_1^2),\ I_{33} = \Sigma m(z_1^2 + z_2^2), \\ I_{23} = I_{32} = -\Sigma m z_2 z_3,\ I_{31} = I_{13} = -\Sigma m z_3 z_1, \\ I_{12} = I_{21} = -\Sigma m z_1 z_2, \end{gathered}$$

which will be recognized as the usual moments and products of inertia, the latter with signs reversed. The symmetric tensor I_{st} is commonly called the *moment of inertia tensor*. It is interesting to see how the three-dimensionality of space effects the reduction from the fourth-order tensor J_{nprq} to the second-order tensor I_{st}.

The equations 5.332 may also be written

5.337. $$dh_s/dt = M_s,$$

where

5.338. $$h_s = \tfrac{1}{2} \epsilon_{smn} h_{mn}.$$

We note that the components J_{nprq} (and I_{st}) depend on the coordinates of the particles of the body, and so are functions of t. This makes the equations of motion difficult to use, and we have recourse to *moving axes*, which we shall now discuss.

5.4. Moving frames of reference. A moving rigid body may be used as a *frame of reference*. Consider a rigid body (S') turning about a fixed point O, and let z'_r be rectangular Cartesian coordinates relative to axes which are fixed in S', but moving in the space S in which axes z_r have been chosen. Then any particle has two sets of coordinates, z_r relative to axes fixed in S and z'_r relative to axes fixed in S'. We choose both coordinate systems to have a common orientation and a common origin at O. Between the two sets of coordinates there exist formulae of transformation, which (with the associated identities) may be written

5.401.
$$z'_r = A_{rm} z_m, \qquad z_r = A_{mr} z'_m,$$
$$A_{mp} A_{mq} = \delta_{pq}, \qquad A_{pm} A_{qm} = \delta_{pq}.$$

The coefficients A_{mn} are the cosines of the angles between the two sets of axes; they are functions of the time t.

A moving particle has two velocities; a velocity relative to S and a velocity relative to S'. The velocity relative to S has components $v_r = \dot{z}_r$ on the z-axes, and the velocity relative to S' has components $v'_r = \dot{z}'_r$ on the z'-axes. We may have occasion to describe velocity relative to S' by its components on the z-axes, or velocity relative to S by its components on the z'-axes, and there is possibility of confusion. This will be avoided by the following notation:

$\dot{z}_r = v_r(S)$ = components on z-axes of velocity relative to S,
$v_r(S')$ = components on z-axes of velocity relative to S',
$v'_r(S)$ = components on z'-axes of velocity relative to S,
$\dot{z}'_r = v'_r(S')$ = components on z'-axes of velocity relative to S'.

In passing from $v_r(S)$ to $v'_r(S)$, we are merely transforming a vector from one set of axes to another. Since vectors transform like coordinates, it follows from 5.401 that

5.402.
$$v'_r(S) = A_{rm} v_m(S) = A_{rm} \dot{z}_m,$$
$$v_r(S') = A_{mr} v'_m(S') = A_{mr} \dot{z}'_m.$$

Indeed, we may, if we like, regard these equations as *defining* $v'_r(S)$ and $v_r(S')$ as functions of t, once A_{mn}, $\dot{z}_r$, and $\dot{z}'_r$ are given as functions of t. These last three functions are, of course, not independent, since the relations 5.401 exist. We have in fact

5.403. $$\dot{z}'_r = A_{rm} \dot{z}_m + \dot{A}_{rm} z_m, \quad \dot{z}_r = A_{mr} \dot{z}'_m + \dot{A}_{mr} z'_m.$$

If we multiply the first of these by A_{rp}, we get

5.404. $$v_p(S') = v_p(S) + A_{rp} \dot{A}_{rm} z_m.$$

Similarly, from the second of 5.403,

5.405. $$v'_p(S) = v'_p(S') + A_{pr} \dot{A}_{mr} z'_m.$$

Equations 5.404 and 5.405 give the transformation of velocity on passage from S to S' (or vice-versa) as frame of reference.

We shall now find out what the coefficients in the above transformations mean in terms of angular velocity.

If we follow a particle fixed in S', its coordinates z_r' are constant and $v_r'(S') = 0$. Hence, by 5.402, $v_r(S') = 0$, and so by 5.404

5.406. $$v_p(S) = -A_{rp}\dot{A}_{rm}z_m.$$

Comparison with 5.310 shows that the angular velocity tensor of S' relative to S may be written

5.407. $$\omega_{pm}(S', S) = A_{rp}\dot{A}_{rm};$$

the notation indicates that we are considering the angular velocity of S' relative to S. This tensor may also be referred to the z'-axes; by the formulae of tensor transformation, we have

5.408. $$\begin{aligned} \omega_{st}'(S', S) &= A_{sp}A_{tm}\omega_{pm}(S', S) \\ &= \delta_{rs}A_{tm}\dot{A}_{rm} \\ &= A_{tm}\dot{A}_{sm}. \end{aligned}$$

We may check the skew-symmetry of the right-hand sides of 5.407 and 5.408 by differentiating the identities in 5.401.

By virtue of the expressions given above for angular velocity, the formulae 5.404 and 5.405 for transformation of velocity on change of frame of reference may be written

5.409. $$\begin{aligned} v_p(S') &= v_p(S) + \omega_{pm}(S', S)z_m, \\ v_p'(S) &= v_p'(S') + \omega_{mp}'(S', S)z_m'. \end{aligned}$$

Relative to S', the original space S may be regarded as a rotating rigid body. Since we have several angular velocities to consider, it will avoid confusion if we set down our notation methodically.

Symbol	Axes of reference	Angular velocity of	relative to
$\omega_{rs}(S', S)$	z	S'	S
$\omega_{rs}'(S', S)$	z'	S'	S
$\omega_{rs}(S, S')$	z	S	S'
$\omega_{rs}'(S, S')$	z'	S	S'

To find the angular velocity of S relative to S', we follow a particle fixed in S. Then we have z_r constant and $v_r(S) = 0$, and consequently $v'_r(S) = 0$. Then, by 5.405 and 5.408,

5.410.
$$\begin{aligned} v'_p(S') &= -A_{pr}\dot{A}_{mr}z'_m \\ &= \omega'_{pm}(S', S)z'_m. \end{aligned}$$

But it follows from 5.310 that for a particle fixed in S and observed by S'

5.411.
$$v'_p(S') = -\omega'_{pm}(S, S')z'_m.$$

Comparing this with 5.410, and remembering that z'_m are arbitrary, we see that

5.412.
$$\omega'_{pm}(S', S) = -\omega'_{pm}(S, S').$$

Similarly

5.413.
$$\omega_{pm}(S', S) = -\omega_{pm}(S, S').$$

In words, *the angular velocity of S' relative to S is the negative of the angular velocity of S relative to S'*. It is understood that both are referred to the same axes, z or z'.

If $N = 3$, we may introduce angular velocity vectors by 5.315; it follows from 5.413 that

5.414.
$$\omega_r(S', S) = -\omega_r(S, S'), \quad \omega'_r(S', S) = -\omega'_r(S, S').$$

We started with a "fixed space" S, and introduced a moving rigid body S'. But, as we have seen, we can regard S' as "fixed," and think of the motion of S relative to it. The two angular velocities are the same except for sign, and there appears to be a complete equivalence in the sense that either frame may be used to describe motions, and neither is to be preferred above the other. It is true that this equivalence is complete as far as *kinematics* is concerned. But the two frames are not equivalent when it comes to dynamics. If we suppose that S is a Newtonian frame, in the sense that the motion of a particle relative to it is governed by Newton's laws of motion, the same will not be true of S'. We proceed to investigate this.

Using S and S' to denote the two frames of reference as above, and assuming S to be Newtonian, the equation of motion of a particle of mass m under the action of a force F_r is

5.415. $$mf_r(S) = F_r,$$

if components are taken on the z-axes. Here f_r denotes acceleration. If we take components on the z'-axes, the equation reads

5.416. $$mf'_r(S) = F'_r.$$

Note that the symbol in parentheses is S, not S', because it is acceleration relative to the Newtonian frame of reference S which must appear in the equation of motion.

We proceed to transform 5.416; by 5.402 and 5.409

5.417.
$$\begin{aligned} f_r(S) &= \dot{v}_r(S) \\ &= \frac{d}{dt}[A_{mr}v'_m(S)] \\ &= \frac{d}{dt}\{A_{mr}\,[v'_m(S') + \omega'_{nm}(S', S)z'_n]\} \\ &= A_{mr}f'_m(S') + \dot{A}_{mr}\,[v'_m(S') + \omega'_{nm}(S', S)z'_n] \\ &\quad + A_{mr}\dot{\omega}'_{nm}(S', S)z'_n + A_{mr}\omega'_{nm}(S', S)v'_n(S'). \end{aligned}$$

We multiply by A_{sr} and obtain by 5.408

5.418.
$$\begin{aligned} A_{sr}f_r(S) = f'_s(S') &+ \omega'_{ms}(S', S)[v'_m(S') + \omega'_{nm}(S', S)z'_n] \\ &+ \dot{\omega}'_{ns}(S', S)z'_n + \omega'_{ns}(S', S)v'_n(S'). \end{aligned}$$

The left-hand side is the same as $f'_s(S)$, i.e. components on z'-axes of acceleration relative to S, and so the equation of motion 5.416 may be written

5.419. $$mf'_s(S') = F'_s + C'_s + G'_s,$$

where the last two symbols are defined by

5.420.
$$\begin{aligned} C'_s &= m\,[\dot{\omega}'_{sn}(S'\,S) + \omega'_{sm}(S', S)\omega'_{nm}(S', S)]\,z'_n, \\ G'_s &= 2m\omega'_{sm}(S', S)v'_m(S'). \end{aligned}$$

Thus Newton's law of motion (mass times acceleration equals force) does not hold when the rotating body S' is used as frame of reference; it is violated by the presence of the two vectors

C'_s and G'_s in 5.419. But, following a common practice, we may regard S' as Newtonian provided we add to the real force F'_s *fictitious forces* C'_s (called *centrifugal force*) and G'_s (called *Coriolis force*).

Since 5.419 is a vector equation, we may take components on the z-axes, and write

5.421. $$mf_s(S') = F_s + C_s + G_s,$$

where C_s and G_s are as in 5.420, but with the primes deleted from ω', v', and z'. The indications of frames of reference remain unchanged.

Exercise. Deduce immediately from 5.420 that the Coriolis force is perpendicular to the velocity.

Exercise. Show that if $N = 3$ and $\dot{\omega}'_r\ (S', S) = 0$, then the centrifugal force may be written

5.422. $$C'_s = m\omega'_n(S', S)\omega'_n(S', S)z'_s - m\omega'_n(S', S)z'_n\omega'_s(S', S).$$

Deduce that C'_s is coplanar with the vectors $\omega'_s(S', S)$ and z'_s and perpendicular to the former.

Let us now consider the dynamics of a rigid body S' turning about a fixed point O under the action of external forces. We have already obtained the equations of motion 5.331, but they are not in useful form because the components J_{nprq} change with time. We shall transform to S'; this will overcome the difficulty, since z'_p are constants, and hence J'_{nprq} are constants, by 5.330. We note that in 5.331 $\omega_{rq} = \omega_{rq}\ (S', S)$.

Let us multiply 5.331 by $A_{an}A_{bp}$; then we have

5.423. $$A_{an}A_{bp}\frac{d}{dt}[A_{cn}A_{dp}J'_{cdrq}\omega'_{rq}(S', S)\,] = M'_{ab}.$$

Remembering the relations in 5.401, and also 5.408, this gives

5.424. $$J'_{abrq}\frac{d}{dt}\omega'_{rq}(S', S) \\ + J'_{cdrq}(\delta_{ac}\delta_{du}\delta_{bv} + \delta_{bd}\delta_{cu}\delta_{av})\omega'_{rq}(S', S)\omega'_{uv}(S', S) = M'_{ab}.$$

These are *the equations of motion of a rigid body in N-dimensions, referred to moving axes.* They are non-linear equations, with constant coefficients.

Exercise. Taking $N = 3$, show that 5.424 may be reduced to the usual Euler equations:

$$5.425.\quad I_{11}\frac{d}{dt}\omega'_1(S', S) - (I'_{22} - I'_{33})\,\omega'_2(S', S)\,\omega'_3(S', S) = M'_1,$$

and two similar equations.

5.5. General dynamical systems. Let us consider a dynamical system the configurations of which are determined by N generalized coordinates $x^r (r = 1, 2, \ldots, N)$. Particular examples are (a) a particle on a surface $(N = 2)$, (b) a rigid body which can turn about a fixed point, as in the preceding section $(N = 3)$, (c) a chain of six rods smoothly hinged together, with one end fixed and all moving on a smooth plane $(N = 6)$. Since the Cartesian coordinates of each particle are functions of the generalized coordinates, the components of velocity of each particle are linear homogeneous functions of the quantities $\dot{x}^r$, the coefficients being functions of x^r. The kinetic energy T of the system is the sum of the kinetic energies of the particles which compose it; hence T is expressible in the form

$$5.501.\qquad T = \tfrac{1}{2}a_{mn}\dot{x}^m\dot{x}^n, \quad (a_{nm} = a_{mn}).$$

The coefficients are functions of the generalized coordinates.

The kinetic energy of a system has the same value at any instant, no matter what generalized coordinates are used. Thus T is an invariant under transformation of the generalized coordinates, and since $\dot{x}^r$ is an arbitrary contravariant vector, the coefficients a_{mn} are the components of a covariant tensor (cf. 1.607).

Let us now think of a *configuration-space* V_N, in which each point corresponds to a configuration of the dynamical system, the correspondence being one-to-one. Since the quantities x^r

specify a configuration, they specify a point in V_N; in fact, they form a coordinate system in V_N.

Two adjacent configurations or points determine an invariant quadratic form

5.502. $$ds^2 = a_{mn} dx^m dx^n.$$

Thus the configuration-space V_N is Riemannian, with the metric form 5.502. The kinetic energy of a particle is positive if the particle is moving, and zero if it is at rest. Hence T is positive if any particle of the dynamical system is moving, and zero if the whole system is at rest. The system is at rest if, and only if, all the quantities $\dot{x}^r$ vanish. It follows that the form 5.502 is *positive definite.* We note that this form may be written

5.503. $$ds^2 = 2T dt^2.$$

To distinguish it from other possible metric forms in the configuration-space, we call 5.502 the *kinematical metric form* or *line element squared*, since it does not depend on the forces acting on the system.

Exercise. Assign convenient generalized coordinates for the three systems (*a*), (*b*), and (*c*) mentioned at the beginning of this section, and calculate the kinematical metric form in each case.

When the system consists of a single particle moving in space or on a surface, the kinematical line element ds is simply the geometrical line element of space or of the surface, multiplied by the square root of the mass of the particle. In these simple cases, the configuration space is not essentially different from the geometrical space in which the particle moves.

With appropriate terminology, the kinematics of a general dynamical system may be made remarkably analogous to the kinematics of a particle. We define the *generalized contravariant velocity vector* by

5.504. $$v^r = \dot{x}^r = \frac{dx^r}{dt}.$$

The covariant components are

5.505. $$v_r = a_{rs}v^s.$$

It is not convenient to define acceleration as the ordinary time-derivative of velocity, because that process does not yield a vector. Instead, we use absolute differentiation as defined in 2.511. We define the *generalized contravariant acceleration vector* by

5.506. $$f^r = \frac{\delta v^r}{\delta t}.$$

The covariant components are

5.507. $$f_r = a_{rs}f^s.$$

As the dynamical system moves, it passes through a sequence of configurations. Correspondingly, the point in configuration-space describes a curve or trajectory, with equations of the form

5.508. $$x^r = g^r(t).$$

We may also use the arc-length of the trajectory as parameter; then the equations are of the form

5.509. $$x^r = h^r(s).$$

Let λ^r be the unit tangent vector to the trajectory, so that

5.510. $$\lambda^r = \frac{dx^r}{ds}.$$

Then

5.511. $$v^r = \frac{dx^r}{dt} = \frac{dx^r}{ds}\frac{ds}{dt} = \lambda^r \frac{ds}{dt}.$$

Let v be the magnitude of the velocity vector, so that

5.512. $$v^2 = a_{mn}v^m v^n, \quad v \geqslant 0.$$

Then, substituting from 5.511,

5.513. $$v^2 = a_{mn}\lambda^m\lambda^n \left(\frac{ds}{dt}\right)^2 = \left(\frac{ds}{dt}\right)^2, \quad v = \frac{ds}{dt},$$

and 5.511 gives

5.514. $$v^r = v\lambda^r.$$

Hence we have the result, familiar in the kinematics of a particle, but now seen to be true for a general dynamical system: *The velocity vector lies along the tangent to the trajectory, and its magnitude is equal to ds/dt.*

Turning now to acceleration, we have from 5.506 and the Frenet formula 2.705

5.515. $$f^r = \frac{\delta v^r}{\delta t} = \frac{\delta}{\delta t}(v\lambda^r) = \frac{dv}{dt}\lambda^r + v\frac{\delta\lambda^r}{\delta t}$$
$$= \frac{dv}{dt}\lambda^r + v^2\frac{\delta\lambda^r}{\delta s} = \frac{dv}{dt}\lambda^r + \kappa v^2 \nu^r,$$

where we have written κ for the first curvature of the trajectory and ν^r for the first normal. Since $dv/dt = (dv/ds)(ds/dt)$, we have the following alternative forms for the acceleration vector:

5.516. $$f^r = \frac{dv}{dt}\lambda^r + \kappa v^2 \nu^r = v\frac{dv}{ds}\lambda^r + \kappa v^2 \nu^r.$$

As in the kinematics of a particle, we may state for a general system: *The acceleration lies in the elementary two-space containing the tangent and the first normal to the trajectory, and has the following components:*

along the tangent: $\dfrac{dv}{dt}$ or $v\dfrac{dv}{ds}$

along the first normal: κv^2.

Let us now consider the dynamics of the system under the action of prescribed forces. The *generalized covariant force vector* X_r is defined by the equation

5.517. $$X_r dx^r = dW,$$

where dW is the work done by the forces in an arbitrary infinitesimal displacement dx^r. Since dW is invariant, and dx^r contravariant, it is evident that X_r is indeed covariant, as anticipated in the definition. The contravariant components are given by

5.518.
$$X^r = a^{rs}X_s.$$

If the system possesses a potential energy V, then $dW = -dV$, and

5.519.
$$X_r = -\frac{\partial V}{\partial x^r}.$$

It is interesting to note that the *contravariant* components of velocity and the *covariant* components of force appear most naturally in setting up the terminology for a general dynamical system.

We shall now find the equations of motion of a general dynamical system in terms of the generalized coordinates and generalized forces. Suppose that the system consists of P particles. Denote the mass of the first particle by $m_1 = m_2 = m_3$, and its rectangular Cartesian coordinates by z_1, z_2, z_3. For the second particle, denote the mass by $m_4 = m_5 = m_6$, and the rectangular Cartesian coordinates by z_4, z_5, z_6. And so on. Denote by Z_1, Z_2, Z_3 the components of force acting on the first particle, by Z_4, Z_5, Z_6 the components of force acting on the second particle, and so on. Then, withholding the summation convention for Greek suffixes, the equations of motion of all the particles are contained in the formula

5.520.
$$m_\alpha \ddot{z}_\alpha = Z_\alpha, \quad \alpha = 1, 2, \ldots, 3P.$$

Now z_α are functions of the generalized coordinates x^r, and so the kinetic energy of the system is

5.521.
$$T = \tfrac{1}{2} \sum_\alpha m_\alpha \dot{z}_\alpha^2 = \tfrac{1}{2} \sum_\alpha m_\alpha \frac{\partial z_\alpha}{\partial x^r} \frac{\partial z_\alpha}{\partial x^s} \dot{x}^r \dot{x}^s.$$

Comparing this with 5.501, we have for the metric tensor

5.522.
$$a_{rs} = \sum_\alpha m_\alpha \frac{\partial z_\alpha}{\partial x^r} \frac{\partial z_\alpha}{\partial x^s}.$$

The symbol $\sum_\alpha$ means summation with respect to α for $\alpha = 1, 2, \ldots, 3P$. Applying the definition 2.421, we find for the Christoffel symbols of the first kind

5.523. $$[rs, t] = \sum_\alpha m_\alpha \frac{\partial^2 z_\alpha}{\partial x^r \partial x^s} \frac{\partial z_\alpha}{\partial x^t}.$$

We have

5.524. $$\dot{z}_\alpha = \frac{\partial z_\alpha}{\partial x^s} v^s, \quad \ddot{z}_\alpha = \frac{\partial z_\alpha}{\partial x^s} \dot{v}^s + \frac{\partial^2 z_\alpha}{\partial x^s \partial x^t} v^s v^t,$$

and hence by 5.523

5.525. $$\begin{aligned} \sum_\alpha m_\alpha \frac{\partial z_\alpha}{\partial x^r} \ddot{z}_\alpha &= a_{rs} \dot{v}^s + [st, r] v^s v^t \\ &= a_{rs} \left(\dot{v}^s + \left\{ \begin{matrix} s \\ m\, n \end{matrix} \right\} v^m v^n \right) \\ &= a_{rs} f^s = f_r. \end{aligned}$$

If we multiply the equations of motion 5.520 by $\partial z^\alpha / \partial x^r$ and sum with respect to α, we get

5.526. $$\sum_\alpha m_\alpha \frac{\partial z_\alpha}{\partial x^r} \ddot{z}_\alpha = \sum_\alpha \frac{\partial z_\alpha}{\partial x^r} Z_\alpha.$$

From 5.525 the left-hand side is equal to the acceleration f_r. As for the right-hand side, the work done in an infinitesimal displacement corresponding to increments dx^r in the generalized coordinates is

5.527. $$dW = \sum_\alpha Z_\alpha dz_\alpha = \sum_\alpha Z_\alpha \frac{\partial z_\alpha}{\partial x^r} dx^r.$$

Comparing this with 5.517 and remembering that dx^r are arbitrary, we get

5.528. $$\sum_\alpha Z_\alpha \frac{\partial z_\alpha}{\partial x^r} = X_r.$$

Hence *the equations of motion of a general dynamical system are*

5.529. $$f_r = X_r \text{ or } f^r = X^r;$$

in words, *acceleration equals force.*

It follows from 2.431 and 2.438 that the acceleration may also be written in the form

5.530. $$f_r = \frac{d}{dt}\left(\frac{\partial T}{\partial \dot{x}^r}\right) - \frac{\partial T}{\partial x^r};$$

consequently the equations of motion may be written

5.531. $$\frac{d}{dt}\left(\frac{\partial T}{\partial \dot{x}^r}\right) - \frac{\partial T}{\partial x^r} = X_r.$$

These are *Lagrange's equations of motion for a general dynamical system*. If the system is conservative, so that $X_r = -\partial V/\partial x^r$, than 5.531 may be written in the form

5.532. $$\frac{d}{dt}\left(\frac{\partial L}{\partial \dot{x}^r}\right) - \frac{\partial L}{\partial x^r} = 0, \quad L = T - V.$$

Exercise. Establish the general results

5.533. $$v\frac{dv}{ds} = X_r\lambda^r, \quad \kappa v^2 = X_r\nu^r.$$

Deduce that, if no forces act on a system, the trajectory is a geodesic in configuration-space and the magnitude of the velocity is constant.

Lines of force in configuration-space are defined as curves which have at each point the direction of the generalized force vector X^r. Their differential equations are

5.534. $$dx^r = \theta X^r,$$

where θ is an indeterminate infinitesimal. We assign a positive sense to the line of force by making θ positive. If the system possesses a potential energy V, it is easy to see that the lines of force are the orthogonal trajectories of the equipotential surfaces $V =$ constant; the positive sense on the line of force is the sense for which V decreases.

It is easy to throw 5.517 into the form

5.535. $$dW = X\, ds \cos\phi,$$

where X is the magnitude of the generalized force, ds the magnitude of the displacement, and ϕ the angle between the displacement and the positive sense of the line of force. It is then evident that, if we consider all infinitesimal displacements of constant magnitude ds, the work dW is a maximum if the displacement is taken in the direction of the positive sense of the line of force, and a minimum if the displacement is taken in the opposite sense.

In dynamical systems of physical interest it often happens that the lines of force are geodesics in configuration-space. When this is the case, the system, if started properly, will travel along a line of force. We shall now prove that *if the system is started with velocity tangent to a geodesic line of force, the trajectory will lie along that line of force.*

To prove this, let us look again at the equation of motion 5.529. Usually we think of the force X_r as prescribed, and regard the problem as that of determining the trajectory by solving a set of differential equations. We recall that f_r is an abbreviation for a function of the coordinates and their first and second derivatives with respect to the time. But we can look at the equation the other way round. We can think of the motion as prescribed, and the equation as one which determines the force under which this motion takes place. If motion and force are both prescribed, and the equation is not satisfied, it will be possible to satisfy it by introducing an extra force Y_r, given by

5.536. $$Y_r = f_r - X_r,$$

for then we shall have

5.537. $$f_r = X_r + Y_r,$$

which is of the form 5.529 with X_r replaced by $X_r + Y_r$.

Consider now motion along a geodesic line of force. The extra force necessary to maintain the motion is

5.538. $$Y_r = f_r - X_r = v\frac{dv}{ds}\lambda_r + \kappa v^2 \nu_r - X_r.$$

So far the magnitude of the velocity has not been assigned; let it be given by

5.539. $$v^2 = v_0^2 + 2\int_0^s X\,ds,$$

where X is the magnitude of the force. Then

5.540. $$v\frac{dv}{ds} = X.$$

Since the trajectory is a line of force, we have $X_r = X\lambda_r$, and since it is a geodesic, $\kappa = 0$. From these facts and 5.540, we see that the right-hand side of 5.538 is zero. This means that no extra force is needed to make the system move along the geodesic line of force with a velocity of magnitude given by 5.539. The theorem is proved.

Exercise. For a spherical pendulum show that the lines of force are geodesics on the sphere on which the particle is constrained to move. What does the theorem stated above tell us in this case?

Exercise. A system starts from rest at a configuration O. Prove that the trajectory at O is tangent to the line of force through O, and that the first curvature of the trajectory is one-third of the first curvature of the line of force.

So far, in dealing with configuration-space, we have used only the kinematical metric 5.502. There is another metric of importance, but it exists only for a system possessing a potential energy. For such a system we have

$$T = \tfrac{1}{2}a_{mn}v^m v^n,$$

$$\dot{T} = a_{mn}v^m f^n = a_{mn}v^m X^n = X_m v^m = -\frac{\partial V}{\partial x^m}v^m = -\dot{V},$$

and so

5.541. $$\frac{d}{dt}(T + V) = 0,$$

or

5.542. $$T + V = E;$$

thus the *sum of kinetic and potential energies is a constant* (E).

The metric which we are about to introduce is based on the concept of *action*, which is defined by the formula

5.543. $$A = \int T\,dt,$$

the integral being taken along any trajectory in configuration-space, not necessarily one which satisfies the equation of motion. However, it is understood that the motion satisfies 5.542, E being an assigned constant of total energy. Then, with ds as in 5.502, we have

5.544. $$T = \tfrac{1}{2}v^2 = \tfrac{1}{2}\left(\frac{ds}{dt}\right)^2, \quad dt = \frac{ds}{\sqrt{2T}},$$

and so the action is

5.545. $$A = \int \frac{T\,ds}{\sqrt{2T}} = \frac{1}{\sqrt{2}}\int \sqrt{T}\,ds = \frac{1}{\sqrt{2}}\int \sqrt{E - V}\,ds.$$

We define the *action line element* $d\sigma$ by

5.546. $$d\sigma = \sqrt{E - V}\,ds,$$

or, equivalently, the *action metric form* by

5.547. $$d\sigma^2 = b_{mn}dx^m dx^n = (E - V)\,a_{mn}dx^m dx^n.$$

The basic theorem which makes the action metric of interest is the following: *A dynamical system possessing potential energy V and moving with total energy E describes in configuration-space a geodesic with respect to the action metric.* This is commonly called the Principle of Least Action, or, more correctly, the Principle of Stationary Action.

In proving this theorem, we must avoid confusion between the two metric tensors, a_{mn} and b_{mn}; we shall distinguish Christoffel symbols by suffixes a and b. We note the following relations, which are easily established:

$$b_{mn} = (E - V)a_{mn}, \quad b^{mn} = \frac{a^{mn}}{E - V},$$

5.548. $$\left\{ {r \atop m\,n} \right\}_a = \left\{ {r \atop m\,n} \right\}_b + \frac{1}{2(E - V)}\left(\delta^r_m \frac{\partial V}{\partial x^n} + \delta^r_n \frac{\partial V}{\partial x^m} - a_{mn}a^{rs}\frac{\partial V}{\partial x^s}\right).$$

In order to prove the stated theorem, we start from the equation of motion 5.529, which may be written

5.549. $$\frac{d^2x^r}{dt^2}+\left\{\begin{matrix} r \\ m\,n \end{matrix}\right\}_a \frac{dx^m}{dt}\frac{dx^n}{dt} = -a^{rs}\frac{\partial V}{\partial x^s}.$$

Substituting from 5.548, this becomes

5.550. $$\frac{d^2x^r}{dt^2}+\left\{\begin{matrix} r \\ m\,n \end{matrix}\right\}_b \frac{dx^m}{dt}\frac{dx^n}{dt}+\frac{1}{E-V}\frac{\partial V}{\partial x^k}\frac{dx^k}{dt}\frac{dx^r}{dt}$$
$$= -a^{rs}\frac{\partial V}{\partial x^s}\left[1-\frac{a_{mn}}{2(E-V)}\frac{dx^m}{dt}\frac{dx^n}{dt}\right].$$

Since

$$\frac{a_{mn}}{2(E-V)}\frac{dx^m}{dt}\frac{dx^n}{dt}=\frac{2T}{2T}=1,$$

the expression on the right-hand side of 5.550 vanishes, and the equation of motion reduces to

5.551. $$\frac{d^2x^r}{dt^2}+\left\{\begin{matrix} r \\ m\,n \end{matrix}\right\}_b \frac{dx^m}{dt}\frac{dx^n}{dt}=\phi\frac{dx^r}{dt},$$

where

5.552. $$\phi = -\frac{1}{E-V}\frac{\partial V}{\partial x^k}\frac{dx^k}{dt}=-\frac{1}{E-V}\frac{dV}{dt}.$$

Equation 5.551 is the general equation of a geodesic (for metric tensor b_{mn}) in terms of an arbitrary parameter (cf. 2.427). This proves the theorem.

From the general theory of Chapter II we know that if the action arc-length σ is introduced as independent variable, the equation of motion, i.e., of the geodesic, simplifies to

5.553. $$\frac{d^2x^r}{d\sigma^2}+\left\{\begin{matrix} r \\ m\,n \end{matrix}\right\}_b \frac{dx^m}{d\sigma}\frac{dx^n}{d\sigma}=0.$$

From the manner in which we have developed the geometry of a Riemannian space from its metric form, it might be thought that the metric determined all the properties of the space. But we have already seen in 4.1 that there exist two distinct spaces of constant positive curvature with the same metric. In general we may say that the properties of a space "in the small" are determined by the metric, but that the properties "in the large" are not completely so determined. A simple illustration is given by a plane and a cylinder immersed in Euclidean 3-space. They are both flat manifolds, with the same metric forms for suitable choice of coordinates. But they have not the same properties in the large. On the cylinder (but not on the plane) there exist closed circuits which cannot be contracted continuously to a point. It is not possible to put the points of the cylinder into continuous one-to-one correspondence with the points of the plane; otherwise stated, supposing the surface of the cylinder to be a thin elastic sheet, it is not possible to stretch it to cover the plane without tearing it. These are intrinsic properties of the manifolds.

When the points of two spaces can be put into continuous one-to-one correspondence with one another, they are said to be *homeomorphic*. For example, the surfaces of a sphere and an ellipsoid are homeomorphic; the surfaces of a sphere and a torus are not homeomorphic. If this correspondence can be made without change of distance between adjacent points, the spaces are said to be *completely applicable*. Thus a portion of a cylinder which does not go right round it is completely applicable to a portion of a plane.

Dynamical systems provide interesting illustrations of topological properties of this sort. Consider a system which consists of a flywheel which can turn about its axis; the axis maintains a fixed direction, but can move in a direction perpendicular to itself. This is a system with two degrees of freedom, and the kinetic energy is

5.554. $$T = \tfrac{1}{2}m\dot{x}^2 + \tfrac{1}{2}I\dot{\theta}^2,$$

where m is the mass of the flywheel, I its moment of inertia, x the displacement of the axis, and θ the angle through which it is turned. The kinematical metric is

5.555. $$ds^2 = m\,dx^2 + I\,d\theta^2 = (dx^1)^2 + (dx^2)^2,$$

where $x^1 = \sqrt{m}\,x$ and $x^2 = \sqrt{I}\,\theta$. Thus the configuration-space is flat. But since an increase of 2π in θ restores the original configuration, the configuration-space is homeomorphic to a cylinder, not to a plane. In fact, the configuration-space is completely applicable to a cylinder in Euclidean 3-space. It is easily seen that the trajectories under no forces correspond to helices on this cylinder.

Consider now a system consisting of two flywheels rotating independently about an axis. The kinetic energy and kinematical metric are

5.556. $$T = \tfrac{1}{2}I_1\dot{\theta}_1^2 + \tfrac{1}{2}I_2\dot{\theta}_2^2, \quad ds^2 = I_1 d\theta_1^2 + I_2 d\theta_2^2,$$

where I_1, I_2 are the moments of inertia and θ_1, θ_2 the angles of rotation of the flywheels. Obviously the configuration-space is flat. It is homeomorphic to the surface of a torus, because the configuration is restored by an increase of 2π in either θ_1 or θ_2. In Euclidean 3-space there exists no surface completely applicable to this configuration-space.

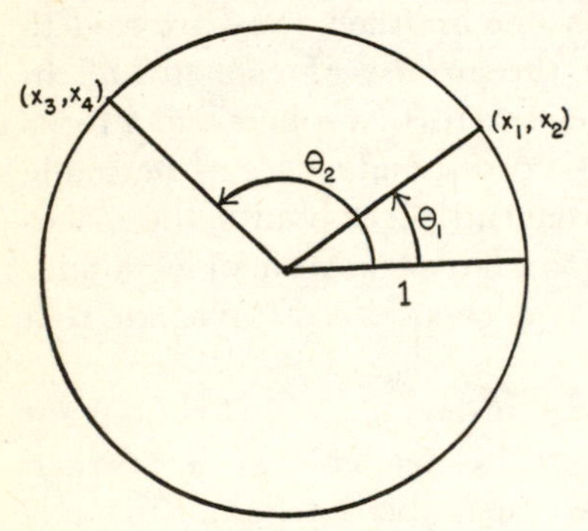

FIG. 12. Model of a flat 2-space homeomorphic to a torus.

If we resolve a dynamical system into its constituent particles (say P of them), we have a flat space of $3P$ dimensions. The configuration-space of the generalized coordinates is a subspace immersed in this flat space. Suppose that in the example just considered we replace the two flywheels by two particles of unit mass, constrained to move on a circle of unit radius (Fig. 12). If θ_1 and θ_2 are angles determining the positions of the particles on the circle, we have

5.557. $$ds^2 = d\theta_1^2 + d\theta_2^2.$$

Again we have a flat configuration-space with the connectivity of a torus. If we introduce the rectangular Cartesian coordinates of the particles, x_1, x_2 for the first particle, and x_3, x_4 for the second, we have

5.558.
$$x_1 = \cos\theta_1, \quad x_2 = \sin\theta_1,$$
$$x_3 = \cos\theta_2, \quad x_4 = \sin\theta_2.$$

If we suppose the particles to be free, the configuration-space has four dimensions and a kinematical metric

5.559.
$$ds^2 = dx_1^2 + dx_2^2 + dx_3^2 + dx_4^2.$$

When the two adjacent configurations correspond to keeping the particles on the circles, this metric reduces to 5.557, as we can see at once on differentiating 5.558. It appears therefore that in a Euclidean space of four dimensions there exist finite closed flat subspaces homeomorphic to a torus.

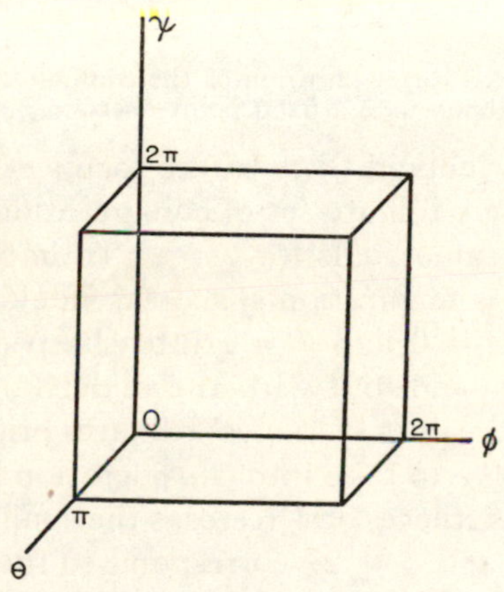

FIG. 13. Representation of the configurations of a rigid body with a fixed point—first phase.

Another interesting dynamical system is a rigid body turning about a fixed point. A configuration is determined by the values of three Eulerian angles θ, ϕ, ψ, and so the configuration-space has three dimensions. All possible configurations are included if we vary the coordinates in the ranges

5.560.
$$0 \leqslant \theta \leqslant \pi, \quad 0 \leqslant \phi < 2\pi, \quad 0 \leqslant \psi < 2\pi.$$

To get an idea of the topology of the configuration-space, let us take θ, ϕ, ψ as rectangular Cartesian coordinates in Euclidean 3-space (Fig. 13); the ranges 5.560 define a rectangular

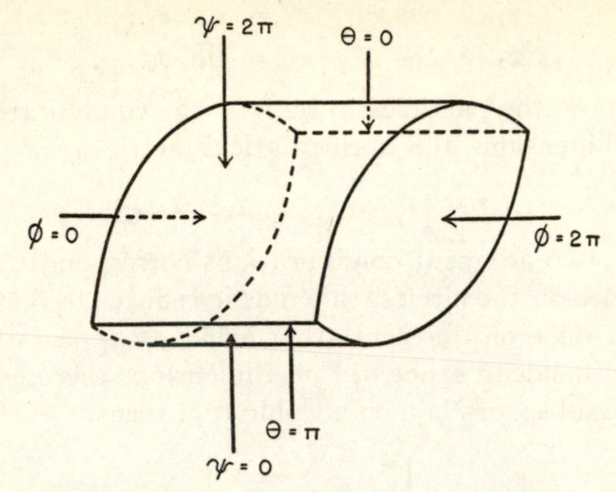

FIG. 14. Representation of the configurations of a rigid body with a fixed point—second phase.

parallelepiped or cuboid. But to the face $\theta = 0$ there corresponds only a singly infinite set of configurations, not a doubly infinite set; the same holds for $\theta = \pi$. To improve our representation of the configuration-space we should therefore compress the faces $\theta = 0$ and $\theta = \pi$ into sharp edges (Fig. 14), deforming the cuboid, but with the understanding that each point retains the values of the coordinates originally assigned to it. Now we have to take into consideration the fact that an increase of 2π in either ϕ or ψ restores the configuration. Thus the faces $\phi = 0$ and $\phi = 2\pi$ correspond to the same configurations, and these two faces should therefore be brought into

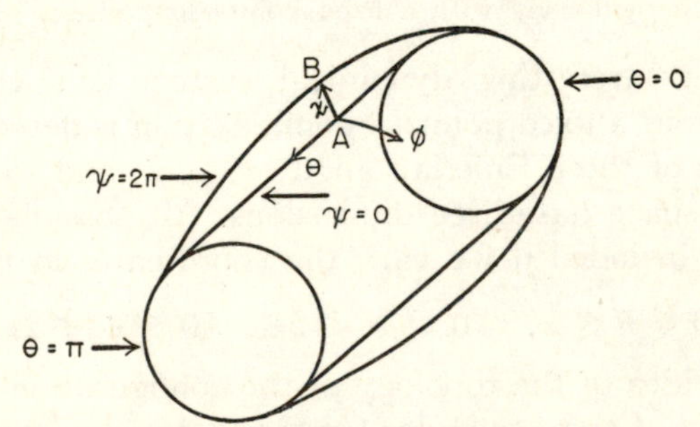

FIG. 15. Representation of the configurations of a rigid body with a fixed point—third phase.

coincidence. If this is done, the original cuboid is changed into a ring with sharp edges (Fig. 15). The parametric lines of ψ run (from 0 to 2π) through the thickness of the ring. To complete the representation, we would have to deform again so that points for which $\psi = 0$ and $\psi = 2\pi$ coincide; a point such as A should be brought into coincidence with a point such as B. But here our attempt to represent the configuration-space in ordinary 3-space breaks down. The best we can do is to leave our model in the last form, and remember the coincidence of the inside with the outside of the ring.

So far no metric has been mentioned. For simplicity, let us suppose that the momental ellipsoid of the body, relative to the fixed point, is a sphere. Then there is only one moment of inertia, which we shall denote by I. The kinetic energy is

5.561. $$T = \tfrac{1}{2} I (\dot{\theta}^2 + \dot{\phi}^2 + \dot{\psi}^2 + 2\dot{\phi}\dot{\psi} \cos \theta),$$

and the kinematical metric

5.562. $$ds^2 = I (d\theta^2 + d\phi^2 + d\psi^2 + 2\, d\phi\, d\psi \cos \theta).$$

By a short cut, we can show that the configuration-space is of constant curvature. If the three principal moments of inertia were not all equal it would be possible to give instructions about moving the body, without mentioning the coordinate system. For example, we could say that the body was to be rotated so much about the principal axis of greatest moment of inertia. But in our case that cannot be done. Except for magnitude, as given by 5.562, all displacements are intrinsically indistinguishable. In other words, configuration-space is isotropic. It is therefore of constant curvature (K), by Schur's theorem, 4.1. We can find K by another short cut. Motion under no forces takes place along a geodesic in configuration-space. For the body we are considering, motion under no forces consists of rotation about a fixed axis with constant angular velocity. In completing a revolution specified by

$$\theta = \text{const.}, \quad \phi = \text{const.}, \quad 0 \leqslant \psi \leqslant 2\pi,$$

the representative point in configuration-space describes a closed geodesic of length

5.563. $$L = \int ds = \int_0^{2\pi} \sqrt{I}\, d\psi = 2\pi\sqrt{I}.$$

Now if, starting from an assigned configuration, we compare two rotations about adjacent axes, we see that the two sequences of configurations so obtained have no configuration in common except the initial one. Thus in configuration-space two geodesics drawn from a point P in adjacent directions do not meet again until they meet at P. This means that configuration-space is a space of constant curvature of the polar type, and we know from 4.123 that the length of each closed geodesic is

5.564. $$L = \frac{\pi}{\sqrt{K}}.$$

Equating this value to L as given in 5.563, we get for the constant curvature the value

5.565. $$K = \frac{1}{4I}.$$

SUMMARY V

Physical components:

$$X^n \lambda_n, \quad X^{mn} \lambda_m \mu_n;$$
$$(\lambda_n \lambda^n = \mu_n \mu^n = 1).$$

Physical components for orthogonal curvilinear coordinates:

$$(X_1/h_1,\ X_2/h_2,\ X_3/h_3),$$

or

$$(h_1 X^1,\ h_2 X^2,\ h_3 X^3);$$
$$ds^2 = (h_1 dx^1)^2 + (h_2 dx^2)^2 + (h_3 dx^3)^2.$$

Dynamics of a particle:

$$m \frac{\delta v^r}{\delta t} = F^r \quad , \quad v^r = \frac{dx^r}{dt} \quad ,$$

or

$$m\, v \frac{dv}{ds} \lambda^r + m \kappa v^2 \nu^r = F^r,$$

or

$$\frac{d}{dt}\frac{\partial T}{\partial \dot{x}^r} - \frac{\partial T}{\partial x^r} = F_r, \quad T = \tfrac{1}{2}\, m a_{pq}\dot{x}^p \dot{x}^q, \quad \dot{x}^p = \frac{dx^p}{dt}.$$

Dynamics of a rigid body in N-dimensions:

$$v_r = -\omega_{rn} z_n, \quad \omega_{rn} = -\omega_{nr},$$

$$\frac{d}{dt}(\sum m v_r) = \sum F_r, \quad \frac{d}{dt} h_{np} = M_{np},$$

$$M_{np} = \sum (z_n F_p - z_p F_n),$$

$$h_{np} = \sum m(z_n v_p - z_p v_n) = J_{nprq}\omega_{rq},$$

$$J_{nprq} = \sum m(\delta_{nr} z_p z_q - \delta_{pr} z_n z_q).$$

For $N = 3$:

$$v_r = \epsilon_{rmn}\omega_m z_n, \quad \omega_r = \tfrac{1}{2}\,\epsilon_{rmn}\omega_{mn},$$

$$\frac{d}{dt}(I_{sn}\omega_n) = M_s,$$

$$I_{sn} = \delta_{sn}\sum m\, z_q z_q - \sum m z_s z_n, \quad M_s = \sum \epsilon_{spq} z_p F_q.$$

Moving frames of reference:

$$z'_r = A_{rm} z_m, \quad z_r = A_{mr} z'_m,$$

$$A_{mp} A_{mq} = \delta_{pq}, \quad A_{pm} A_{qm} = \delta_{pq},$$

$$\omega_{pq}(S', S) = -\omega_{pq}(S, S') = A_{mp}\dot{A}_{mq} = -A_{mq}\dot{A}_{mp},$$

$$\omega'_{pq}(S', S) = -\omega'_{pq}(S, S') = -A_{pm}\dot{A}_{qm} = A_{qm}\dot{A}_{pm}.$$

Kinematical line element:

$$ds^2 = 2T\, dt^2 = a_{mn} dx^m dx^n;$$

$$v^r = \frac{dx^r}{dt} = \dot{x}^r, \quad v_r = a_{rs} v^s,$$

$$f^r = \frac{\delta v^r}{\delta t} = \frac{d}{dt}\left(\frac{\partial T}{\partial \dot{x}^r}\right) - \frac{\partial T}{\partial x^r} = v\frac{dv}{ds}\lambda^r + \kappa v^2 \nu^r, \quad f_r = a_{rs} f^s.$$

Generalized force:

$$X_r = \sum_a Z_a \frac{\partial z^a}{\partial x^r}$$

Conservative system:

$$X_r = -\frac{\partial V}{\partial x^r}.$$

Dynamics of a system:

$$f_r = X_r.$$

Action line element:

$$d\sigma^2 = b_{mn}dx^m dx^n, \quad b_{mn} = (E - V)\, a_{mn},$$

$$d\sigma = \sqrt{E - V}\, ds.$$

Dynamics of a conservative system:

$$\frac{d^2x^r}{d\sigma^2} + \left\{ \begin{matrix} r \\ m\,n \end{matrix} \right\}_b \frac{dx^m}{d\sigma}\frac{dx^n}{d\sigma} = 0.$$

EXERCISES V

1. If a vector at the point with coordinates (1, 1, 1) in Euclidean 3-space has components (3, −1, 2), find the contravariant, covariant, and physical components in spherical polar coordinates.

2. In cylindrical coordinates (r, ϕ, z) in Euclidean 3-space, a vector field is such that the vector at each point points along the parametric line of ϕ, in the sense of ϕ increasing, and its magnitude is kr, where k is a constant. Find the contravariant, covariant, and physical components of this vector field.

3. Find the physical components of velocity and acceleration along the parametric lines of cylindrical coordinates in terms of the coordinates and their derivatives with respect to the time.

4. A particle moves on a sphere under the action of gravity. Find the covariant and contravariant components of force, using colatitude and azimuth, and write down the equations of motion.

5. Consider the motion of a particle on a smooth torus under no forces except normal reaction. The geometrical line element may be written

$$ds^2 = (a - b\cos\theta)^2 d\phi^2 + b^2 d\theta^2,$$

where ϕ is an azimuthal angle and θ an angular displacement from the equatorial plane. Show that the path of the particle satisfies the following two differential equations in which h is a constant:

(a) $$(a - b\cos\theta)^2 \frac{d\phi}{ds} = h,$$

(b) $$b^2 \left(\frac{d\theta}{d\phi}\right)^2 = (a - b\cos\theta)^4/h^2 - (a - b\cos\theta)^2.$$

6. Consider the motion of a particle under gravity on the smooth torus of the previous problem, the equatorial plane of the torus being horizontal. Taking the mass of the particle to be unity, so that $V = bg\sin\theta$, show that the path of the particle satisfies the following two differential equations:

(a) $$(E - V)(a - b\cos\theta)^2 d\phi/d\sigma = h,$$

(b) $$b^2 \left(\frac{d\theta}{d\phi}\right)^2 = (E - V)(a - b\cos\theta)^4/h^2 - (a - b\cos\theta)^2,$$

where E is the total energy, h is a constant, and $d\sigma$ is the action line element.

7. A dynamical system consists of a thin straight smooth tube which can rotate in a horizontal plane about one end O, together with a bead B inside the tube connected to O by a spring. Taking as coordinates $r = OB$ and $\theta =$ angle of rotation of the tube about O, the potential energy V is a function of r only. Show that, in configuration-space, all the lines of force are geodesics for the kinematical line element.

8. Show that if a line of force is a geodesic for the kinematical line element, it is also a geodesic for the action line element.

9. Using the methods of Chapter II and 5.532, show that the trajectories of a dynamical system with kinetic energy T and potential energy V satisfy the variational equation

$$\delta \int_{t_1}^{t_2} (T - V)\, dt = 0,$$

where the family of trajectories considered have common end points at $t = t_1$ and $t = t_2$. (This is known as Hamilton's Principle.)

10. Using the definition 5.335 for I_{rs}, prove that if X_r is any non-zero vector, then $I_{rs}X_rX_s \geqslant 0$, and that the equality occurs only if all particles of the system are distributed on a single line.

11. Let $Oz_1z_2z_3$ and $O'z'_1z'_2z'_3$ be two sets of Cartesian axes, parallel to one another. Consider a mass distribution and let I_{rs}, I'_{rs} be its moment of inertia tensors calculated for these two axes in accordance with 5.335. Writing $I'_{rs} = I_{rs} + K_{rs}$, evaluate K_{rs}.

12. A rigid body is turning about a fixed point. Referred to right-handed axes $Oz_1z_2z_3$, its angular velocity tensor has components

$$\omega_{23} = 1, \quad \omega_{31} = 2, \quad \omega_{12} = 3.$$

What are the components of its angular velocity vector ω_r? If we refer the same motion to axes $Oz'_1z'_2z'_3$, such that the axis Oz'_1 is Oz_1 reversed, while Oz_2z_3 coincide with $Oz'_2z'_3$, what are ω'_{rs} and ω'_r?

13. Consider three rigid bodies, S, S', S'', turning about a common point. If all angular velocities are referred to common axes, show that the angular velocity tensor of S'' relative to S is the sum of the angular velocity tensors of S' relative to S and of S'' relative to S'.

14. A freely moving particle is observed from a platform S' which rotates with angular velocity $\omega_r = n\delta_{r3}$, where n is a constant, relative to a Newtonian frame S in which z_r are rectangular Cartesians. Use 5.421 to find the equations of motion relative to S' in terms of coordinates z'_r in S', such that the axis of z'_3 coincides permanently with the axis of z_3.

15. If the tensor I_{st} is defined by 5.335 for N dimensions, and J_{nprq} is defined by 5.330, establish the following relations:

$$J_{nprq} = (N-1)^{-1}I_{ss}(\delta_{nr}\delta_{pq} - \delta_{nq}\delta_{pr}) - \delta_{nr}I_{pq} + \delta_{pr}I_{nq},$$

$$J_{nppn} = -I_{ss},$$

$$I_{nq} = (N-1)^{-1}(J_{nppq} - \delta_{nq}J_{spps}) = J_{nprp}.$$

16. The motion of a dynamical system is represented by a curve in configuration-space. Using the kinematical line element, express the curvature of the trajectory as a function of

its total energy E, and deduce that as E tends to infinity, the trajectory tends to become a geodesic. Illustrate by considering a particle moving under gravity on a smooth sphere.

17. A particle moves on a smooth sphere under the action of gravity. Using the action line element, calculate the Gaussian curvature of configuration-space as a function of total energy E and height z above the centre of the sphere. Show that if the total energy is not sufficient to raise the particle to the top of the sphere, but only to a level $z = h$, then the Gaussian curvature tends to infinity as z approaches h from below.

18. Show that the equations of motion of a rigid body with a fixed point may be written in either of the forms

$$(a) \qquad \dot{h}'_r + \omega'_{mr}(S', S)\, h'_m = M'_r,$$

$$(b) \qquad \dot{h}'_r - K'_{rmn}\, h'_m\, h'_n = M'_r,$$

where h'_r are the components on z'-axes (moving with the body) of angular momentum as given in 5.338 and K'_{rmn} is a certain moment of inertia tensor. Evaluate the components K'_{rmn} in terms of the moments and products of inertia.

19. A rigid body turns about a fixed O in a flat space of N dimensions. Prove that if N is odd, there exists at any instant a line OP of particles instantaneously at rest, but that, if N is even, no point other than O is, in general, instantaneously at rest. Show that if $N = 4$, there are points other than O instantaneously at rest if, and only if,

$$\omega_{23}\omega_{14} + \omega_{31}\omega_{24} + \omega_{12}\omega_{34} = 0.$$

20. The equations 5.329 do not determine J_{nprq} uniquely. Why? As an alternative to 5.330, we can require J_{nprq} to be skew-symmetric in the last two suffixes. Show that this defines J_{nprq} uniquely as follows:

$$J_{nprq} = \tfrac{1}{2} \sum m\, (\delta_{nr}\, z_p z_q + \delta_{pq}\, z_n z_r - \delta_{nq}\, z_p z_r - \delta_{pr}\, z_n z_q)\,.$$

Prove that J_{nprq}, as defined here, has the same symmetries as the covariant curvature tensor (cf. 3.115, 3.116) and that, for N = 3, we have

$$I_{st} = \tfrac{1}{2}\, \epsilon_{snp}\, \epsilon_{trq}\, J_{nprq}, \quad J_{nprq} = \tfrac{1}{2}\, \epsilon_{snp}\, \epsilon_{trq}\, I_{st}\,.$$

CHAPTER VI

APPLICATIONS TO HYDRODYNAMICS, ELASTICITY, AND ELECTROMAGNETIC RADIATION

6.1. Hydrodynamics. The mathematical fluid of hydrodynamics is a continuum of moving particles, each of which remains identifiable, so that we can speak of following a particle of the fluid. The space in which the motion takes place is Euclidean 3-space. We shall use rectangular Cartesians z_r, and later curvilinear coordinates x^r. In general, it is easiest to use the Cartesians to establish formulae, and then translate these (with a little judicious guessing) into a form valid for curvilinear coordinates.

The history of any particle is described by equations of the form $z_r = z_r(t)$, where t is the time. But we have a continuum of particles in the fluid, and so we introduce labels to distinguish one particle from another. These labels (which we shall denote by a_r) may be the coordinates of the individual particle at time $t = 0$, or they may be arbitrary functions of these initial coordinates. The complete history of the whole fluid may then be described by equations of the form

6.101. $$z_r = z_r(a, t),$$

where a stands for the set of three labels. The components of velocity v_r of a particle of the fluid are then given by taking derivatives with respect to t, holding the a's fixed, since we are interested in the rates of change of the coordinates of an individual particle:

6.102. $$v_r = \frac{\partial z_r}{\partial t}.$$

The above method of describing the motion of a fluid is called the *Lagrangian method.*

Another (and generally more useful) way of describing the motion of a fluid is the *Eulerian method.* We disregard the labels we attach to the particles, and start from the fact that, at a definite point and at a definite instant, there exists a definite velocity in the fluid; in other words, velocity v_r is a function of coordinates z_n and time t, and we may express this by writing

6.103. $$v_r = v_r(z, t).$$

Exercise. A fluid rotates as a rigid body about the axis of z_3 with variable angular velocity $\omega(t)$. Write out explicitly the three Lagrangian equations 6.101 and the three Eulerian equations 6.103.

Of course, it remains true in the Eulerian method that the co-ordinates z_r of an individual particle are functions of t, and that the velocity is obtained by differentiation, as in 6.102. But since we have no occasion in the Eulerian method to refer to the labels a_r, we shall, for future reference, write 6.102 in the form

6.104. $$v_r = \frac{dz_r}{dt}.$$

The *acceleration* of a particle is the rate of change of v_r, provided that, in differentiating, we follow a definite particle. Thus, if we differentiate 6.103 and use 6.104, we obtain for the acceleration f_r

6.105. $$f_r = \frac{dv_r}{dt} = \frac{\partial v_r}{\partial t} + \frac{\partial v_r}{\partial z_s}\frac{dz_s}{dt} = \frac{\partial v_r}{\partial t} + v_{r,s}\, v_s,$$

where $,s$ denotes $\partial/\partial z_s$.

Exercise. Compute the components of acceleration for the motion described in the preceding exercise.

What we have calculated above, in deriving 6.105, may be called the *comoving time-derivative*; it is "rate of change moving with the fluid." Sometimes this is emphasized by using the symbol D/Dt instead of d/dt, to remind the reader that a special procedure is involved, but we shall not find this necessary. For any function $F(z, t)$, the comoving time-derivative is

6.106. $$\frac{dF}{dt} = \frac{\partial F}{\partial t} + F_{,s} v_s.$$

The *density* ρ of a fluid is defined as mass per unit volume. If the ratio of mass to volume is the same for all volumes in the fluid, then the fluid is of constant density. But if this is not the case, the definition actually involves a limiting process. We take a sequence of volumes enclosing a point P, the sequence shrinking in on P in the limit. The density at P is then the limit of the ratio of mass to volume for this sequence of volumes. Density is in general a function of position and time, and so we write $\rho = \rho(z, t)$. The comoving time-derivative of ρ is of course

6.107. $$\frac{d\rho}{dt} = \frac{\partial \rho}{\partial t} + \rho_{,s} v_s.$$

So far we have been working with rectangular Cartesians. Let us introduce curvilinear coordinates x^r. Now we have a *contravariant velocity vector*

6.108. $$v^r = \frac{dx^r}{dt},$$

and a *covariant velocity vector.**

6.109. $$v_r = a_{rs} v^s,$$

where a_{rs} is the metric tensor. The simplest way to get the acceleration is not to transform 6.105, but to guess the covariant and contravariant forms

*To avoid complicating the notation, we use the same symbols v_r, f_r to denote components for rectangular Cartesian coordinates, and covariant components for curvilinear coordinates. The context should remove any possible ambiguity.

6.110. $$f_r = \frac{\partial v_r}{\partial t} + v_{r|s}v^s, \quad f^r = \frac{\partial v^r}{\partial t} + v^r{}_{|s}v^s,$$

where the stroke indicates covariant differentiation, as in 2.520 or 2.521. We verify that these equations are correct, because they are vector equations, and they reduce to 6.105 when the curvilinear coordinates reduce to rectangular Cartesians.

For the comoving time-derivative of an invariant, F or ρ, we have

6.111. $$\frac{dF}{dt} = \frac{\partial F}{\partial t} + F_{,s}v^s, \quad \frac{d\rho}{dt} = \frac{\partial \rho}{\partial t} + \rho_{,s}v^s.$$

Exercise. Verify that the operator $\partial/\partial t$ does not alter tensor character.

In what follows we shall make use of Green's theorem (also known as Gauss' theorem, or the divergence theorem). Consider a volume V, bounded by a surface S. Let us use rectangular Cartesians z_r. Let n_r be the direction cosines of the normal to S, drawn outward. Let F be a function of the coordinates, which is continuous and possesses continuous partial derivatives of the first order throughout V. Then Green's theorem states that

6.112. $$\int F n_r dS = \int F_{,r} dV.$$

The integral on the left is a surface integral taken over S, and the integral on the right is a volume integral taken throughout V. Only a single sign of integration is used for economy in notation, because it is obvious from the differential what multiplicity of integral is meant. As usual, $F_{,r} = \partial F/\partial z_r$. We assume that the reader is familiar with the proof of 6.112.*

The equation 6.112 is often written with a vector function in the integrand. Indeed, it follows from 6.112 that

6.113. $$\int F_r n_r dS = \int F_{r,r} dV.$$

*See any book on advanced calculus, or, in particular, R. Courant, *Differential and Integral Calculus*, London and Glasgow, Blackie, 1936, p. 384. For a proof of Green's theorem, in generalized form, see chap. VII, in particular 7.610.

This may be called a *divergence theorem*, since $F_{r,r}$ is the divergence of the vector F_r. Actually, however, Green's theorem is a theorem in analysis, and the question of the transformation of coordinates is not necessarily involved. If we like, we can replace F in 6.112 by a set of quantities which may be tensor components, and write

6.114. $$\int F_{st} n_r dS = \int F_{st,r}\, dV.$$

But the tensor character of the integrand becomes important if we wish to transform Green's theorem to curvilinear coordinates x^r. It is then essential that the integrand be an invariant, and the transformation of 6.112 (where F is an invariant) or 6.114 (where F_{st} is a tensor of the second order) presents difficulties. But 6.113 transforms directly. It is true that

6.115. $$\int F_r n^r ds = \int F^r{}_{|r}\, dV,$$

for each integrand is an invariant, and so the statement is true for all coordinate systems if true for one—and obviously 6.115 reduces to 6.113 for rectangular Cartesians. In 6.115 n^r are the contravariant components of the outward unit normal, i.e., $n^r = dx^r/ds$, where ds is an element of the normal to S.

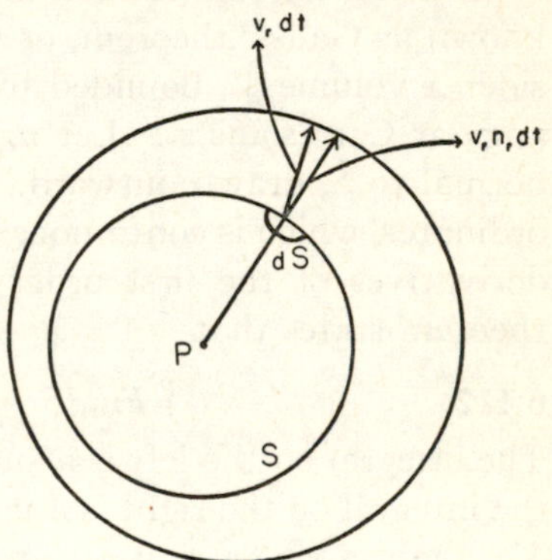

FIG. 16. Expansion of a fluid.

We shall now consider the *expansion* of a fluid. Let V be a volume of the fluid, bounded by a surface S. Let S move with the fluid, so that it is always formed of the same particles. Let us use rectangular Cartesians. Let v_r be the velocity. Then $v_r n_r$ is the component of velocity along the outward normal, n_r being the direction cosines of that normal. As the fluid moves, a particle on the surface is displaced obliquely (Fig. 16); the normal component of its displacement in infinitesimal time dt is $v_r n_r dt$. Hence the volume of the thin shell between the surface S at time t and the surface formed by the same particles at time $t + dt$ is

$$6.116. \qquad dt \int v_r n_r dS.$$

But this is the increase in the volume V, and so we have

$$6.117. \qquad \frac{dV}{dt} = \int v_r n_r dS = \int v_{r,r} dV,$$

by Green's theorem. We divide by the volume V, and then consider a sequence of volumes, all enclosing the point P and shrinking down on this point. Then it is easy to see that

$$6.118. \qquad \lim_{V\to 0} \frac{1}{V}\frac{dV}{dt} = \lim_{V\to 0} \frac{1}{V}\int v_{r,r} dV = v_{r,r},$$

evaluated at the point P. We define this limit to be the *expansion* of the fluid at the point P. Denoting it by θ, we have

$$6.119. \qquad \theta = v_{r,r};$$

in vector language, this is the divergence of the velocity vector.

No logical contradiction is implied by choosing the velocity v_r and the density ρ as arbitrary functions of position and time. But such arbitrary choice will, in general, violate a basic physical principle—*the conservation of mass*. It is, in fact, a basic hypothesis of Newtonian mechanics that the mass of any system is constant, provided the system is always composed of the same particles. (In relativity, the conservation of mass is abandoned—hence the source of atomic energy.)

Let us now investigate the connection between velocity and density arising from the conservation of mass. The mass of a fluid element is ρdV. As the fluid moves, the volume dV may change, but the mass ρdV does not. Let us write for the volume of a portion of the fluid

$$6.120. \qquad V = \int dV = \int \frac{1}{\rho}\rho dV.$$

Following this portion of the fluid, we have

$$6.121. \qquad \frac{dV}{dt} = \int \frac{d}{dt}\left(\frac{1}{\rho}\right)\rho dV,$$

the derivatives being comoving derivatives. Dividing by V and letting the volume shrink to zero as in 6.118, we get

6.122. $$\theta = \lim_{V \to 0} \frac{1}{V}\frac{dV}{dt} = \rho \frac{d}{dt}\left(\frac{1}{\rho}\right).$$

But

6.123. $$\frac{d}{dt}\left(\frac{1}{\rho}\right) = -\frac{1}{\rho^2}\frac{d\rho}{dt},$$

and hence 6.122 may be written

6.124. $$\frac{d\rho}{dt} + \rho\theta = 0.$$

This is the *equation of conservation of mass*, traditionally called the *equation of continuity*. By virtue of 6.119, it can also be written in the following forms:

6.125a. $$\frac{d\rho}{dt} + \rho v_{r,r} = 0,$$

6.125b. $$\frac{\partial \rho}{\partial t} + (\rho v_r)_{,r} = 0.$$

The above equations are easily transformed to curvilinear coordinates by the usual process of guessing and verifying. We have

6.126. $$\theta = v^r{}_{|r},$$

6.127a. $$\frac{d\rho}{dt} + \rho v^r{}_{|r} = 0,$$

6.127b. $$\frac{\partial \rho}{\partial t} + (\rho v^r)_{|r} = 0.$$

Exercise. Write out 6.126 and 6.127b explicitly for spherical polar coordinates.

The *vorticity tensor* at a point in a fluid is defined as

6.128. $$\omega_{rs} = \tfrac{1}{2}(v_{s,r} - v_{r,s})$$

for rectangular Cartesian coordinates. Obviously, it is skew-symmetric ($\omega_{sr} = -\omega_{rs}$). We can at once transform 6.128 to curvilinear coordinates by replacing the comma by a stroke. However, it follows from 2.521 that, for any covariant vector V,

6.129. $$V_{s|r} - V_{r|s} = V_{s,r} - V_{r,s}.$$

Therefore 6.128, as written, represents the vorticity tensor in curvilinear coordinates as well as in rectangular Cartesian coordinates.

For rectangular Cartesian coordinates, the *vorticity vector* is defined as

6.130. $$\omega_r = \tfrac{1}{2}\epsilon_{rmn}\omega_{mn},$$

so that, explicitly,

6.131. $$2\omega_1 = v_{3,2} - v_{2,3}, \quad 2\omega_2 = v_{1,3} - v_{3,1}, \quad 2\omega_3 = v_{2,1} - v_{1,2}.$$

The right-hand side of 6.130 is an oriented Cartesian vector (cf. 4.3). To transform to curvilinear coordinates, we must investigate the tensor character of the permutation symbol ϵ_{rmn} for general transformations.

The Jacobian of a transformation from coordinates x to coordinates x' may be written

6.132. $$J = \left|\frac{\partial x^p}{\partial x'^q}\right| = \epsilon_{rmn}\frac{\partial x^r}{\partial x'^1}\frac{\partial x^m}{\partial x'^2}\frac{\partial x^n}{\partial x'^3}.$$

If we interchange two of the numerical suffixes, say 1 and 2, we change the sign of the expression on the right, without altering its absolute value. If we permute 123 into uvw, the effect is the same as multiplication by ϵ_{uvw}, or, if we prefer to write it so, ϵ'_{uvw}. Thus

6.133. $$J\epsilon'_{uvw} = \epsilon_{rmn}\frac{\partial x^r}{\partial x'^u}\frac{\partial x^m}{\partial x'^v}\frac{\partial x^n}{\partial x'^w},$$

or

6.134. $$\epsilon'_{uvw} = J^{-1}\epsilon_{rmn}\frac{\partial x^r}{\partial x'^u}\frac{\partial x^m}{\partial x'^v}\frac{\partial x^n}{\partial x'^w}.$$

If the Jacobian is unity, this becomes the usual transformation for a covariant tensor of the third order. In general, a set of quantities transforming as in 6.134 are said to be the components of a *relative tensor* of the third order of *weight* -1. If J^P occurred instead of J^{-1}, we would say that the weight was P.*

Exercise. If ϵ^{rmn} is defined in precisely the same way as ϵ_{rmn}, prove that

6.135. $$\epsilon'^{uvw} = J\epsilon^{rmn}\frac{\partial x'^u}{\partial x^r}\frac{\partial x'^v}{\partial x^m}\frac{\partial x'^w}{\partial x^n},$$

so that ϵ^{rmn} is a relative tensor of weight $+1$.

Now consider the transformation of the determinant a of the metric tensor. We have

6.136. $$a' = \left| a'_{rs} \right| = \left| a_{mn}\frac{\partial x^m}{\partial x'^r}\frac{\partial x^n}{\partial x'^s} \right| = \left| a_{mn} \right| \left| \frac{\partial x^p}{\partial x'^r} \right| \left| \frac{\partial x^q}{\partial x'_s} \right|,$$

by the rule for the multiplication of determinants. Hence

6.137. $$a' = J^2 a,$$

so that *a is a relative invariant of weight* 2.

To avoid complications, let us think only of transformations for which the Jacobian is positive. Then we have

6.138. $$\sqrt{a'} = J\sqrt{a},$$

so that *$\sqrt{a}$ is a relative invariant of weight* 1.

Combining 6.138 with 6.134, we get

6.139. $$\epsilon'_{uvw}\sqrt{a'} = (\epsilon_{rmn}\sqrt{a})\frac{\partial x^r}{\partial x'^u}\frac{\partial x^m}{\partial x'^v}\frac{\partial x^n}{\partial x'^w}.$$

Hence $\epsilon_{rmn}\sqrt{a}$ is a tensor (not relative—we may call it *absolute* for emphasis), covariant and of the third order.

Exercise. Prove that $\epsilon^{rmn}/\sqrt{a}$ is an (absolute) contravariant tensor of the third order.

*For the general theory of relative tensors, see 7.1 and 7.2.

We are now ready to write 6.130 in curvilinear coordinates. Tentatively, we put

6.140. $$\omega^r = \frac{1}{2\sqrt{a}} \epsilon^{rmn} \omega_{mn}.$$

Clearly, this is an (absolute) contravariant vector. If we take rectangular Cartesian coordinates, we have $a = 1$, and so 6.140 reduces to 6.130. Therefore, in view of its tensor character, 6.140 is the contravariant transform of 6.130 in curvilinear coordinates.

A fluid motion is said to be *irrotational* if there exists an invariant function ϕ such that

6.141. $$v_r = -\phi_{,r}.$$

This equation is unchanged in form if we pass to curvilinear coordinates. It is evident from 6.128 that $\omega_{rs} = 0$ for an irrotational motion; $\omega_r = 0$ also.

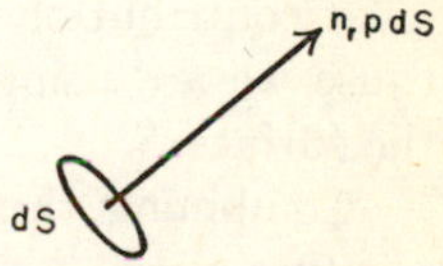

FIG. 17. Pressure in a perfect fluid.

So far we have considered only the kinematics of a fluid. Let us now discuss the internal forces. We shall here consider only *perfect fluids*.

A perfect fluid is defined as one in which the force transmitted across any plane element is perpendicular to that element (Fig. 17). The force per unit area is called the *pressure*, and will be denoted by p. This definition does not in itself imply that, if two plane elements with different normals are taken at a point, the value of p is the same for both of them. However, the application of Newton's law of motion (mass times acceleration equals force) to a small tetrahedron of fluid does in fact establish the fact that the value of the pressure at a point is independent of the element across which it is measured.* Thus the force across an element of area dS and unit normal n_r is the vector $n_r p \, dS$ in rectangular Cartesian coordinates, where p is a function of position only.

*Sir H. Lamb, *Hydrodynamics*, Cambridge, 1932, p. 2.

In addition to the pressure, there may be present a *body force*, such as gravity. We shall denote by X_r the body force per unit mass, so that the body force on an element of volume dV is $X_r \rho dV$.

To determine the equations of motion of a fluid, we have available the principle of linear momentum, which states that for any system of particles the rate of change of linear momentum is equal to the total *external* force. For our system, let us take the fluid contained in a surface S which moves with the fluid. Its linear momentum is $\int \rho v_r dV$. To the total external force, the body forces contribute $\int \rho X_r dV$. The pressures across plane elements in the fluid give internal, not external, forces. It is only the pressure across the bounding surface S that gives a contribution to the total external force. That contribution is $-\int p n_r dS$, the minus sign occurring because we are using n_r to denote the outward unit normal to the surface S.

Combining these expressions, the principle of linear momentum gives

6.142. $$\frac{d}{dt}\int \rho v_r dV = \int \rho X_r dV - \int p n_r dS.$$

We bring the differentiation under the sign of integration, remembering that ρdV is constant on account of the conservation of mass; at the same time we transform the surface integral by Green's theorem, and so obtain

6.143. $$\int \rho \frac{dv_r}{dt} dV = \int \rho X_r dV - \int p_{,r} dV.$$

Combining the integrals into a single integral, we get

6.144. $$\int \left(\rho \frac{dv_r}{dt} - \rho X_r + p_{,r} \right) dV = 0.$$

Since this integral vanishes for *every* volume taken in the fluid, the integrand must vanish, and so we get

6.145. $$\rho \frac{dv_r}{dt} - \rho X_r + p_{,r} = 0,$$

or

6.146. $$\frac{\partial v_r}{\partial t} + v_s v_{r,s} = X_r - \rho^{-1} p_{,r}.$$

These are *the general equations of motion of a perfect fluid in rectangular Cartesian coordinates.* In curvilinear coordinates they read

6.147. $$\frac{\partial v_r}{\partial t} + v^s v_{r|s} = X_r - \rho^{-1} p_{,r},$$

in covariant form.

Exercise. Write down the contravariant form of 6.147.

In the case of irrotational motion (which is by far the most important type of motion in the applications of hydrodynamics) we have, by 6.141 and 6.146, in rectangular Cartesian coordinates,

6.148. $$-\frac{\partial}{\partial t}(\phi_{,r}) + \phi_{,s}\phi_{,rs} = X_r - \rho^{-1} p_{,r}.$$

It is generally assumed that the density ρ is a function of the pressure p. (The case ρ = const. is a special case of this.) Then, if we define

6.149. $$P = \int \frac{dp}{\rho},$$

P is a function of p, and so (through p) a function of the coordinates. Hence

6.150. $$P_{,r} = \frac{dP}{dp} p_{,r} = \rho^{-1} p_{,r}.$$

Let us also assume that the body forces X_r are conservative and can be derived from a potential U such that

6.151. $$X_r = -U_{,r}.$$

Then 6.148 may be written

6.152. $$\left(-\frac{\partial \phi}{\partial t} + \tfrac{1}{2}\phi_{,s}\phi_{,s} + P + U \right)_{,r} = 0,$$

so that the quantity in parenthesis is independent of the coordinates. It can be a function of the time t only; so we write

6.153. $$-\frac{\partial\phi}{\partial t}+\tfrac{1}{2}\phi_{,s}\phi_{,s}+P+U=F(t).$$

This is *the integral of Bernoulli for irrotational motion.* This form holds only for rectangular Cartesian coordinates; for curvilinear coordinates we have

6.154. $$-\frac{\partial\phi}{\partial t}+\tfrac{1}{2}\,a^{mn}\phi_{,m}\phi_{,n}+P+U=F(t).$$

The simplest and most important type of fluid motion is that in which the density is constant (homogeneous incompressible fluid). For irrotational motion of this type, the equation of conservation of mass 6.125b becomes

6.155. $$\phi_{,rr}=0$$

for rectangular Cartesian coordinates; in curvilinear coordinates it reads

6.156. $$a^{mn}\phi_{|mn}=0.$$

Written explicitly, 6.155 is Laplace's equation

$$\frac{\partial^2\phi}{\partial z_1^2}+\frac{\partial^2\phi}{\partial z_2^2}+\frac{\partial^2\phi}{\partial z_3^2}=0,$$

and 6.156 is the transformation of this equation to curvilinear coordinates. However, the second-order covariant derivative in this expression is sometimes tedious to compute, and 6.156 is more conveniently written

6.157. $$(\sqrt{a}a^{mn}\phi_{,m})_{,n}=0.$$

Exercise. Verify by means of 3.204 that 6.157 and 6.156 are the same equation.

6.2. Elasticity. The theory of elasticity involves the concepts of *strain* and *stress* in an elastic solid. The theory of strain belongs to geometry; it consists of a systematic mathematical description of the deformations which can occur in a continuous medium. The theory of stress involves a study of the internal reactions which can occur in a continuous medium. We shall start by considering strain and stress separately, and later link them together by Hooke's law.

Consider a continuous medium, consisting of particles which retain their identities. First, the medium is at rest in what we shall call the *unstrained state.* Using rectangular Cartesian coordinates z_r, we shall suppose that the particle situated at the point z_r receives a displacement $u_r(z)$, the notation indicating that each particle receives a displacement which is a function of its coordinates. If the functions $u_r(z)$ are constants, the whole medium is translated without deformation. For other special choices of these functions, the displacement of the medium may be that of a rigid body. But, in general, for arbitrary displacements $u_r(z)$, the medium will be *deformed* or *strained.*

It is characteristic of a rigid body displacement that the distance between any two particles remains unchanged by the displacement. It is therefore natural that we should analyse a strain in terms of the changes in length of the lines joining particles. Consider two particles with coordinates z_r, z'_r in the unstrained state. After strain, their coordinates are respectively

$$z_r + u_r(z), \quad z'_r + u_r(z').$$

Let L_0 be the distance between the particles in the unstrained state and L_1 the distance between them after strain. Then the *extension* e of the line joining the particles is defined to be

6.201. $$e = (L_1 - L_0)/L_0,$$

i.e., the increase in length per unit length. Let us now see how the extension is to be evaluated.

We have

6.202. $$L_0^2 = (z'_r - z_r)(z'_r - z_r),$$

$$L_1^2 = [z'_r - z_r + u_r(z') - u_r(z)][z'_r - z_r + u_r(z') - u_r(z)],$$

and hence

6.203. $$L_1^2 - L_0^2 = 2(z'_r - z_r)[u_r(z') - u_r(z)] + [u_r(z') - u_r(z)][u_r(z') - u_r(z)].$$

It is clear that we could derive from 6.202 a formal expression for e. But it would be very complicated. Let us rather investigate the extensions of infinitesimal elements drawn from the point z_r. To do this, we keep z_r fixed, and allow z'_r to approach it along a curve, the unit tangent vector to this curve at z_r being denoted by λ_r. Since, in this process, $z'_r - z_r$ becomes small, we have approximately

6.204. $$u_r(z') - u_r(z) = (z'_s - z_s)\, u_{r,s}(z),$$

and so 6.204 gives approximately

6.205. $$L_1^2 - L_0^2 = 2u_{r,s}(z)(z'_r - z_r)(z'_s - z_s) + (z'_s - z_s)(z'_t - z_t)u_{r,s}(z)u_{r,t}(z).$$

But $z'_r - z_r = L_0\lambda_r$ approximately. So, dividing 6.205 by L_0^2 and letting z'_r approach z_r along the curve, we get after a change of suffixes,

6.206. $$\lim \frac{L_1^2 - L_0^2}{L_0^2} = 2u_{r,s}(z)\lambda_r\lambda_s + u_{m,r}(z)u_{m,s}(z)\lambda_r\lambda_s.$$

But even this has not given us the desired value of 6.201.

To get a simple expression for the extension it is necessary to introduce a limitation on the character of the strain. We shall consider only *small strains*, a small strain being one in which the derivatives $u_{r,s}$ are small. Then the last term on the right-hand side of 6.206 is small of the second order, and will be dropped. Moreover, we have identically

6.207. $$\frac{L_1^2 - L_0^2}{L_0^2} = \frac{L_1 + L_0}{L_0} \cdot \frac{L_1 - L_0}{L_0} = \left(2 + \frac{L_1 - L_0}{L_0}\right)\frac{L_1 - L_0}{L_0}.$$

On account of the smallness of the strain, the left-hand side of 6.206 is small because the right-hand side is small. Hence the limit of the term on the left of 6.207 is small. Since $(L_1 + L_0)/L_0$ is greater than unity, it follows that the limit of $(L_1 - L_0)/L_0$ is small, and 6.207 gives approximately

6.208. $$\lim \frac{L_1^2 - L_0^2}{L_0^2} = 2 \lim \frac{L_1 - L_0}{L_0} = 2e,$$

e being the extension of the element in question, i.e. the element at z_r pointing in the direction λ_r. Combining this result with 6.206, we have

6.209. $$e = u_{r,s}(z)\lambda_r\lambda_s.$$

We now define the *components of strain* e_{rs} by

6.210. $$e_{rs} = \tfrac{1}{2}(u_{r,s} + u_{s,r}).$$

It is obvious that these are the components of a Cartesian tensor. Equation 6.209 for the extension e may be written

6.211. $$e = e_{rs}\lambda_r\lambda_s.$$

Thus, if the tensor e_{rs} is given as a function of position throughout the medium, the extension of every element is determined. But it must be remembered that 6.211 is valid only for small strain.

Exercise. Show that a small strain is a rigid body displacement if, and only if, $e_{rs} = 0$. In the case of finite strain, deduce from 6.206 the conditions which must be satisfied by the partial derivatives of the displacement in order that it may be a rigid body displacement.

If we use curvilinear coordinates x^r, the extension of an element with direction determined by a unit vector λ^r is (for small strain)

6.212. $$e = e_{rs}\lambda^r\lambda^s,$$

where

6.213. $$e_{rs} = \tfrac{1}{2}(u_{r|s} + u_{s|r}).$$

This statement is verified at once from the invariant character of 6.212.

In future it will be understood that the strain is small in every case.

Stress is a generalization of the concept of pressure introduced in 6.1. We consider the force transmitted across a plane element dS, but no longer insist that the force shall be perpendicular to the element. These ideas apply both to an elastic solid and to a viscous fluid.

Let us use rectangular Cartesian coordinates z_r. Let n_r be the direction cosines of the normal to the element dS, and $T_r\,dS$ the force across the element. (We naturally take the force proportional to the area dS.) We call T_r the *stress* across the plane element, so that stress is force per unit area. In general, T_r is a function of the coordinates z_r which determine the position of the element dS, as well as of the direction cosines n_r which determine the direction of the element. We may indicate this dependence by writing

6.214. $$T_r = F_r(z, n).$$

We shall next see how this function involves the direction cosines.

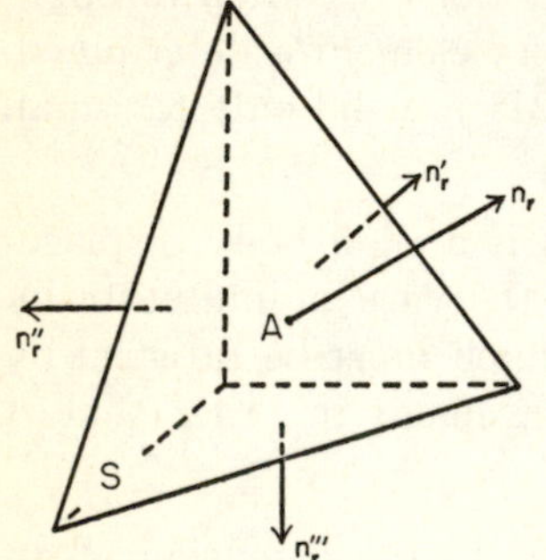

FIG. 18. Analysis of stress.

Consider any point A in the medium and a plane P passing through A. Take three mutually perpendicular planes, P', P'', P''', cutting the plane P in a triangle containing A (Fig. 18), so that a tetrahedron is formed with A situated in one of its faces. Let the direction cosine of the normals to P, P', P'', P''', all drawn outwards from the tetrahedron be n_r, n_r', n_r'', n_r''', respectively. The areas, S, S', S'', S''' of the faces of the tetrahedron are easily seen to be connected by the relations

6.215. $$S' = -Sn_s n_s', \quad S'' = -Sn_s n_s'', \quad S''' = -Sn_s n_s'''.$$

Let us suppose that the medium is subject to a body force such as gravity, the force on a volume dV being $X_r\,dV$. Now the rate of change of the linear momentum of the material contained in the tetrahedron is equal to the total external force acting on it. Therefore

6.216.

$$\frac{dM_r}{dt} = \int X_r dV + \int T_r dS + \int T_r' dS' + \int T_r'' dS'' + \int T_r''' dS''',$$

where M_r is the total linear momentum of the tetrahedron and the surface integrals are taken over the several faces of the tetrahedron.

Let us now make the tetrahedron contract towards A by a uniform contraction. This means that the directions of all four planes are maintained, and the relations 6.215 remain unchanged. Let a denote the longest edge of the tetrahedron in this process of contraction. Then, clearly, since the volume of the tetrahedron is less than a^3,

6.217. $$\lim_{a\to 0} \frac{1}{a^2}\frac{dM_r}{dt} = 0, \qquad \lim_{a\to 0} \frac{1}{a^2}\int X_r\, dV = 0,$$

and so by 6.216

6.218. $$\lim_{a\to 0} \frac{1}{a^2}\left(\int T_r dS + \int T_r' dS' + \int T_r'' dS'' + \int T_r''' dS'''\right) = 0.$$

Assuming the function F_r of 6.214 to be continuous in z, the value of the above limit will not be changed if we replace the integrands by constants as follows:

6.219. $$T_r \to F_r(z, n), \quad T_r' \to F_r(z, n'), \quad T_r'' \to F_r(z, n''),$$
$$T_r''' \to F_r(z, n'''),$$

where z stands for the coordinates of A. It then follows that 6.218 may be written

6.220. $$[F_r(z, n) - F_r(z, n')n_s n_s' - F_r(z, n'')n_s n_s''$$
$$- F_r(z, n''')n_s n_s'''] \lim_{a\to 0} (S/a^2) = 0.$$

But $\lim_{a\to 0} S/a^2$ is not zero, and so

6.221. $$F_r(z, n) = E_{rs}(z)n_s,$$

where

6.222. $$E_{rs} = F_r(z, n')n_s' + F_r(z, n'')n_s'' + F_r(z, n''')n_s'''.$$

We note that E_{rs} is independent of the direction cosines n_t.

We may write 6.221 equivalently in the form

6.223. $$T_r = E_{rs}n_s.$$

Since T_r and n_r are Cartesian vectors, it follows that E_{rs} is a Cartesian tensor; we call it the *stress tensor*. Equation 6.223 gives the stress across every plane element in the medium if E_{rs} is given as a function of position. So far there is no evidence that E_{rs} is symmetric, but we shall see later that it is.

Exercise. Show that the stress across a plane $z_1 =$ const. has components E_{11}, E_{21}, E_{31}. What are the components across planes $z_2 =$ const. and $z_3 =$ const.?

The concept of stress which we have just developed applies to any continuous medium, solid or fluid. We shall now find the equations of motion for a continuous medium by applying to a portion of it the principle of linear momentum. We have already done this for a perfect fluid, starting with 6.142 and ending with 6.145 or 6.146. To generalize 6.142 to the case of a general medium, we have merely to replace the last integral on the right by an expression for the total force due to stress across the bounding surface of the portion of the medium under consideration. The appropriate expression is

6.224. $$\int T_r \, dS = \int E_{rs} n_s dS,$$

by 6.223, the integral being taken over the bounding surface and n_s denoting the unit normal to this surface, drawn outward. Using Green's theorem to replace this surface integral by a volume integral we easily obtain the generalization of 6.145 in the form

6.225. $$\rho f_r = \rho X_r + E_{rs,s},$$

where f_r denotes the acceleration of a particle. *These are the equations of motion of a continuous medium. The equations of equilibrium are obtained by putting* $f_r = 0$.

We shall now establish the symmetry of the tensor E_{rs} by means of the principle of angular momentum, which states that the rate of change of angular momentum of any system about a fixed point is equal to the moment of external forces acting on the system about that point. As fixed point let us take the origin $z_r = 0$. The angular momentum for a rigid body was given in 5.327; actually that expression holds for any system of particles, rigid or not, and the appropriate expression for a continuous medium is

$$h_{rs} = \int \rho (z_r v_s - z_s v_r)\, dV, \tag{6.226}$$

and the moment of external forces is

$$M_{rs} = \int \rho (z_r X_s - z_s X_r)\, dV + \int (z_r T_s - z_s T_r)\, dS. \tag{6.227}$$

The principle of angular momentum then gives

$$\frac{d}{dt} h_{rs} = M_{rs}. \tag{6.228}$$

This yields, by 6.223,

$$\begin{aligned} \int \rho (z_r f_s - z_s f_r)\, dV &= \int \rho (z_r X_s - z_s X_r)\, dV \\ &\quad + \int (z_r E_{sm} - z_s E_{rm}) n_m\, dS. \end{aligned} \tag{6.229}$$

By Green's theorem

$$\begin{aligned} \int (z_r E_{sm} - z_s E_{rm}) n_m dS &= \int (z_r E_{sm,m} - z_s E_{rm,m})\, dV \\ &\quad + \int (E_{sr} - E_{rs})\, dV. \end{aligned} \tag{6.230}$$

When this is substituted in 6.229, all the integrals cancel out on account of 6.225, except one, and we are left with

$$\int (E_{sr} - E_{rs})\, dV = 0. \tag{6.231}$$

Since this holds for every volume, we deduce that

$$E_{sr} = E_{rs}. \tag{6.232}$$

Thus the symmetry of the stress tensor is established.

Having dealt separately with strain and stress, we shall now connect them, and so establish the theory of elasticity. The basic assumption is the *generalized Hooke's law*, which states that stress is a linear homogeneous function of strain. This means that there exists a stress-strain equation of the form

$$E_{rs} = c_{rsmn} e_{mn}. \tag{6.233}$$

In the case of a heterogeneous body, with elastic properties varying from point to point, the coefficients will be functions of position; we shall consider here only homogeneous bodies, for which the coefficients are constants.

Since $e_{mn} = e_{nm}$, there is obviously no loss of generality in taking

6.234. $$c_{rsmn} = c_{rsnm}.$$

Moreover, since $E_{rs} = E_{sr}$, we must have

6.235. $$c_{rsmn} = c_{srmn}.$$

By applying the tests for tensor character, 1.6, it is easily seen that c_{rsmn} are the components of a Cartesian tensor; we shall call it the first *elasticity tensor*. On account of 6.234 and 6.235, the number of independent components is 36.

By appeal to thermodynamic arguments, into which we shall not enter here, it can be shown that

6.236. $$c_{rsmn} = c_{mnrs},$$

in the limiting cases of isothermal and adiabatic states. This relation reduces the number of independent components of the elasticity tensor to 21. This is generally accepted as the maximum number of independent coefficients.

Exercise. Show that if 6.233 is solved for strain, so as to read

6.237. $$e_{rs} = C_{rsmn}E_{mn},$$

then the symmetry conditions 6.234, 6.235 and 6.236 imply similar conditions on C_{rsmn}. (The tensor C_{rsmn} is the second elasticity tensor.)

The simplest type of body is *isotropic*. This word implies that all systems of rectangular Cartesian coordinates are equivalent as far as the description of elastic properties is concerned. But the elastic properties of a body are completely described by the elasticity tensor c_{rsmn}, and so this tensor must have the same components for all sets of axes of coordinates. Thus, under the orthogonal transformation

6.238. $$z_r = A_{sr}z'_s, \quad (A_{rm}A_{sm} = \delta_{rs}),$$

the general transformation

6.239. $$c'_{rsmn} = c_{pquv}A_{rp}A_{sq}A_{mu}A_{nv}$$

must reduce to

6.240. $$c'_{rsmn} = c_{rsmn}.$$

We have already met a tensor that transforms identically like this—the Kronecker delta δ_{rs}. This suggests that the elasticity tensor must be built up out of Kronecker deltas. It can in fact be shown that this is the case.* But we shall content ourselves here by building up a tensor which satisfies 6.240 and the symmetry conditions, without proving that it is the most general possible.

The most general tensor of the fourth order that we can construct from the Kronecker delta is

6.241. $$c_{rsmn} = \lambda\delta_{rs}\delta_{mn} + \mu\delta_{rm}\delta_{sn} + \nu\delta_{rn}\delta_{sm},$$

where λ, μ, ν are invariants. Since $\delta'_{pq} = \delta_{pq}$, 6.240 is satisfied. On account of 6.234, we have

6.242. $$\mu\delta_{rm}\delta_{sn} + \nu\delta_{rn}\delta_{sm} = \mu\delta_{rn}\delta_{sm} + \nu\delta_{rm}\delta_{sn},$$

or

6.243. $$(\mu - \nu)(\delta_{rm}\delta_{sn} - \delta_{rn}\delta_{sm}) = 0,$$

so that $\mu = \nu$, and 6.241 becomes

6.244. $$c_{rsmn} = \lambda\delta_{rs}\delta_{mn} + \mu(\delta_{rm}\delta_{sn} + \delta_{rn}\delta_{sm}).$$

The other symmetry conditions 6.235 and 6.236 are automatically satisfied. *We accept* 6.244 *as the elasticity tensor of an isotropic body;* it contains just two elastic constants, λ and μ.

Substituting 6.244, we obtain from 6.233, as stress-strain equation for an isotropic body,

6.245. $$E_{rs} = \lambda\delta_{rs}\theta + 2\mu e_{rs},$$

where

6.246. $$\theta = e_{nn};$$

θ is called the *expansion* or *dilatation.*

It is easy to solve 6.245 for strain in terms of stress. The result is usually written in the form

*H. Jeffreys, *Cartesian Tensors*, Cambridge, Cambridge University Press, 1931, p. 66.

6.247. $$e_{rs} = \frac{1}{E}\left\{ (1+\sigma)E_{rs} - \sigma\delta_{rs}E_{nn} \right\},$$

where the constants are given by

6.248. $$E = \frac{\mu(3\lambda + 2\mu)}{\lambda + \mu}, \quad \sigma = \frac{\lambda}{2(\lambda + \mu)},$$

or, written the other way round,

6.249. $$\lambda = \frac{\sigma E}{(1+\sigma)(1-2\sigma)}, \quad \mu = \frac{E}{2(1+\sigma)}.$$

Specific names are given to these constants:

E is Young's modulus,
σ is Poisson's ratio,
λ is Lamé's constant,
μ is the rigidity.

If we substitute from 6.245 in 6.225, and use 6.210, we obtain the equations of motion of an elastic body in the form

6.250. $$\rho f_r = \rho X_r + (\lambda + \mu)\theta_{,r} + \mu\Delta u_r,$$

where Δ is the Laplacian differential operator. On account of the assumed smallness of the displacement, we may write $f_r = \partial^2 u_r/\partial t^2$. Then 6.250 gives a set of three partial differential equations for three quantities, u_r.

Exercise. Deduce from 6.250 that if an isotropic elastic body is in equilibrium under no body forces, then the expansion θ is a harmonic function ($\Delta\theta = 0$).

Let us now translate our results from rectangular Cartesian coordinates z_r to curvilinear coordinates x^r. This is easy to do by guessing, and the verification is immediate in each case.

From 6.223 we have

6.251. $$T_r = E_{rs}n^s, \quad T^r = E^{rs}n_s.$$

The equations of motion 6.225 read, in contravariant form,

6.252. $$\rho f^r = \rho X^r + E^{rs}{}_{|s},$$

or, in covariant form,

6.253. $$\rho f_r = \rho X_r + E^s_{r|s},$$

where E^s_r is the mixed stress tensor. The symmetry property 6.232 is invariant under general transformations of coordinates, and so it holds for curvilinear coordinates since it holds for rectangular Cartesians. On account of this symmetry, we can write E^s_r instead of $E_r{}^{\cdot s}$, since $E_r{}^{\cdot s} = E^s{}_{\cdot r}$, so that there is no risk of confusion in omitting the dot.

Hooke's law 6.233 reads in covariant form

6.254. $$E_{rs} = c_{rsmn}e^{mn},$$

the coefficients being the covariant components of the first elasticity tensor. The symmetry relations 6.234, 6.235, and 6.236 hold for curvilinear coordinates. The isotropic stress-strain relation 6.245 reads, in covariant form,

6.255. $$E_{rs} = \lambda a_{rs}\theta + 2\mu e_{rs},$$

where a_{rs} is the metric tensor and

6.256. $$\theta = a^{mn}e_{mn}.$$

Exercise. Express the equations of motion 6.250 in curvilinear coordinates.

6.3. Electromagnetic radiation. We shall consider only electromagnetic fields in vacuo. This means that we omit the application of tensors to electromagnetic circuits and machines, and to the propagation of electromagnetic waves through material media. However, for most practical purposes, air may be regarded as a vacuum, and so our theory applies to the ordinary propagation through air of radio waves, radar waves, heat waves, light waves, and X-rays.

An electromagnetic field in vacuo is characterized by two vectors—an *electric vector* and a *magnetic vector*. For the present we shall use rectangular Cartesian coordinates z_r; the electric vector will be denoted by E_r and the magnetic vector by H_r. In any given field, the values of the components E_r and H_r depend on the choice of units in which they are measured. We shall use Heaviside or rational units, which are most convenient for our purposes.

The basic hypothesis from which we start consists of a set of partial differential equations, *Maxwell's equations*. For right-handed axes, these read

$$\text{6.301.} \qquad \frac{1}{c}\frac{\partial E_r}{\partial t} = \epsilon_{rmn} H_{n,m}, \quad \frac{1}{c}\frac{\partial H_r}{\partial t} = -\epsilon_{rmn} E_{n,m},$$

$$\text{6.302.} \qquad E_{n,n} = 0, \quad H_{n,n} = 0.$$

The constant c is a universal constant; if the units of length and time are the centimeter and the second, its numerical value is approximately $c = 3 \times 10^{10}$.

There are six equations in 6.301 and two in 6.302. The equations 6.302 cannot be deduced from 6.301, but they are connected with them. If we differentiate 6.301 with respect to z_r and note that

$$\text{6.303.} \qquad \epsilon_{rmn} H_{n,mr} = 0, \quad \epsilon_{rmn} E_{n,mr} = 0,$$

on account of the skew-symmetry of the permutation symbols, we obtain

$$\text{6.304.} \qquad \frac{\partial}{\partial t} E_{r,r} = 0, \quad \frac{\partial}{\partial t} H_{r,r} = 0.$$

Thus 6.301 tells us that $E_{r,r}$ and $H_{r,r}$ are independent of t. This conclusion is consistent with 6.302. This interlocking of 6.301 with 6.302 enables us to obtain solutions of what at first sight appears to be an overdetermined problem—namely, to find six quantities satisfying eight equations.

If we differentiate the first of 6.301 with respect to t, and then use the second equation, we get, with the aid of 4.329,

$$\text{6.305.} \qquad \begin{aligned} \frac{1}{c}\frac{\partial^2 E_r}{\partial t^2} &= \epsilon_{rmn} \frac{\partial}{\partial z_m}\frac{\partial H_n}{\partial t} \\ &= -c\, \epsilon_{rmn} \frac{\partial}{\partial z_m} (\epsilon_{npq} E_{q,p}) \\ &= -c\, \epsilon_{nrm}\epsilon_{npq} E_{q,pm} \\ &= -c\, (\delta_{rp}\delta_{mq} - \delta_{rq}\delta_{pm}) E_{q,pm} \\ &= -c\, E_{m,mr} + c\, E_{r,mm}. \end{aligned}$$

The first term on the right-hand side vanishes on account of 6.302, and so

6.306. $$\frac{1}{c^2}\frac{\partial^2 E_r}{\partial t^2} - E_{r,mm} = 0.$$

Similarly,

6.307. $$\frac{1}{c^2}\frac{\partial^2 H_r}{\partial t^2} - H_{r,mm} = 0.$$

This form of partial differential equation is called the *wave equation*. Thus *the electric and magnetic vectors satisfy the wave equation*.

Exercise. Verify that $E_r = z_r$, $H_r = 0$ satisfy the wave equations, but not Maxwell's equations.

It will be noted that Maxwell's equations 6.301 and 6.302 are linear homogeneous partial differential equations. This is an extremely important property, since it enables us to superimpose solutions. If $(E_r^{(1)}, H_r^{(1)})$ satisfy Maxwell's equations, and if $(E_r^{(2)}, H_r^{(2)})$ satisfy Maxwell's equations, then the field given by

$$E_r = E_r^{(1)} + E_r^{(2)},\ H_r = H_r^{(1)} + H_r^{(2)}$$

also satisfies Maxwell's equations. Further, if $(E_r^{(1)}, H_r^{(1)})$ satisfy Maxwell's equations, then so also do

$$E_r = kE_r^{(1)},\ H_r = kH_r^{(1)},$$

where k is any constant.

This technique of superposition is useful in connection with complex solutions of Maxwell's equations. Physically, we are interested only in real electromagnetic vectors. However, if (E_r, H_r) is a complex field satisfying Maxwell's equations, it is clear that the complex conjugate field $(\overline{E}_r, \overline{H}_r)$ also satisfies the equations. Hence, the real field

$$E_r^* = \tfrac{1}{2}(E_r + \overline{E}_r),\ H_r^* = \tfrac{1}{2}(H_r + \overline{H}_r)$$

satisfies Maxwell's equations.

We shall now study complex solutions of Maxwell's equations of the form

6.308. $$E_r = E_r^{(0)} e^{iS}, \quad H_r = H_r^{(0)} e^{iS}.$$

Here $E_r^{(0)}$ and $H_r^{(0)}$ are complex vectors independent of time but, in general, functions of position; S is given by

6.309. $$S = \frac{2\pi}{\lambda}[V - ct],$$

where λ is a real constant and V a real function of position only; c is the constant occurring in Maxwell's equations 6.301. Since c has the dimensions of a velocity, λ has the dimensions of a length.

Since E_r and H_r involve the time through the factor $e^{-i(2\pi/\lambda)ct}$ only, equations 6.304 imply 6.302. Thus the only conditions on $E_r^{(0)}$, $H_r^{(0)}$, V, λ, are imposed by Maxwell's equations 6.301. Substitution in these equations leads to

6.310. $$E_r^{(0)} = i\frac{\lambda}{2\pi}\epsilon_{rmn}H^{(0)}_{n,m} - \epsilon_{rmn}H_n^{(0)} V_{,m},$$

6.311. $$H_r^{(0)} = -i\frac{\lambda}{2\pi}\epsilon_{rmn}E^{(0)}_{n,m} + \epsilon_{rmn} E_n^{(0)} V_{,m}.$$

We now introduce the following approximation. We shall assume that λ is small, or, more precisely, that $\lambda E^{(0)}_{n,m}$ and $\lambda H^{(0)}_{n,m}$ are small in comparison with $E_r^{(0)}$ and $H_r^{(0)}$. The omission of these small terms is the essential step in passing from physical or electromagnetic optics to geometrical optics. In the approximation considered, 6.310 and 6.311 read

6.312. $$E_r^{(0)} = -\epsilon_{rmn} H_n^{(0)} V_{,m},$$

6.313. $$H_r^{(0)} = \epsilon_{rmn} E_n^{(0)} V_{,m}.$$

We immediately deduce the following relations

6.314. $$E_r^{(0)} V_{,r} = 0, \quad H_r^{(0)} V_{,r} = 0, \quad E_r^{(0)} H_r^{(0)} = 0.$$

Equations 6.312 and 6.313 are algebraic equations, linear and homogeneous, in $E_r^{(0)}$ and $H_r^{(0)}$. We can eliminate $E_r^{(0)}$ and $H_r^{(0)}$ by substituting from the second equation into the first, thus:

6.315.
$$\begin{aligned} E_r^{(0)} &= -\epsilon_{rmn}\epsilon_{nuv}E_v^{(0)} V_{,u} V_{,m} \\ &= (\delta_{rv}\delta_{mu} - \delta_{ru}\delta_{mv})E_v^{(0)} V_{,u} V_{,m} \\ &= E_r^{(0)} V_{,m}V_{,m} - E_m^{(0)} V_{,m} V_{,r} \\ &= E_r^{(0)} V_{,m} V_{,m}. \end{aligned}$$

We immediately obtain Hamilton's famous partial differential equation of geometrical optics:

6.316.
$$V_{,m}V_{,m} = 1.$$

From 6.312, we have

$$\begin{aligned} E_r^{(0)}E_r^{(0)} &= \epsilon_{rmn}\epsilon_{rpq}H_n^{(0)}H_q^{(0)} V_{,m} V_{,p} \\ &= H_n^{(0)}H_n^{(0)} V_{,m}V_{,m} - H_n^{(0)} V_{,n} H_m^{(0)} V_{,m}. \end{aligned}$$

By 6.316 and 6.314, this becomes

6.317.
$$E_r^{(0)}E_r^{(0)} = H_r^{(0)}H_r^{(0)}.$$

It is now easy to verify the following statement: *The complex field* 6.308 *satisfies Maxwell's equations, to the approximation considered, if* V, $E_r^{(0)}$, $H_r^{(0)}$, λ *are chosen as follows:*

(i) V is an arbitrary solution of the partial differential equation 6.316.

(ii) $E_r^{(0)}$ is arbitrary except for the first equation 6.314.

(iii) $H_r^{(0)}$ is given by 6.313.

(iv) λ is an arbitrary small constant.

It remains to verify that 6.312 is satisfied. But, with 6.313 established by (iii), equation 6.312 is equivalent to 6.315; and this is satisfied by virtue of (i), i.e., by virtue of 6.316.

We have now obtained fields of the form 6.308 which satisfy Maxwell's equations approximately when λ is small. Let us examine these fields more closely.

Consider the surfaces V = constant, where V is a solution of 6.316. The vector $V_{,m}$ is normal to V = constant, and points in the direction of V increasing. By virtue of 6.316, $V_{,m}$ is a *unit normal.* If we proceed along a curve cutting the surfaces V = constant orthogonally (i.e., having the unit tangent $V_{,m}$ at each point), we obtain

6.318. $$ds = V_{,m}dz_m = dV,$$

and

6.319. $$\frac{d}{ds} V_{,m} = V_{,mn} \frac{dz_n}{ds} = V_{,mn}V_{,n} = \tfrac{1}{2} \frac{\partial}{\partial z_m} (V_{,n}V_{,n}) = 0.$$

The first of these equations tells us that two of the surfaces $V =$ constant, say $V = V_1$ and $V = V_2$, cut off equal intercepts on all orthogonal trajectories to the system of surfaces; the length of each intercept is $V_2 - V_1$. Equation 6.319 shows that each normal trajectory to the system of surfaces $V =$ constant is a straight line, since its tangent vector is a constant. We summarize by saying that $V =$ constant is a system of parallel surfaces with rectilinear orthogonal trajectories.

By 6.314, the real and imaginary parts of the vectors $E_r^{(0)}$ and $H_r^{(0)}$ are tangential to the surfaces $V =$ constant. The same statement obviously applies to E_r and H_r.

The equation $S =$ constant, or

6.320. $$V - ct = \text{constant},$$

is the equation of a moving surface, and $V_{,m}$ is the unit normal to this surface. To find the normal velocity v with which the surface moves, we follow a moving point which always lies in the surface and whose velocity is always normal to the surface. For such a point we have

6.321. $$\frac{dz_r}{dt} = vV_{,r}.$$

But since the point always lies on the surface, we have from 6.320

$$\frac{dV}{dt} = V_{,r}\frac{dz_r}{dt} = c,$$

and thus, by 6.321 and 6.316,

6.322. $$v = c.$$

The moving surfaces 6.320 are called *phase waves*, and we have just established the fact that *phase waves are propagated with normal velocity c, where c is the constant occurring in Maxwell's equations.* Since visible light is a type of electromagnetic radiation, c is also known as *the velocity of light in vacuo.*

We observe that E_r and H_r vary as slowly as $E_r^{(0)}$ and $H_r^{(0)}$ over a phase wave S=constant. Over a region of this surface whose linear dimensions are of the order of λ, and for a time interval of the order of λ/c, E_r and H_r are constants to the approximation considered. Moreover, to our approximation, we get the same value again if S is increased by 2π, or any small multiple of 2π.

It is clear from 6.309 that, at any fixed point in space, S is changed by 2π if t is changed by an amount λ/c. This is called the *period* τ, and the *frequency* ν is its reciprocal. Hence

6.323. $$\tau = \lambda/c, \quad \nu = c/\lambda.$$

Since λ is small, we have on moving a distance λ along the normal to a phase wave, without changing the value of t, the following approximate result:

$$\Delta S = \frac{2\pi}{\lambda}\Delta V = \frac{2\pi}{\lambda}\lambda = 2\pi.$$

Thus, at any instant, wave surfaces of equal "phase" e^{iS} are separated by a normal distance λ; λ is therefore called the *wave-length.*

In exploring the nature of the field, the situation is slightly complicated by the fact that $E_r^{(0)}$ and $H_r^{(0)}$ are complex vectors.

If we multiply 6.312 and 6.313 across by e^{iS}, we get

6.324. $$E_r = -\epsilon_{rmn}H_n V_{,m}, \quad H_r = \epsilon_{rmn}E_n V_{,m}.$$

These relations between the complex vectors E_r, H_r (involving twelve real components) can be replaced by relations between the real vectors

6.325. $$E_r^* = \tfrac{1}{2}(E_r + \overline{E}_r), \quad H_r^* = \tfrac{1}{2}(H_r + \overline{H}_r),$$

6.326. $$E_r^{**} = \tfrac{1}{2}i(E_r - \overline{E}_r), \quad H_r^{**} = \tfrac{1}{2}i(H_r - \overline{H}_r).$$

The vectors E_r^* and H_r^* will be taken as the *physical components* of the electromagnetic field. The vectors E_r^{**}, H_r^{**} also satisfy Maxwell's equations, and may be regarded as the physical components of a complementary electromagnetic field.

If we take the complex conjugates of 6.324 and add, we obtain

$$\textbf{6.327.}\qquad E_r^* = -\epsilon_{rmn}H_n^*V_{,m}, \quad H_r^* = \epsilon_{rmn}E_n^*V_{,m}.$$

Hence the vectors $V_{,r}$, E_r^*, H_r^* form a right-handed orthogonal triad, and we obtain the relation

$$\textbf{6.328.}\qquad E_r^*E_r^* = H_r^*H_r^*$$

in the same way as we obtained 6.317. Thus, *the physical electric and magnetic vectors are mutually perpendicular and lie in the phase waves; their magnitudes are equal.* (Fig. 19.)

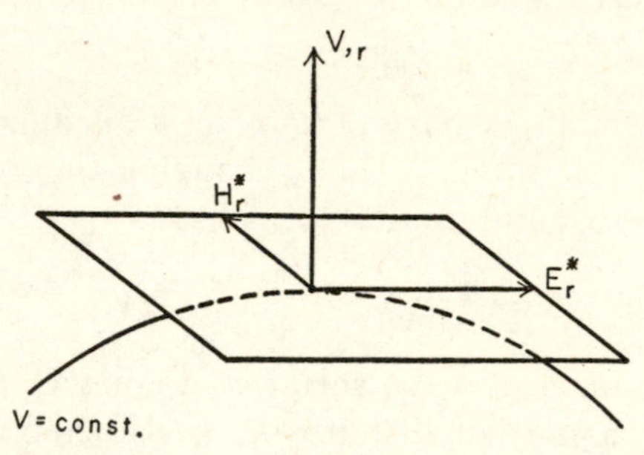

FIG. 19. Orthogonality relations in electromagnetic radiation.

Exercise. Prove a similar statement for the electric and magnetic vectors of the complementary electromagnetic field.

Each of the vectors $E_r^{(0)}$, $H_r^{(0)}$ defines a pair of fixed directions at any point in space, but the directions of the physical vectors E_r^*, H_r^* are not fixed. It is true that, if we move with a phase wave, E_r and H_r, and consequently also E_r^* and H_r^*, are approximately constant for time intervals of the order of the period τ. But if we remain at a fixed point in space, we have, by 6.308 and 6.309,

$$\textbf{6.329.}\qquad \frac{\partial E_r}{\partial t} = -\frac{2\pi ic}{\lambda}E_r, \quad \frac{\partial H_r}{\partial t} = -\frac{2\pi ic}{\lambda}H_r,$$

and so

6.330.
$$\frac{\partial E^*_r}{\partial t} = -\frac{2\pi c}{\lambda} E^{**}_r, \quad \frac{\partial H^*_r}{\partial t} = -\frac{2\pi c}{\lambda} H^{**}_r,$$
$$\frac{\partial E^{**}_r}{\partial t} = \frac{2\pi c}{\lambda} E^*_r, \quad \frac{\partial H^{**}_r}{\partial t} = \frac{2\pi c}{\lambda} H^*_r.$$

Hence

6.331.
$$\frac{\partial^2 E^*_r}{\partial t^2} + \left(\frac{2\pi c}{\lambda}\right)^2 E^*_r = 0, \quad \frac{\partial^2 H^*_r}{\partial t^2} + \left(\frac{2\pi c}{\lambda}\right)^2 H^*_r = 0.$$

Thus the extremities of the vectors E^*_r and H^*_r, which lie in the fixed plane tangent to the surface $V=$ constant, follow the behaviour of a particle attracted to a fixed point by a force proportional to the distance from that point. It is well known that the path of such a particle is an ellipse, and therefore the extremities of E^*_r and H^*_r describe ellipses, these vectors being at all times perpendicular to one another. We say that the radiation is *elliptically polarized* (Fig. 20).

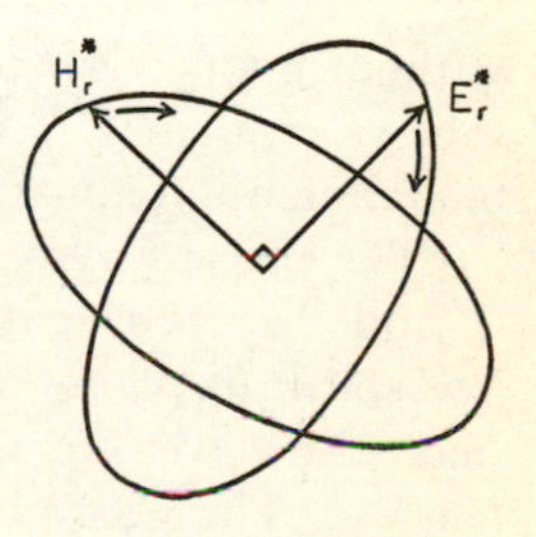

FIG. 20. Elliptically polarized wave travelling up from paper.

A case of particular interest is that in which the vectors E^*_r and E^{**}_r have the same direction, or opposite directions. It follows from 6.330 that, in this case, the ellipse degenerates to a straight line, and the vector E^*_r maintains a fixed direction, except for reversals which occur twice in each period. The same is true for H^*_r, on account of the orthogonality relation. We speak then of a *plane-polarized wave*.

Exercise. What conditions must be imposed on the fixed complex vectors $E^{(0)}_r$ and $H^{(0)}_r$ in order that the wave may be plane-polarized?

Our discussion of electromagnetic waves has involved an approximation and is valid only for a small wave-length λ. There is, however, an important special case for which our theory is exact. Consider the linear function

6.332. $$V = a_m z_m + b,$$

where a_m and b are real constants. We immediately have

6.333. $$V_{,m} = a_m,$$

and thus 6.316 is equivalent to

6.334. $$a_m a_m = 1.$$

This means that a_m is a unit vector.

Equations 6.312 and 6.313 become

6.335. $$E_r^{(0)} = -\epsilon_{rmn} H_n^{(0)} a_m, \quad H_r^{(0)} = \epsilon_{rmn} E_n^{(0)} a_m.$$

These are algebraic equations with *constant* coefficients. We may therefore take $E_r^{(0)}$ and $H_r^{(0)}$ to be *constant complex vectors* without giving rise to inconsistencies. But in this case 6.335 follows *rigorously* from 6.308 and from Maxwell's equations 6.301; no approximation is required.

The phase waves are planes orthogonal to the unit vector a_r and they propagate with velocity c in the direction of a_r. We speak of *plane electromagnetic* waves. For plane electromagnetic waves our theory is exact for arbitrarily large wavelength. The vectors E and H are rigorously constant over a phase wave and are also constant in time if we move with the wave.

Let us now return to the general Maxwell equations 6.301 and 6.302. We shall consider some useful types of solution.

Let ψ be an invariant and ϕ_r a vector, both functions of position and time. Let us write tentatively

6.336. $$E_r = -\frac{1}{c}\frac{\partial \phi_r}{\partial t} - \psi_{,r}, \qquad H_r = \epsilon_{rpq}\phi_{q,p}.$$

Let us see what conditions must be imposed on ψ and ϕ_r in order that Maxwell's equations may be satisfied.

The second of 6.301 is satisfied automatically, and the first gives

6.337. $$\frac{1}{c^2}\frac{\partial^2 \phi_r}{\partial t^2} + \frac{1}{c}\frac{\partial \psi_{,r}}{\partial t} = \phi_{r,mm} - \phi_{m,mr}.$$

The second of 6.302 is satisfied automatically, and the first gives

6.338. $$\frac{1}{c}\frac{\partial}{\partial t}\phi_{m,m} + \psi_{,mm} = 0.$$

If, then, we subject ψ and ϕ_r to the partial differential equations

6.339. $$\frac{1}{c^2}\frac{\partial^2\phi_r}{\partial t^2} - \phi_{r,mm} = 0, \quad \frac{1}{c^2}\frac{\partial^2\psi}{\partial t^2} - \psi_{,mm} = 0,$$

6.340. $$\frac{1}{c}\frac{\partial\psi}{\partial t} + \phi_{m,m} = 0,$$

we have in 6.336 an electromagnetic field which satisfies Maxwell's equations. The invariant ψ is called the *scalar potential* and the vector ϕ_r is called the *vector potential.*

If we are given an electromagnetic field satisfying Maxwell's equations, it can be shown that functions ψ and ϕ_r exist so that 6.336 are satisfied. It follows that these functions must satisfy 6.337 and 6.338. It does not follow that they necessarily satisfy 6.339 and 6.340. However, other functions ψ and ϕ_r can be found to satisfy 6.336, 6.339, and 6.340. (Cf. Exercise 16 at end of chapter.)

There is another useful type of solution in terms of the *Hertz vector* Π_r. Let Π_r be a vector function of position and time, and let us write tentatively

6.341. $$E_r = \Pi_{m,mr} - \frac{1}{c^2}\frac{\partial^2\Pi_r}{\partial t^2},$$
$$H_r = \frac{1}{c}\epsilon_{rpq}\frac{\partial}{\partial t}\Pi_{q,p}.$$

On substitution in 6.301 and 6.302, we find that these equations are satisfied provided the Hertz vector Π_r satisfies the wave equation

6.342. $$\frac{1}{c^2}\frac{\partial^2\Pi_r}{\partial t^2} - \Pi_{r,mm} = 0.$$

It must not of course be assumed that any field satisfying Maxwell's equations can be represented in the form 6.341. However the Hertz vector enables us to build up a fairly general type of solution, to be discussed below.

Let z_r and ζ_r be any two points in space and let R be the distance between them, so that

$$\textbf{6.343.} \qquad R^2 = (z_m - \zeta_m)(z_m - \zeta_m).$$

Let k be any constant, and let us write

$$\textbf{6.344.} \qquad F(z, \zeta) = \frac{e^{ikR}}{R},$$

the notation indicating that we have here a function of the six coordinates z_r, ζ_r; but of course these variables are involved only in the form R. Let us denote derivatives with respect to R by primes, and partial derivatives with respect to z_r by a comma. By 6.343 we have

$$\textbf{6.345.} \qquad R_{,m} = \frac{z_m - \zeta_m}{R},$$

and

$$\textbf{6.346.} \qquad F_{,m} = \frac{(z_m - \zeta_m)\, F'}{R}.$$

Hence, since $z_{m,m} = 3$, we obtain

$$\textbf{6.347.} \qquad F_{,mm} = F'' + \frac{2F'}{R}.$$

But, directly from 6.344,

$$\textbf{6.348.} \qquad F' = \left(\frac{ik}{R} - \frac{1}{R^2}\right) e^{ikR},$$

$$F'' = \left(-\frac{k^2}{R} - \frac{2ik}{R^2} + \frac{2}{R^3}\right) e^{ikR},$$

and so

$$\textbf{6.349.} \qquad F'' + \frac{2F'}{R} = -\frac{k^2}{R} e^{ikR} = -k^2 F.$$

On substituting this in 6.347, we find that F satisfies the partial differential equation

6.350. $$F_{,mm} + k^2F = 0.$$

It will be recalled that the comma indicates differentiation with respect to z_r. It is very easy to see that the same equation is also satisfied if the comma indicates differentiation with respect to ζ_r.

Let us define $G(z, \zeta, t)$ by

6.351. $$G(z, \zeta, t) = e^{-ikct}F(z, \zeta) = \frac{1}{R}e^{ik(R-ct)}.$$

Then

6.352. $$G_{,mm} = e^{-ikct}F_{,mm}, \quad \frac{\partial^2 G}{\partial t^2} = -k^2c^2e^{-ikct}F,$$

and so, by 6.350, G satisfies the partial differential equation

6.353. $$G_{,mm} - \frac{1}{c^2}\frac{\partial^2 G}{\partial t^2} = 0,$$

which is the wave equation. The function G may be called *the fundamental solution of the wave equation*, the constant k having any value.

We are now in a position to build up a fairly general class of solutions of Maxwell's equations. We shall confine our attention to fields with a simple harmonic variation in time, so that the physical electric and magnetic vectors have the forms

6.354. $$\begin{aligned} E_r^* &= A_r \cos kct + B_r \sin kct, \\ H_r^* &= C_r \cos kct + D_r \sin kct, \end{aligned}$$

where the coefficients are real functions of position and k is a real constant. This field corresponds to a complex field of the form

6.355. $$E_r = E_r^{(0)}e^{-ikct}, \quad H_r = H_r^{(0)}e^{-ikct},$$

where $E_r^{(0)}$ and $H_r^{(0)}$ are complex vector functions of position only. (The use of the negative form of the exponential is merely a notational convenience.) When we substitute 6.355 in Maxwell's equations 6.301, we get

6.356. $-ikE_r^{(0)} = \epsilon_{rmn}H^{(0)}_{n,m}, \quad ikH_r^{(0)} = \epsilon_{rmn}E^{(0)}_{n,m}.$

The equations 6.302 are satisfied identically.

Now we know that Maxwell's equations are satisfied by expressions of the form 6.341, provided Π_r satisfies 6.342. Hence 6.356 will be satisfied if we put

6.357. $$\Pi_r = \Pi_r^{(0)} e^{-ikct},$$

where $\Pi_r^{(0)}$ is a vector function of position satisfying

6.358. $$\Pi^{(0)}_{r,mm} + k^2\Pi_r^{(0)} = 0.$$

By 6.341 the vectors $E_r^{(0)}$ and $H_r^{(0)}$ are given in terms of $\Pi_r^{(0)}$ by

6.359. $$E_r^{(0)} = \Pi^{(0)}_{m,mr} + k^2\Pi_r^{(0)},$$
$$H^{(0)} = -ik\,\epsilon_{rpq}\Pi^{(0)}_{q,p}.$$

Thus we can build a Maxwellian field out of any vector field satisfying 6.358. By 6.350 we know that 6.358 is satisfied by a vector

6.360. $$\Pi_r^{(0)} = B_r \frac{e^{ikR}}{R},$$

where B_r is any constant vector, and R is the distance from the variable point z_r to a fixed point ζ_r, as in 6.343. But we can immediately construct a much more general solution of 6.358 by writing

6.361. $$\Pi_r^{(0)}(z) = \int_{V_\zeta} P_r(\zeta) F(z, \zeta) dV_\zeta$$

where $F(z, \zeta) = e^{ikR}/R$, $P_r(\zeta)$ is an arbitrary vector function of the coordinates ζ, and dV_ζ is the volume element of $d\zeta_1 d\zeta_2 d\zeta_3$. The integration is carried through a volume V_ζ which does not contain the point z_r under consideration. If z_r did lie inside the volume V_ζ, we would have to consider carefully the possibility of differentiating under the sign of integration with respect to z_r. However, this is not the case by hypothesis, and differentiation under the sign of integration is permissible. It follows that 6.358 is satisfied. The corresponding Maxwellian field is given by 6.355, where

6.362.
$$E_r^{(0)} = \int_{V_\zeta} P_m(\zeta)\,(F_{,mr} + k^2 F\delta_{mr})\,dV_\zeta,$$
$$H_r^{(0)} = -ik\epsilon_{rpq}\int_{V_\zeta} P_q(\zeta) F_{,p}\,dV_\zeta,$$

where the comma indicates partial differentiation with respect to z. It will be noted that the vector field $P_r(\zeta)$ remains completely arbitrary. It need not even be continuous throughout V_ζ.

It is not essential that the integration be through a volume; integration over a surface or along a curve would serve equally well.

Exercise. Show that Maxwell's equations in the form 6.356 are satisfied by

6.363.
$$E_r^{(0)} = ik\epsilon_{rpq}\int_{V_\zeta} Q_q(\zeta) F_{,p}\,dV_\zeta,$$
$$H_r^{(0)} = \int_{V_\zeta} Q_m(\zeta)\,(F_{,mr} + k^2 F\delta_{mr})\,dV_\zeta,$$

where $Q_r(\zeta)$ is an arbitrary vector field, and F is as in 6.344.

So far we have used rectangular Cartesians in discussing electromagnetic fields. We shall now introduce *curvilinear coordinates*. One method of obtaining Maxwell's equations in covariant form is by introducing

6.364.
$$\eta_{rmn} = \sqrt{a}\,\epsilon_{rmn},\quad \eta^{rmn} = \frac{1}{\sqrt{a}}\,\epsilon^{rmn}.$$

We recall from 6.139 that the η's are absolute tensors, at least for transformations with positive Jacobian. It is now easy to guess and verify that Maxwell's equations 6.301 read in tensor form

6.365.
$$\frac{1}{c}\frac{\partial E^r}{\partial t} = \eta^{rmn}H_{n\,|\,m},\quad \frac{1}{c}\frac{\partial H^r}{\partial t} = -\,\eta^{rmn}E_{n\,|\,m},$$

and 6.302 read

6.366.
$$E^n{}_{|\,n} = 0,\quad H^n{}_{|\,n} = 0.$$

The stroke indicates covariant differentiation.

Due to the behaviour of the η's, Maxwell's equations in the form 6.365 and 6.366 only hold in right-handed curvilinear coordinate systems; this means that the Jacobian of the transformation to right-handed Cartesians must be positive.

However, there is another covariant form of Maxwell's equations valid in both right- and left-handed coordinate systems. For the moment we retain rectangular Cartesian coordinates. From its usual definition as force on a unit charge at rest, it follows that E_r is a Cartesian vector. Hence, by 6.301, H_r must be oriented since ϵ_{rmn} is an oriented tensor, 4.3. If we therefore introduce a skew-symmetric *magnetic tensor* by the equations

6.367. $$H_{rm} = \epsilon_{rmn}H_n, \quad H_r = \tfrac{1}{2}\,\epsilon_{rmn}H_{mn},$$

then both E_r and H_{rm} are unoriented. It is easily verified that Maxwell's equations may be written in the following form:

6.368. $$\frac{1}{c}\frac{\partial E_r}{\partial t} = H_{rm,m}, \quad \frac{1}{c}\frac{\partial H_{rm}}{\partial t} = E_{r,m} - E_{m,r}\,,$$
$$E_{n,n} = 0, \quad H_{rm,n} + H_{mn,r} + H_{nr,m} = 0.$$

These equations are equally valid in right- and left-handed Cartesian coordinates.

Exercise. Write out Maxwell's equations in terms of a magnetic vector and a skew-symmetric electric tensor.

We have succeeded in getting rid of the permutation symbols in Maxwell's equations for rectangular Cartesians, and are now in a position to write them in the covariant form, valid for arbitrary curvilinear coordinates (right- or left-handed):

6.369. $$\frac{1}{c}\frac{\partial E_r}{\partial t} = a^{mn}H_{rm\,|\,n}, \quad \frac{1}{c}\frac{\partial H_{rm}}{\partial t} = E_{r,m} - E_{m,r},$$
$$a^{nm}E_{n\,|\,m} = 0, \quad H_{rm,n} + H_{mn,r} + H_{nr,m} = 0.$$

In two of these equations the partial derivatives have been retained. This is justified since $E_{r,m} - E_{m,r}$ and $H_{rm,n} + H_{mn,r} + H_{nr,m}$ are both tensors (cf. Exercises I, No. 8 and 1.707).

There is a third and most compact form into which Maxwell's equations can be brought. This entails the introduction of a fourth dimension, but gives the most fundamental form of Maxwell's equations from the point of view of relativity.

Introducing Greek suffixes for the range 1, 2, 3, while reserving Latin suffixes for the range 1, 2, 3, 4, we may write two of Maxwell's equations 6.369 in the form

$$H_{\alpha\beta,\gamma} + H_{\beta\gamma,\alpha} + H_{\gamma\alpha,\beta} = 0,$$

6.370.

$$E_{\alpha.,\gamma} - E_{.\gamma,\alpha} + \frac{\partial}{\partial(ct)} H_{\gamma\alpha} = 0.$$

The coordinates are curvilinear. The dot in $E_{\alpha.}$ and $E_{.\gamma}$ is merely a device to bring into focus the similarity between the two equations. This similarity suggests the following procedure: we introduce a fourth coordinate $x^4 = ct$ and the skew-symmetric quantities F_{mn}, defined by

6.371. $$F_{\alpha\beta} = H_{\alpha\beta},\ F_{\alpha 4} = -F_{4\alpha} = E_\alpha,\ F_{44} = 0.$$

With respect to transformation of the curvilinear coordinates x^α, it is clear that $F_{\alpha\beta}$ is a skew-symmetric tensor and $F_{\alpha 4}$ (or $F_{4\alpha}$) is a covariant vector. Then the equations 6.370 can be combined into the single equation

6.372. $$F_{rm,n} + F_{mn,r} + F_{nr,m} = 0.$$

The remaining two of Maxwell's equations can now be written as follows:

$$a^{\beta\gamma} H_{\alpha\beta||\gamma} - E_{\alpha,4} = 0, \quad a^{\beta\gamma} E_{\beta||\gamma} = 0,$$

or, equivalently,

6.373. $$a^{\beta\gamma} F_{\alpha\beta||\gamma} - F_{\alpha 4,4} = 0, \quad a^{\beta\gamma} F_{4\beta||\gamma} = 0.$$

The double stroke indicates covariant differentiation with respect to the 3-metric $a_{\alpha\beta}$.

Let us now consider a 4-dimensional manifold, *space-time*, in which the four coordinates are x^1, x^2, x^3, x^4, where $x^4 =$ ct. Let us adopt in space-time the metric g_{mn} defined by

6.374. $$g_{\alpha\beta} = a_{\alpha\beta},\ g_{\alpha 4} = 0,\ g_{44} = -1,$$

so that the metric form is

6.375. $$\Phi = g_{mn}dx^m dx^n = a_{\alpha\beta}dx^\alpha dx^\beta - (dx^4)^2, \; x^4 = ct.$$

From 6.375 it follows that

6.376. $$g^{\alpha\beta} = a^{\alpha\beta}, \quad g^{\alpha 4} = 0, \quad g^{44} = -1.$$

We have already defined the quantities F_{mn}. We shall understand the symbol $F_{rm|n}$ to mean

6.377. $$F_{rm|n} = F_{rm,n} - \left\{ \begin{matrix} s \\ r\,n \end{matrix} \right\} F_{sm} - \left\{ \begin{matrix} s \\ m\,n \end{matrix} \right\} F_{rs},$$

which is the formula for the covariant derivative of a tensor in 4-space, the Christoffel symbols being calculated from the g's. So far the only transformations of coordinates in space-time have been transformations of the coordinates x^1, x^2, x^3 among themselves, with x^4 unchanged. For such transformations $F_{rm|n}$ has tensor character, as is easily verified (cf. the discussion of normal coordinate systems in 2.6).

Consider now the equations

6.378. $$F_{rm,n} + F_{mn,r} + F_{nr,m} = 0, \qquad g^{mn}F_{rm|n} = 0.$$

The first was established in 6.372. As for the second, the four equations contained in it are precisely the four equations in 6.373. Equations 6.378 express Maxwell's equations in the form suitable for the theory of relativity.

Should we wish to use *arbitrary* coordinates in space-time, i.e., any functions of x^1, x^2, x^3, x^4, we have in 6.378 a valid form for Maxwell's equations provided that the following rule is observed: under the space-time transformation

6.379. $$x'^r = f^r(x^1, x^2, x^3, x^4)\,,$$

g_{mn} and F_{mn} are to be transformed as covariant tensors. The fundamental nature of equations 6.378 becomes apparent only when coordinate transformations of the general type 6.379 are considered which involve all four coordinates on an equal footing. These are the transformations of the theory of relativity.

If rectangular Cartesian coordinates are used in space, 6.375 becomes

6.380. $$\Phi = (dx^1)^2 + (dx^2)^2 + (dx^3)^2 - (dx^4)^2.$$

The linear transformations of the coordinates in space-time which preserve the form of this line element are called the *Lorentz transformations*. They form the basis of the special theory of relativity.

Physically, a Lorentz transformation connects, in general, the spatial and temporal frames of reference of two observers moving with uniform velocity relative to each other. The electromagnetic field is characterized for each observer by F_{mn} which, under a Lorentz transformation, behaves like a four-dimensional tensor. Since F_{mn} embodies both the electric and magnetic fields, these will in general not transform as separate entities. Physically this means that if, for example, an observer finds an electric but no magnetic field present, then a second observer, moving relative to the first, may find both electric and magnetic fields present. It is obvious that Maxwell's equations preserve their explicit form under Lorentz transformations. They read as in equation 6.378, with the simplification that the stroke may be replaced by a comma, and all the g^{mn} vanish except $g^{11} = g^{22} = g^{33} = -g^{44} = 1$.

Exercise. Show that with homogeneous coordinates z_r (z_1, z_2, z_3 being rectangular Cartesians in space and $z_4 = ict = ix^4$) Maxwell's equations read

$$F_{rm,n} + F_{mn,r} + F_{nr,m} = 0, \quad F_{rm,m} = 0.$$

Write out the components of F_{mn} in terms of the real electric and magnetic vectors, noting which components are real and which are imaginary.

In the general theory of relativity space-time is a curved Riemannian 4-space with line element

6.381. $$\Phi = g_{mn}dx^m dx^n.$$

The coordinate transformations considered are of the general type 6.379. We can no longer distinguish between an electric and a magnetic field, even if we restrict ourselves to a single coordinate system. Only the fusion of the electric and magnetic fields in the tensor F_{mn} has physical significance. The tensor F_{mn} satisfies Maxwell's equations in the covariant form 6.378 in space free of charges.

SUMMARY VI

HYDRODYNAMICS

Eulerian description of fluid motion:

$$v_r = v_r(x, t).$$

Equation of continuity:

$$\frac{\partial \rho}{\partial t} + (\rho v^r)_{|r} = 0.$$

Equations of motion of perfect fluid:

$$\frac{\partial v_r}{\partial t} + v^s v_{r|s} = X_r - \rho^{-1} p_{,r}\,.$$

Vorticity:

$$\omega_{rs} = \tfrac{1}{2}(v_{s,r} - v_{r,s}),\quad \omega^r = \frac{1}{2\sqrt{a}}\,\epsilon^{rmn}\omega_{mn}.$$

Irrotational flow:

$$\omega_{rs} = 0,\quad v_r = -\phi_{,r}\,.$$

Bernoulli's integral for irrotational flow:

$$-\frac{\partial \phi}{\partial t} + \tfrac{1}{2}a^{mn}\phi_{,m}\phi_{,n} + P + U = F(t),$$

$$P = \int \frac{dp}{\rho},\quad X_r = -U_{,r}\,.$$

ELASTICITY

Displacement in elastic medium:

$$u_r(x)$$

Extension and strain tensor:

$$e = e_{rs}\lambda^r\lambda^s, \quad \lambda^r\lambda_r = 1,$$
$$e_{rs} = \tfrac{1}{2}\,(u_{r|s} + u_{s|r}).$$

Stress tensor:

$$E_{rs} = E_{sr}, \quad T_r = E_{rs}n^s, \quad n^s n_s = 1.$$

Equations of motion of continuous medium:

$$\rho f^r = \rho X^r + E^{rs}{}_{|s}.$$

Generalized Hooke's law:

$$E_{rs} = c_{rsmn}e^{mn},$$
$$c_{rsmn} = c_{rsnm} = c_{srmn} = c_{mnrs}.$$

Isotropic stress-strain relation:

$$c_{rsmn} = \lambda a_{rs}a_{mn} + \mu(a_{rm}a_{sn} + a_{rn}a_{sm}),$$
$$E_{rs} = \lambda a_{rs}\theta + 2\mu e_{rs}, \quad \theta = a^{mn}e_{mn}.$$

ELECTROMAGNETIC RADIATION

Maxwell's equations (Cartesian coordinates):

$$\frac{1}{c}\frac{\partial E_r}{\partial t} = \epsilon_{rmn}H_{n,m}, \quad \frac{1}{c}\frac{\partial H_r}{\partial t} = -\,\epsilon_{rmn}E_{n,m},$$
$$E_{n,n} = 0, \qquad H_{n,n} = 0.$$

Wave equation:

$$\frac{1}{c^2}\frac{\partial^2 E_r}{\partial t^2} - E_{r,mm} = 0, \quad \frac{1}{c^2}\frac{\partial^2 H_r}{\partial t^2} - H_{r,mm} = 0.$$

Electromagnetic waves (small wave-length λ):

$$E_r = E_r^{(0)}e^{iS}, \quad H_r = H_r^{(0)}e^{iS}, \quad S = \frac{2\pi}{\lambda}\,(V - ct);$$
$$V_{,m}V_{,m} = 1\,;$$
$$E_r^{(0)}V_{,r} = H_r^{(0)}V_{,r} = E_r^{(0)}H_r^{(0)} = 0, \quad E_r^{(0)}E_r^{(0)} = H_r^{(0)}H_r^{(0)};$$
$$\tau = \lambda/c, \quad \nu = c/\lambda.$$

Plane waves:

$$V = a_m z_m + b, \quad a_m a_m = 1.$$

Electromagnetic potentials:

$$E_r = -\frac{1}{c}\frac{\partial \phi_r}{\partial t} - \psi_{,r},\ H_r = \epsilon_{rpq}\phi_{q,p};$$

$$\frac{1}{c^2}\frac{\partial^2 \phi_r}{\partial t^2} - \phi_{r,mm} = 0\quad,\ \frac{1}{c^2}\frac{\partial^2 \psi}{\partial t^2} - \psi_{,mm} = 0,$$

$$\frac{1}{c}\frac{\partial \psi}{\partial t} + \phi_{m,m} = 0.$$

Hertz vector:

$$E_r = \Pi_{m,mr} - \frac{1}{c^2}\frac{\partial^2 \Pi_r}{\partial t^2},\ H_r = \frac{1}{c}\epsilon_{rpq}\frac{\partial}{\partial t}\Pi_{q,p};$$

$$\frac{1}{c^2}\frac{\partial^2 \Pi_r}{\partial t^2} - \Pi_{r,mm} = 0.$$

Fundamental solution:

$$\Pi_r = \Pi_r^{(0)} e^{-ikct},\quad \Pi_r^{(0)} = B_r\frac{e^{ikR}}{R}.$$

$$R^2 = (z_m - \zeta_m)(z_m - \zeta_m).$$

Maxwell's equations (curvilinear coordinates):

$$H_{rm} = \sqrt{a}\epsilon_{rmn}H^n,\ H^r = \frac{1}{2\sqrt{a}}\epsilon^{rmn}H_{mn};$$

$$\frac{1}{c}\frac{\partial E_r}{\partial t} = a^{mn}H_{rm|n},\ \frac{1}{c}\frac{\partial H_{rm}}{\partial t} = E_{r,m} - E_{m,r},$$

$$a^{nm}E_{n|m} = 0,\ H_{rm,n} + H_{mn,r} + H_{nr,m} = 0.$$

Maxwell's equations in 4-dimensional (space-time) notation:

$$\Phi = g_{mn}dx^m dx^n = a_{\alpha\beta}dx^\alpha dx^\beta - (dx^4)^2,\ x^4 = ct;$$

$$F_{\alpha\beta} = H_{\alpha\beta},\ F_{\alpha 4} = -F_{4\alpha} = E_\alpha,\ F_{44} = 0;$$

$$F_{rm,n} + F_{mn,r} + F_{nr,m} = 0,\ g^{mn}F_{rm|n} = 0.$$

EXERCISES VI

1. For a fluid in motion referred to curvilinear coordinates the kinetic energy of the fluid in any region R is

$$T = \tfrac{1}{2}\int_R \rho v_r v^r dV.$$

Use the equations of motion 6.147 to show that, if we follow the particles which compose R, we have

$$\frac{dT}{dt} = -\int_S pn_r v^r dS + \int_R \theta p dV + \int_R \rho v_r X^r dV,$$

where S is the bounding surface of R, and n_r the unit vector normal to S and drawn outward. Show further that if, instead of following the particles, we calculate the rate of change of T for a fixed portion of space, we get the above expression with the following additional term:

$$-\tfrac{1}{2}\int_S \rho n_r v^r v_s v^s dS.$$

2. Consider a fluid in which ρ is a function of p, moving under a conservative body force. Show that if the motion is steady, but not necessarily irrotational, then the following quantity is constant along each stream line*:

$$\tfrac{1}{2} v_r v^r + P + U.$$

Compare and contrast this result with 6.154.

3. For the general motion of the fluid described in Exercise 2, prove that

$$\frac{d}{dt}\int_C v_r dx^r = 0,$$

where the integral is taken round any closed curve, and d/dt is the co-moving time derivative.

4. Curves having at each point the direction of the vorticity vector ω^r are called "vortex lines." Prove that $\int_C v_r dx^r$ has the same value for all closed curves C which lie on the surface of a tube of vortex lines, and go once round the tube in the same sense. (Use Stokes' theorem; cf. 7.502.)

5. Prove that for the type of fluid described in Exercise 2, the vorticity tensor satisfies the differential equations

$$\frac{d}{dt}\omega_{rs} = \omega_{pr} v_{p,s} - \omega_{ps} v_{p,r},$$

*A stream line is a curve which, at each point, has the direction of the velocity vector v^r.

the coordinates being rectangular Cartesians. Write these equations for curvilinear coordinates.

Deduce from these equations that, if $\omega_{rs} = 0$ initially at some point P in the fluid, these quantities will remain zero permanently for the particle which was initially at P.

6. By eliminating the three components of displacement u_r from the six equations 6.210, obtain the following Cartesian equations of compatibility:

$$e_{rs,mn} + e_{mn,rs} - e_{rm,sn} - e_{sn,rm} = 0.$$

Show that there are only six independent equations here. Write the equations of compatibility in general tensor form.

7. For rectangular Cartesian coordinates z_r, a state of simple tension is represented by $E_{11} = C$ (a constant), all the other components of stress being zero. Find all six covariant components of stress for spherical polar coordinates r, θ, ϕ.

8. By substitution from 6.247 in the Cartesian equations of compatibility given in Exercise 6, deduce that in a homogeneous isotropic body in equilibrium under body forces X_r, the invariant $\Theta = E_{nn}$ satisfies the following partial differential equation:

$$(1 - \sigma)\Theta_{,rr} = (1 + \sigma)\rho X_{r,r}.$$

9. In a state of plane stress we have $E_{\alpha 3} = 0$, $E_{33} = 0$, the coordinates being Cartesian, and Greek suffixes taking the values 1, 2. Prove that the equations of equilibrium under no body forces are satisfied if we put

$$E_{\alpha\beta} = \epsilon_{\alpha\rho}\epsilon_{\beta\sigma}\psi_{,\rho\sigma},$$

where ψ is an arbitrary function. Show that this gives

$$E_{11} = \psi_{,22}, \; E_{12} = -\psi_{,12}, \; E_{22} = \psi_{,11}.$$

10. An isotropic elastic body is in equilibrium under no body forces. Show that, for rectangular Cartesian coordinates, the displacement satisfies the partial differential equations

$$(1 - 2\sigma)\Delta u_r + \theta_{,r} = 0.$$

Deduce that θ is a harmonic function.

Show that the above equations are satisfied if we put

$$u_r = \psi_r - \frac{1}{4(1-\sigma)}(z_s\psi_s + \phi)_{,r},$$

provided $\Delta\phi = 0$, $\Delta\psi_r = 0$. (Papcovich-Neuber)

11. If, for rectangular Cartesian coordinates z_r, χ_{rs} is any symmetric tensor, show that the tensor E_{mn}, defined by

$$E_{mn} = \epsilon_{mpr}\epsilon_{nqs}\chi_{rs,pq},$$

is symmetric, and satisfies the equations $E_{mn,n} = 0$. (Finzi)
Show that if we choose $\chi_{rs} = z_r z_s$, then $E_{mn} = -2\delta_{mn}$.

12. The determinantal equation $|\lambda\delta_{mn} - E_{mn}| = 0$ is important in elasticity because it gives the three principal stresses at a point. Show that if we introduce the three Cartesian invariants

$$A = E_{mm},\quad B = E_{mn}E_{mn},\quad C = E_{mn}E_{np}E_{pm},$$

this cubic equation may be written in the form

$$\lambda^3 - A\lambda^2 + \tfrac{1}{2}(A^2 - B)\lambda - (\tfrac{1}{6}A^3 - \tfrac{1}{2}AB + \tfrac{1}{3}C) = 0.$$

[Hint: Note the Cartesian invariance of this expression, and use coordinates which make $E_{rs} = 0$ for $r \neq s$.]

13. A plane electromagnetic wave in complex form is given, for rectangular Cartesians z_r, by the formulae

$$E_\alpha = A_\alpha e^{iS},\ E_3 = 0,\ H_\alpha = -\epsilon_{\alpha\beta}A_\beta e^{iS},\ H_3 = 0,$$

$$S = \frac{2\pi}{\lambda}(z_3 - ct),$$

where A_α is a constant complex vector, and Greek suffixes take the values 1, 2. Verify that Maxwell's equations are satisfied, and that the wave is propagated in the positive z_3-direction. The wave meets a perfectly conducting wall $z_3 = 0$, and is reflected. Given that the condition on such a wall is that the tangential component of the electric vector for the total field vanishes, show that the reflected wave is given by

$$E'_\alpha = -A_\alpha e^{iS'},\ E'_3 = 0,\ H'_\alpha = -\epsilon_{\alpha\beta}A_\beta e^{iS'},\ H'_3 = 0,$$

$$S' = -\frac{2\pi}{\lambda}(z_3 + ct).$$

14. Taking for the Hertz vector the fundamental solution of the wave equation 6.342

$$\Pi_r = B_r \frac{e^{-ik(ct-R)}}{R}, \quad R^2 = z_m z_m,$$

where B_r is a constant vector, show that, for R much less than $\lambda = 2\pi/k$, we have approximately

$$E_r = -\frac{\partial}{\partial z^r}\left[B_m z_m \frac{e^{-ikct}}{R^3} \right],$$

$$H_r = -ik\epsilon_{rmn} B_m z_n \frac{e^{-ikct}}{R^3}.$$

Show also that, for R much greater than $\lambda = 2\pi/k$ (wave-zone), we have approximately

$$E_r = k^2 (B_r R^2 - B_m z_m z_r) \frac{e^{-ik(ct-R)}}{R^3},$$

$$H_r = -k^2 \epsilon_{rmn} B_m z_n \frac{e^{-ik(ct-R)}}{R^2}.$$

(This is the electromagnetic field of the Hertzian dipole oscillator, which is the simplest model of a radio antenna.)

15. In terms of the magnetic tensor H_{mn}, defined in 6.367, show that, in curvilinear coordinates in space,

(a) The condition that λ^n be parallel to the magnetic field is

$$H_{mn}\lambda^n = 0.$$

(b) The condition that μ_n be perpendicular to the magnetic field is

$$H_{mn}\mu_r + H_{nr}\mu_m + H_{rm}\mu_n = 0.$$

(c) The square of the magnitude of the magnetic vector ($H^2 = H_r H^r$) is

$$H^2 = \tfrac{1}{2} H_{mn} H^{mn}.$$

16. Show that E_r and H_r are unchanged if, in equations 6.336, ϕ_r and ψ are replaced by

$$\phi'_r = \phi_r + v_{,r}, \quad \psi' = \psi - \frac{1}{c}\frac{\partial v}{\partial t},$$

where v is an arbitrary function of position and time. (This transformation of the electromagnetic potentials is called a *gauge transformation.*) If v is any solution of the inhomogeneous wave equation

$$\frac{1}{c^2}\frac{\partial^2 v}{\partial t^2} - v_{,mm} = \frac{1}{c}\frac{\partial\psi}{\partial t} + \phi_{m,m},$$

show that, if ϕ_r and ψ satisfy 6.337 and 6.338, then ϕ'_r and ψ' satisfy equations of the form 6.339 and 6.340.

17. Combining the vector potential ϕ_α ($\alpha = 1, 2, 3$) and the scalar potential ψ into a single 4-vector κ_r, given by

$$\kappa_\alpha = -\phi_\alpha, \quad \kappa_4 = \psi,$$

show that equations 6.336, 6.337, and 6.338, can be written in the relativistic form

$$F_{mn} = \kappa_{m,n} - \kappa_{n,m},$$
$$g^{mn}\kappa_{r\,|\,mn} - g^{mn}\kappa_{m\,|\,rn} = 0,$$

where F_{mn} is as in 6.371, and g^{mn} as in 6.376. Show further that equations 6.339 and 6.340 become

$$g^{mn}\kappa_{r\,|\,mn} = 0, \quad g^{mn}\kappa_{m\,|\,n} = 0.$$

Show also that the gauge transformation of the preceding exercise can be written

$$\kappa'_r = \kappa_r - v_{,r}.$$

18. Using homogeneous coordinates z_r (z_1, z_2, z_3 being rectangular Cartesians in space and $z_4 = ict = ix^4$), and defining

$$\hat{F}_{mn} = \tfrac{1}{2}\,\epsilon_{mnrs}F_{rs},$$

show that the complete set of Maxwell's equations 6.378 reads

$$F_{rm,m} = 0, \quad \hat{F}_{rm,m} = 0.$$

CHAPTER VII

RELATIVE TENSORS, IDEAS OF VOLUME, GREEN-STOKES THEOREMS

7.1. Relative tensors, generalized Kronecker delta, permutation symbol. In Chapter I we defined tensors by their transformation properties. The characteristics to which the tensor concept owes its importance may be summarized as follows:

A. The tensor transformation is linear and homogeneous. Hence if all the components of a tensor vanish in one coordinate system, they vanish in every coordinate system. It follows that a tensor equation, if true in one system of coordinates, holds in all systems of coordinates.

B. The tensor transformation is transitive.

We shall now study in detail a new set of geometrical objects which share with tensors both the above properties. These are the relative tensors which we have already met in 6.1.

As before, we denote the Jacobian of the transformation from coordinates x^r to x'^r by

7.101. $$J = \left| \frac{\partial x^k}{\partial x'^s} \right|,$$

where the vertical bars denote a determinant. We now define relative tensors as follows:

A set of quantities $T^r{}_s{}^{\cdot\,\cdot\,\cdot}_{\cdot\,\cdot\,\cdot}$ *are said to be the components of a relative tensor* of weight* W, *contravariant in the superscripts* $r, \ldots,$ *and covariant in the subscripts,* $s, \ldots,$ *if they transform according to the equation*

*Abuse of language, cf. p. 128.

7.102. $$T'^{r}{}_{s}{}^{\cdots}_{\cdots} = J^{W} T^{m}{}_{n}{}^{\cdots}_{\cdots} \frac{\partial x'^{r}}{\partial x^{m}} \cdots \frac{\partial x^{n}}{\partial x'^{s}} \cdots.$$

It is understood that we limit ourselves to transformations for which the Jacobian at the point under consideration is neither zero nor infinite. We shall also assume throughout that the weight W is an integer, since otherwise J^{W} is not single valued.

In accordance with our previous practice, we refer to relative tensors of orders 1 and 0 as relative vectors and relative invariants, respectively. To distinguish, where necessary, the tensors previously considered from relative tensors, we may refer to the former as *absolute* tensors, *absolute* vectors, or *absolute* invariants. These are, in fact, relative tensors of weight 0. Relative tensors of weight 1 are also known as *tensor densities*.

Exercise. If b_{rs} is an absolute covariant tensor, show that the determinant $|b_{rs}|$ is a relative invariant of weight 2. What are the tensor characters of $|c^{rs}|$ and $|f^{r}_{s}|$?

It follows immediately from 7.102 that property A, stated above, holds for relative tensors. In the case of an equation in relative tensors the two sides must be of the same weight.

Property B follows from the transitivity of the tensor transformation and from what we may term the "transitivity of the Jacobian." If x^{r}, x'^{r}, x''^{r} are three systems of coordinates, we have

7.103. $$\frac{\partial x^{r}}{\partial x''^{s}} = \frac{\partial x^{r}}{\partial x'^{m}} \frac{\partial x'^{m}}{\partial x''^{s}}.$$

Hence, by the rule for multiplying determinants

7.104. $$\left|\frac{\partial x^{r}}{\partial x''^{s}}\right| = \left|\frac{\partial x^{r}}{\partial x'^{m}}\right| \cdot \left|\frac{\partial x'^{n}}{\partial x''^{s}}\right|.$$

This establishes the "transitivity of the Jacobian."

Two relative tensors of the same order, type, and weight may be added, the sum being a relative tensor of the same weight. Any two relative tensors may be multiplied, the weight of the product being the sum of the weights of the factors. The process of contraction may be applied to a relative tensor and does not change the weight. All these results follow from 7.102 by using arguments quite analogous to those given in Chapter I for the case of absolute tensors.

We now introduce a set of numerical tensors, the *generalized Kronecker deltas*. In a space of N dimensions, we define (for any positive integer M):

7.105. $\delta^{k_1 k_2 \ldots k_M}_{s_1 s_2 \ldots s_M} = +1$ if $k_1, \ldots, k_M$ are distinct integers selected from the range $1, 2, \ldots, N$, and if $s_1, \ldots, s_M$ is an *even* permutation of $k_1, \ldots, k_M$.

$= -1$ if $k_1, \ldots, k_M$ are distinct integers selected from the range $1, 2, \ldots, N$, and if $s_1, \ldots, s_M$ is an *odd* permutation of $k_1, \ldots, k_M$.

$= 0$ if any two of $k_1, \ldots, k_M$ are equal, or if any two of $s_1, \ldots, s_M$ are equal, or if the set of numbers $k_1, \ldots, k_M$ differs, apart from order, from the set $s_1, \ldots, s_M$.

We immediately notice that for δ^k_s this definition agrees with that given in 1.207. If $M > N$, N being the dimension of our space, then

$$\delta^{k_1 \ldots k_M}_{s_1 \ldots s_M} = 0,$$

since $k_1, \ldots, k_M$ cannot all be different.

Exercise. Show that, in three dimensions, the only non-vanishing components of δ^{kl}_{rs} are

$$\delta^{23}_{23} = \delta^{32}_{32} = \delta^{31}_{31} = \delta^{13}_{13} = \delta^{12}_{12} = \delta^{21}_{21} = 1,$$
$$\delta^{23}_{32} = \delta^{32}_{23} = \delta^{31}_{13} = \delta^{13}_{31} = \delta^{12}_{21} = \delta^{21}_{12} = -1\,.$$

Exercise. Show that equations 5.231 and 6.128 can be written as follows:

$$M_{rs} = \delta^{kl}_{rs} z_k F_l,$$
$$\omega_{rs} = \tfrac{1}{2} \delta^{kl}_{rs} v_{l,k}.$$

To establish the tensor character of the generalized Kronecker delta, let $A_{(1)}{}^{k_1}, A_{(2)}{}^{k_2}, \ldots, A_{(M)}{}^{k_M}$ be M arbitrary contravariant vectors. Then

$$\delta^{k_1 \ldots k_M}_{s_1 \ldots s_M} \; A_{(1)}{}^{s_1} \ldots A_{(M)}{}^{s_M}$$

is a sum of $M!$ outer products of these vectors, viz.,

$$A_{(1)}{}^{k_1} A_{(2)}{}^{k_2} A_{(3)}{}^{k_3} \ldots A_{(M)}{}^{k_M} - A_{(1)}{}^{k_2} A_{(2)}{}^{k_1} A_{(3)}{}^{k_3} \ldots A_{(M)}{}^{k_M} + \ldots.$$

This sum has the tensor character of $T^{k_1 \ldots k_M}$. Hence, by our test for tensor character, 1.6, $\delta^{k_1 \ldots k_M}_{s_1 \ldots s_M}$ *is a mixed absolute tensor of order* $2M$.

For purposes of manipulation we shall find the following formula useful. If $T^{k_1 \ldots k_M}$ is skew-symmetric in all pairs of superscripts, then

7.106. $$\delta^{k_1 \ldots k_M}_{s_1 \ldots s_M} T^{s_1 \ldots s_M} = M! \, T^{k_1 \ldots k_M}.$$

The expression on the left side is, by 7.105, a sum of $M!$ terms. The first of these is $T^{k_1 \ldots k_M}$; the other terms are obtained from it by permuting the superscripts and a minus sign is attached if the permutation is odd. Since $T^{k_1 \ldots k_M}$ is completely skew-symmetric, each of the $M!$ terms in this sum equals $+ T^{k_1 \ldots k_M}$. This proves 7.106.

Exercise. If $T_{k_1 \ldots k_M}$ is completely skew-symmetric, determine

$$\delta^{k_1 \ldots k_M}_{s_1 \ldots s_M} T_{k_1 \ldots k_M}.$$

We now consider the *permutation symbol* $\epsilon_{r_1 r_2 \ldots r_N}$, which was introduced in 4.3 and was there seen to be an oriented Cartesian tensor. Let us recall its definition:

7.107. $\epsilon_{r_1 r_2 \ldots r_N}$ = 0 if any two of the suffixes are equal.
= + 1 if $r_1, r_2, \ldots, r_N$, is an *even* permutation of 1, 2, ..., N.
= − 1 if $r_1, r_2, \ldots, r_N$, is an *odd* permutation of 1, 2, ..., N.

This definition is equivalent to stating

7.108.
$$\epsilon_{r_1 r_2 \ldots r_N} = \delta^{1\ 2\ \ldots\ N}_{r_1 r_2 \ldots r_N}.$$

In order to investigate the tensor character of the permutation symbol under general coordinate transformations, we note that, by the definition of a determinant, we may write the Jacobian

7.109.
$$J = \epsilon_{r_1 r_2 \ldots r_N} \frac{\partial x^{r_1}}{\partial x'^1} \frac{\partial x^{r_2}}{\partial x'^2} \cdots \frac{\partial x^{r_N}}{\partial x'^N}.$$

If, in the right side of this equation, we interchange any two of the suffixes 1, 2, . . . , N, the expression changes sign; if any two of these suffixes are made equal, the expression vanishes; for example

$$\epsilon_{r_1 r_2 \ldots r_N} \frac{\partial x^{r_1}}{\partial x'^2} \frac{\partial x^{r_2}}{\partial x'^1} \cdots \frac{\partial x^{r_N}}{\partial x'^N} = \epsilon_{r_2 r_1 \ldots r_N} \frac{\partial x^{r_1}}{\partial x'^1} \frac{\partial x^{r_2}}{\partial x'^2} \cdots \frac{\partial x^{r_N}}{\partial x'^N} = -J,$$

$$\epsilon_{r_1 r_2 \ldots r_N} \frac{\partial x^{r_1}}{\partial x'^1} \frac{\partial x^{r_2}}{\partial x'^1} \cdots \frac{\partial x^{r_N}}{\partial x'^N} = \epsilon_{r_2 r_1 \ldots r_N} \frac{\partial x^{r_1}}{\partial x'^1} \frac{\partial x^{r_2}}{\partial x'^1} \cdots \frac{\partial x^{r_N}}{\partial x'^N} = -\epsilon_{r_1 r_2 \ldots r_N} \frac{\partial x^{r_1}}{\partial x'^1} \frac{\partial x^{r_2}}{\partial x'^1} \cdots \frac{\partial x^{r_N}}{\partial x'^N}.$$

The last expression is therefore its own negative, and consequently vanishes. Thus, if we change 1, 2, . . . , N in 7.109 into $s_1, s_2, \ldots, s_N$ and divide by J, we get

7.110.
$$\epsilon'_{s_1 s_2 \ldots s_N} = J^{-1} \epsilon_{r_1 r_2 \ldots r_N} \frac{\partial x^{r_1}}{\partial x'^{s_1}} \frac{\partial x^{r_2}}{\partial x'^{s_2}} \cdots \frac{\partial x^{r_N}}{\partial x'^{s_N}}.$$

This proves that the *permutation symbol* $\epsilon_{r_1 \ldots r_N}$ *is a covariant relative tensor of weight* -1.

Similarly to 7.110, we have

7.111.
$$\epsilon'^{s_1 s_2 \ldots s_N} = J \epsilon^{r_1 r_2 \ldots r_N} \frac{\partial x'^{s_1}}{\partial x^{r_1}} \frac{\partial x'^{s_2}}{\partial x^{r_2}} \cdots \frac{\partial x'^{s_N}}{\partial x^{r_N}}.$$

This shows that the *permutation symbol* is also *a contravariant relative tensor of weight* +1. We therefore use the alternative notation $\epsilon^{r_1 \ldots r_N}$.

If the transformation is a positive orthogonal transformation, 4.3, we have $J=1$, and 7.110, 7.111 become the transformation formulae for an absolute tensor, covariant and contravariant respectively. This verifies the fact, indicated in 4.3, that the permutation symbols are components of an oriented Cartesian tensor.

Exercise. Show that

7.112. $$\epsilon^{r_1 r_2 \ldots r_N} \epsilon_{r_1 r_2 \ldots r_N} = N!$$

Since the permutation symbol is completely skew-symmetric, we have, by 7.106, the following formulae

7.113. $$\delta^{k_1 \ldots k_M}_{s_1 \ldots s_M} \epsilon^{s_1 \ldots s_M r_1 \ldots r_{N-M}} = M! \, \epsilon^{k_1 \ldots k_M r_1 \ldots r_{N-M}},$$
$$\delta^{k_1 \ldots k_N}_{s_1 \ldots s_N} \epsilon_{k_1 \ldots k_N} = N! \, \epsilon_{s_1 \ldots s_N}, \text{ etc.}$$

For later work we also require the formulae

7.114. $$\epsilon^{k_1 \ldots k_M r_1 \ldots r_{N-M}} \epsilon_{s_1 \ldots s_M r_1 \ldots r_{N-M}} = (N-M)! \, \delta^{k_1 \ldots k_M}_{s_1 \ldots s_M},$$
$$\epsilon^{k_1 \ldots k_N} \epsilon_{s_1 \ldots s_N} = \delta^{k_1 \ldots k_N}_{s_1 \ldots s_N}.$$

The proof is straightforward, and is left as an exercise.

Having established the tensor character of the permutation symbol for general transformations, not necessarily orthogonal, we can generalize concepts such as that of the vector product and the curl to general 3-spaces and to arbitrary curvilinear coordinates. It is interesting to note that the idea of metric is not involved here, nor indeed throughout this section.

Let X_m, Y_m be absolute covariant vectors in a general 3-space. We put

7.115. $$P^m = \epsilon^{mnr} X_n Y_r = \tfrac{1}{2} \epsilon^{mnr} (X_n Y_r - X_r Y_n).$$

Comparing this with 4.319, we note the similarity between P^m and the vector product of the Cartesian vectors X_m and Y_m.

However, P^m is now a contravariant relative vector of weight 1. Similarly, if X^m and Y^m are absolute vectors,

7.116. $$P_m = \epsilon_{mnr} X^n Y^r = \tfrac{1}{2} \epsilon_{mnr} (X^n Y^r - X^r Y^n)$$

is a covariant relative vector of weight -1.

A generalization of the curl of a covariant vector field is given by the equation

7.117. $$P^m = \epsilon^{mnr} \frac{\partial X_r}{\partial x^n} = \tfrac{1}{2} \epsilon^{mnr} \left(\frac{\partial X_r}{\partial x^n} - \frac{\partial X_n}{\partial x^r} \right).$$

The expression in the parenthesis is an absolute tensor (cf. Exercises I, No. 8).

The permutation symbols may be used to generate relative tensors from other relative tensors, the process in general changing the order. We illustrate with a particular example in four dimensions ($N = 4$). Let T_{mn} be a covariant relative tensor, and let $\hat{T}^{mn}$ be defined by

7.118. $$\hat{T}^{mn} = \tfrac{1}{2} \epsilon^{rtmn} T_{rt}.$$

Then $\hat{T}^{mn}$ is a contravariant relative tensor whose weight exceeds by unity that of T_{mn}, since the weight of ϵ^{rtmn} is $+1$. It is obvious that $\hat{T}^{mn}$ is skew-symmetric.

If T_{mn} is skew-symmetric, we can solve the equations 7.118 for T_{mn} in terms of the components of $\hat{T}^{mn}$. Multiplying 7.118 by $\tfrac{1}{2} \epsilon_{ksmn}$, we obtain, by use of 7.114,

$$\begin{aligned} \tfrac{1}{2} \epsilon_{ksmn} \hat{T}^{mn} &= \tfrac{1}{4} \epsilon_{ksmn} \epsilon^{rtmn} T_{rt} \\ &= \tfrac{1}{2} \delta^{rt}_{ks} T_{rt} \\ &= \tfrac{1}{2} (T_{ks} - T_{sk}) \\ &= T_{ks}. \end{aligned}$$

In the last step the skew-symmetry of T_{mn} has been used. Renaming the free suffixes we can write the result as follows:

7.119. $$T_{rt} = \tfrac{1}{2} \epsilon_{rtmn} \hat{T}^{mn}.$$

The symmetry between the equations 7.118 and 7.119 is quite striking. Two skew-symmetric relative tensors T_{mn} and $\hat{T}^{mn}$ related by these equations are said to be *dual*. The symmetry of the relationship of duality is also exhibited by writing out explicitly all components of 7.118 :

7.120.
$$\hat{T}^{23} = T_{14}, \quad \hat{T}^{31} = T_{24}, \quad \hat{T}^{12} = T_{34},$$
$$\hat{T}^{14} = T_{23}, \quad \hat{T}^{24} = T_{31}, \quad \hat{T}^{34} = T_{12}.$$

Exercise. If T_{rs} is an absolute skew-symmetric tensor in a 4-space, show that $T_{14}T_{23} + T_{24}T_{31} + T_{34}T_{12}$ is a tensor density.

The concept of duality can be generalized. If $T_{r_1 \ldots r_M}$ is a covariant relative tensor which is completely skew-symmetric and if $M \leqslant N$, then we define

7.121.
$$\hat{T}^{r_1 \ldots r_{N-M}} = \frac{1}{M!} \epsilon^{s_1 \ldots s_M r_1 \ldots r_{N-M}} T_{s_1 \ldots s_M}.$$

This equation can be solved for $T_{s_1 \ldots s_M}$ by multiplying by $\epsilon_{k_1 \ldots k_M r_1 \ldots r_{N-M}}$. We then have, using 7.114 and 7.106,

$$\epsilon_{k_1 \ldots k_M r_1 \ldots r_{N-M}} \hat{T}^{r_1 \ldots r_{N-M}} = \frac{(N-M)!}{M!} \delta^{s_1 \ldots s_M}_{k_1 \ldots k_M} T_{s_1 \ldots s_M}$$
$$= (N-M)! \; T_{k_1 \ldots k_M}.$$

We can write this

7.122.
$$T_{s_1 \ldots s_M} = \frac{1}{(N-M)!} \epsilon_{s_1 \ldots s_M r_1 \ldots r_{N-M}} \hat{T}^{r_1 \ldots r_{N-M}}.$$

Two skew-symmetric relative tensors $T_{k_1 \ldots k_M}$ and $\hat{T}^{k_1 \ldots k_{N-M}}$ related by 7.121, or, equivalently by 7.122, are called duals. Equation 7.119 is a particular case of 7.122 for $M = 2$ and $N = 4$. The weight of the contravariant relative tensor in a pair of duals always exceeds by unity the weight of the covariant relative tensor. It is easily seen from 7.122 that if $s_1, \ldots, s_M, k_1, \ldots, k_{N-M}$ is an even permutation of $1, 2, \ldots, N$, then

7.123.
$$T_{s_1 \ldots s_M} = \hat{T}^{k_1 \ldots k_{N-M}}.$$

This result enables us to write out immediately the explicit equations relating the components of a pair of duals.

Exercise. Show that, for rectangular Cartesian coordinates, the vorticity tensor and the vorticity vector of a fluid are duals (cf. 6.130).

7.2. Change of weight. Differentiation. Thus far our discussion of relative tensors has run parallel to the development of absolute tensors as given in Chapter I. All conclusions were based on the transformation properties 7.102. We now consider Riemannian spaces and introduce a metric form

7.201. $$\Phi = a_{mn}dx^m dx^n.$$

In addition to the operations of lowering and raising suffixes by use of the metric tensor, we also have a process of changing the weight of a relative tensor, as will now be explained.

As we wish to cover indefinite as well as definite forms, and transformations with negative Jacobian as well as those with positive Jacobian, certain symbols ($\epsilon(a)$ and $\epsilon(J)$) described below are introduced. For the simpler case of positive-definite metric forms and transformations with positive Jacobian, we may substitute $\epsilon(a) = \epsilon(J) = 1$ in the formulae given below.

Consider the determinant $|a_{mn}|$ to be denoted by a. Under a coordinate transformation, we have

7.202. $$a' = |a'_{mn}| = \left| a_{rs} \frac{\partial x^r}{\partial x'^m} \frac{\partial x^s}{\partial x'^n} \right| = J^2 a.$$

This shows that a is a relative invariant of weight 2. If Φ is positive-definite, then a is positive (cf. Exercises II, No. 13); but if Φ is indefinite, a may be negative, and to take care of that case we introduce a symbol $\epsilon(a) = \pm 1$ such as to make $\epsilon(a)$ positive. We note that, since J^2 is positive, $\epsilon(a)$ does not change sign under the transformation. Thus $\epsilon(a)$ is an absolute invariant and $\epsilon(a)a$ a relative invariant of weight 2:

7.203. $$\epsilon'(a)a' = J^2\epsilon(a)\, a.$$

If $T^{r}_{s}\because$ is a relative tensor of weight W, and w is any positive or negative integer, then $\overline{T}^{r}_{s}\because$, defined by

7.204. $$\overline{T}^{r}_{s}\because = (\epsilon(a)\, a)^w \quad T^{r}_{s}\because,$$

is a relative tensor of weight $W + 2w$. In a Riemannian space we may regard $T^{r}_{s}\because$ and $\overline{T}^{r}_{s}\because$ as two representations of the same geometrical object. We shall now consider how to generalize this result when w is half of an odd integer, so that the weight of a relative tensor may be changed by an odd number.

We have throughout considered only coordinate transformations whose Jacobian does not vanish or become infinite in any region under consideration. It follows that the Jacobian J is everywhere positive or everywhere negative. In the former case we call the transformation positive, in the latter case negative.* Let $\epsilon(J) = +1$ or -1 according as the transformation is positive or negative. We now define oriented relative tensors as follows: A set of quantities $T^r_s{:}{:}$ is said to form an *oriented relative tensor* of weight W, if it transforms according to the equation

$$\textbf{7.205.} \qquad T'^r_s{:}{:} = \epsilon(J) J^W T^m_n{:}{:} \frac{\partial x'^r}{\partial x^m} \cdots \frac{\partial x^n}{\partial x'^s} \cdots .$$

If J is positive, then $\epsilon(J) = 1$, and this transformation law reduces to 7.102.

The product of two oriented relative tensors is a relative tensor which is not oriented. Note that the word "oriented" is used here in a more restricted sense than in 4.3, since 7.205 determines the transformation properties of $T^r_s{:}{:}$ under negative transformations as well.

Taking the positive square root on both sides of equation 7.203, we have

$$\textbf{7.206.} \qquad (\epsilon'(a)a')^{\frac{1}{2}} = \epsilon(J) J \, (\epsilon(a)\, a)^{\frac{1}{2}}.$$

This shows that $(\epsilon(a)a)^{\frac{1}{2}}$ is an oriented relative invariant of weight $+1$. We can now interpret 7.204 without difficulty if w is half an odd integer; $\overline{T}^r_s{:}{:}$ is then an oriented relative tensor. In particular, we can associate with every relative tensor $T^r_s{:}{:}$ of weight W an absolute tensor $\overline{T}^r_s{:}{:}$, which is oriented if W is odd and not oriented if W is even; it is given by

$$\textbf{7.207.} \qquad \overline{T}^r_s{:}{:} = (\epsilon(a)\, a)^{-\frac{1}{2}W} \; T^r_s{:}{:}.$$

The permutation symbol $\epsilon_{r_1 \ldots r_N}$, or $\epsilon^{r_1 \cdots r_N}$, gives rise to two oriented absolute tensors:

$$\textbf{7.208.} \qquad \eta_{r_1 \ldots r_N} = (\epsilon(a)a)^{\frac{1}{2}} \, \epsilon_{r_1 \ldots r_N},$$

*Cf. 4.3.

7.209. $$\eta^{r_1 \dots r_N} = (\epsilon(a)a)^{-\frac{1}{2}} \epsilon^{r_1 \dots r_N}.$$

Using these, rather than the permutation symbols ϵ, in equations 7.115 and 7.117, we can, in three dimensions, define the vector product and the curl as oriented absolute vectors. In a similar manner we may define the dual of a skew-symmetric absolute tensor as an oriented absolute tensor.

In one respect our notational conventions are violated by the permutation symbols, when there is a metric. This is not serious as long as the following caution is kept in mind: $\epsilon^{r_1 \dots r_N}$ and $\epsilon_{r_1 \dots r_N}$ are *not* obtained from one another by raising or lowering suffixes with the metric tensor. Instead, we have the relations:

7.210. $$\epsilon_{r_1 \dots r_N} = \frac{1}{a} a_{r_1 s_1} \dots a_{r_N s_N} \epsilon^{s_1 \dots s_N},$$

7.211. $$\epsilon^{r_1 \dots r_N} = a\, a^{r_1 s_1} \dots a^{r_N s_N} \epsilon_{s_1 \dots s_N}.$$

These equations are established by an argument completely analogous to that used in deriving 7.110.

Exercise. Show that

$$\eta_{r_1 \dots r_N} = \epsilon(a)\ a_{r_1 s_1} \dots a_{r_N s_N}\ \eta^{s_1 \dots s_N},$$
$$\eta^{r_1 \dots r_N} = \epsilon(a)\ a^{r_1 s_1} \dots a^{r_N s_N}\ \eta_{s_1 \dots s_N}.$$

We now turn to the problem of defining the absolute and covariant derivatives of a relative tensor. We can proceed as we like, provided that the definition agrees with the old definition of 2.5 when the weight of the relative tensor is zero. The absolute derivative of the metric tensor a_{mn} is zero (cf. 2.526); we shall now make sure that the absolute derivative of the determinant $a = |a_{mn}|$ is also zero. This suggests the following definition of *the absolute derivative of a relative tensor* $T^{r \dots}_{s \dots}$ *of weight* W:

7.212. $$\frac{\delta}{\delta u} T^{r \dots}_{s \dots} = (\epsilon(a)a)^{\frac{1}{2}W} \frac{\delta}{\delta u} [(\epsilon(a)a)^{-\frac{1}{2}W} T^{r \dots}_{s \dots}].$$

The operation of absolute differentiation on the right side of this equation can be carried out by the process of 2.5 since the bracket expression $(\epsilon(a)a)^{-\frac{1}{2}W}\, T^{r\cdots}_{s\cdots}$ is an absolute tensor (which is, however, oriented if W is an odd integer). Note that $\dfrac{\delta}{\delta u} T^{r\cdots}_{s\cdots}$ is a relative tensor which is not oriented, even if W is odd.

The absolute derivative of a, for which $W = 2$, is, by 7.212,

$$\textbf{7.213.} \qquad \begin{aligned} \frac{\delta a}{\delta u} &= \epsilon(a)a\frac{\delta}{\delta u}[(\epsilon(a)a)^{-1}a] \\ &= \epsilon(a)a\frac{\delta}{\delta u}\epsilon(a) = 0, \end{aligned}$$

since $\epsilon(a)$ is an absolute invariant, and a constant. Also in the case of the invariant density $(\epsilon(a)a)^{\frac{1}{2}}$, for which $W = 1$, it is easily seen that the absolute derivative is zero.

The definition of the covariant derivative of a relative tensor is now obvious:

$$\textbf{7.214.} \qquad T^{r\cdots}_{s\cdots|k} = (\epsilon(a)a)^{\frac{1}{2}W}[(\epsilon(a)a)^{-\frac{1}{2}W}\, T^{r\cdots}_{s\cdots}]_{|k}\,.$$

Neither absolute nor covariant differentiation alters the weight of a relative tensor.*

We have already seen that the operation of absolute (or covariant) differentiation, when applied to a product of absolute tensors, proceeds according to the laws of ordinary (or partial) differentiation. Thus, for example,

$$\textbf{7.215.} \qquad \frac{\delta}{\delta u}(S^{k}_{.r}\, T^{mn}) = \frac{\delta S^{k}_{.r}}{\delta u} T^{mn} + S^{k}_{.r}\frac{\delta T^{mn}}{\delta u}\,.$$

The same holds good for relative tensors. This is easily proved from the definition 7.212. Alternatively, we may take Riemannian coordinates with an arbitrary point O as origin. At O, all Christoffel symbols vanish and further $\partial(\epsilon(a)a)/\partial x^r = 0$, since $\partial a_{mn}/\partial x^r = 0$. Thus, at O, the absolute derivative of a relative tensor, as defined by 7.212, reduces to an ordinary derivative. We then know from elementary calculus that 7.215 holds for relative tensors at the origin of a Riemannian coordinate system. But this being an equation in relative

*For an alternative (but equivalent) definition of absolute and covariant differentiation of relative tensors, see Exercises VIII, Nos. 18, 20, 21.

tensors, it must be true in all systems of coordinates; and O being an arbitrary point, it must hold generally.

Exercise. Using Riemannian coordinates, prove that

7.216.
$$\epsilon_{r_1 \ldots r_N|k} = \epsilon^{r_1 \ldots r_N}{}_{|k} = 0,$$
$$\eta_{r_1 \ldots r_N|k} = \eta^{r_1 \ldots r_N}{}_{|k} = 0.$$

We shall now establish a simple but useful formula for the divergence of a contravariant vector density. If T^n is a vector density, i.e., a relative vector of weight 1, then

7.217.
$$t^n = (\epsilon(a)a)^{-\frac{1}{2}} T^n$$

is an absolute vector (oriented). Then, by equation 2.545, generalized slightly to include the case of a negative determinant a, we have

7.218.
$$t^n{}_{|n} = (\epsilon(a)a)^{-\frac{1}{2}} \frac{\partial}{\partial x^n} [(\epsilon(a)a)^{\frac{1}{2}} t^n] .$$

Multiplying across by $(\epsilon(a)a)^{\frac{1}{2}}$ and substituting from 7.217, we get

$$(\epsilon(a)a)^{\frac{1}{2}} [(\epsilon(a)a)^{-\frac{1}{2}} T^n]_{|n} = \frac{\partial T^n}{\partial x^n} .$$

Since the covariant derivative of $(\epsilon(a)a)^{\frac{1}{2}}$ is zero, this reduces to the simple result

7.219.
$$T^n{}_{|n} = \frac{\partial T^n}{\partial x^n} .$$

Note that this striking result holds only in the case of a contravariant vector density.

Exercise. Prove that if T^n is a relative vector of weight W then

7.220.
$$T^n{}_{|n} = (\epsilon(a)a)^{\frac{1}{2}(W-1)} \frac{\partial}{\partial x^n} [(\epsilon(a)a)^{\frac{1}{2}(1-W)} T^n] .$$

7.3. Extension. In 1.3 the infinitesimal displacement was studied as the prototype of a contravariant vector. The infinitesimal displacement was characterized by the vector dx^r. Only later, in 2.1, did we introduce a metric and the concept of a length ds associated with the displacement dx^r. This indicates that the vector displacement is more fundamental than its length. Both magnitude (in a sense) and direction of the displacement are determined by the vector dx^r; for example, it is possible to say that one displacement ($d'x^r$) is twice another (dx^r) if $d'x^r = 2dx^r$. It is only when we wish to compare the magnitudes of infinitesimal displacements at different points or in different directions that we require the concepts of metric and length.

We are all familiar with the ideas of area (2-volume) and ordinary volume (3-volume). In this section we shall be concerned with the more fundamental *non-metrical* concept of *extension** which bears the same relationship to volume as the vector displacement bears to length. We shall work from the beginning with a general number of dimensions, since the tensor notation is such that little is gained by restriction to a special number of dimensions.

In a space of N dimensions, in which no metric is assigned, consider a subspace V_M of M dimensions ($M \leqslant N$) defined by the parametric equations

7.301. $$x^k = x^k(y^1, y^2, \ldots, y^M).$$

In this V_M consider a certain region R_M. Let us divide R_M into a system of cells by M families of surfaces

7.302. $$f^{(\alpha)}(y) = c^{(\alpha)}, \quad \alpha = 1, 2, \ldots, M,$$

where $c^{(\alpha)}$ are constants, each taking on a number of discrete values and so forming a family of surfaces (Fig. 21). Here and throughout the remainder of this chapter the Greek letters α, β, γ assume values from 1 to M and the range and summation convention will apply to them unless an explicit statement to the contrary is made. Note that this applies only to the first few letters of the Greek alphabet. Others, such as θ, ρ, σ, τ will be reserved for special ranges of values as required.

It is easy to attach a precise meaning to the statement that a point P of V_M lies in a certain cell; for P may be said to lie between two surfaces of the same family, say family α, if the expression $f^{(\alpha)}(y) - c^{(\alpha)}$ changes sign when we change $c^{(\alpha)}$ from

*Historically, this concept is quite old. It is due to Grassmann (1842).

the value belonging to one of these surfaces to the value belonging to the other: here y refers to the parameter values at the point P.

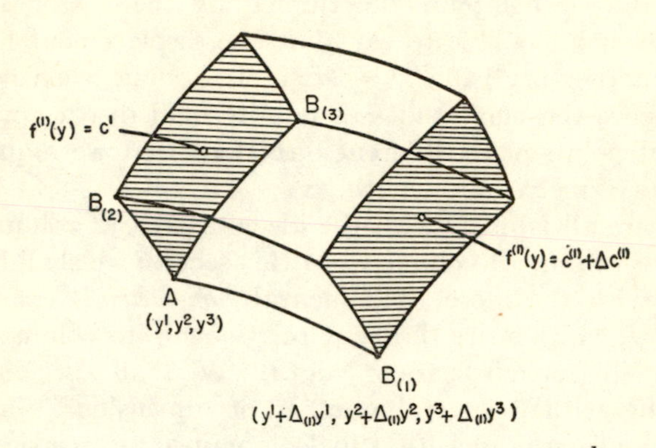

FIG. 21.—A cell in subspace V_3; the coordinates of A are x^k, and those of $B_{(1)}$ are $x^k + \Delta_{(1)}x^k$.

Let A be a corner of a cell (Fig. 21). Through A there passes one surface of each family. Let the corresponding values of the constants be $c^{(1)}, c^{(2)}, \ldots, c^{(M)}$. Passing along the edge for which $c^{(1)}$ alone changes, we arrive at another corner $B_{(1)}$, for which the constants have values $c^{(1)} + \Delta c^{(1)}, c^{(2)}, \ldots, c^{(M)}$. In passing from A to $B_{(1)}$ the parameters change from, say, y^a at A to $y^a + \Delta_{(1)}y^a$ at $B_{(1)}$, and the coordinates from x^k to $x^k + \Delta_{(1)}x^k$. Similarly the parameter values and the coordinates of all the corners $B_{(1)}, B_{(2)}, \ldots, B_{(M)}$, of the cell which are reached by passing from A along an edge of the cell may respectively be written

$$y^a + \Delta_{(\beta)}y^a;$$

and

$$x^k + \Delta_{(\beta)}x^k.$$

We now form the determinants

$$\textbf{7.303.} \qquad \Delta^{k_1 \cdots k_M} = \begin{vmatrix} \Delta_{(1)}x^{k_1}, & \Delta_{(1)}x^{k_2}, & \ldots & , \Delta_{(1)}x^{k_M} \\ \Delta_{(2)}x^{k_1}, & \Delta_{(2)}x^{k_2}, & \ldots & , \Delta_{(2)}x^{k_M} \\ \cdot & \cdot & \cdot & \cdot \\ \cdot & \cdot & \cdot & \cdot \\ \Delta_{(M)}x^{k_1}, & \Delta_{(M)}x^{k_2}, & \ldots & , \Delta_{(M)}x^{k_M} \end{vmatrix}.$$

Introducing the generalized Kronecker delta $\delta^{k_1 \dots k_M}_{s_1 \dots s_M}$, these determinants can be written compactly as follows:

7.304. $$\Delta^{k_1 \dots k_M} = \delta^{k_1 \dots k_M}_{s_1 \dots s_M} \Delta_{(1)} x^{s_1} \dots \Delta_{(M)} x^{s_M}.$$

Had the edges of the cell been taken in a different order, the determinants might all have had the opposite signs; apart from this common change in sign, the determinants are independent of the order of selection.

If we employ the same cell, but a different corner to start from, the determinants formed in the same manner would in general have different values. But if the cell is infinitesimal, that is, if the $\Delta c^{(\beta)}$ are infinitesimals, the differences will be infinitesimals of a higher order than the determinants 7.304, which we shall now write

7.305. $$d\tau_{(M)}{}^{k_1 \dots k_M} = \delta^{k_1 \dots k_M}_{s_1 \dots s_M} d_{(1)} x^{s_1} \dots d_{(M)} x^{s_M}.$$

We are therefore to regard this set of determinants as something which, except for the possibility of a common change in sign, is uniquely determined by an infinitesimal M-cell whose βth edge is $d_{(\beta)} x^k$. *The set of quantities* $d\tau_{(M)}{}^{k_1 \dots k_M}$ *is called the extension of the infinitesimal M-cell.* This tensor is the nearest approach to the concept of elementary volume possible in a non-metrical space.

Exercise. Show that 7.305 may be written in the equivalent form

7.306. $$d\tau_{(M)}{}^{k_1 \dots k_M} = \epsilon^{\beta_1 \dots \beta_M} d_{(\beta_1)} x^{k_1} \dots d_{(\beta_M)} x^{k_M}.$$

In this equation, what is the tensor character of the permutation symbol $\epsilon^{\beta_1 \dots \beta_M}$?

The edges of an infinitesimal M-cell of V_M can obviously be written in the form

7.307. $$d_{(\beta)} x^k = \frac{\partial x^k}{\partial y^\alpha} \quad d_{(\beta)} y^\alpha.$$

We therefore obtain the following alternative expression for the extension of the M-cell

7.308. $$d\tau_{(M)}{}^{k_1 \ldots k_M} = \delta^{k_1 \ldots k_M}_{s_1 \ldots s_M} \frac{\partial x^{s_1}}{\partial y^{\alpha_1}} \ldots \frac{\partial x^{s_M}}{\partial y^{\alpha_M}} d_{(1)}y^{\alpha_1} \ldots d_{(M)}y^{\alpha_M}.$$

The expression on the right side of this equation is essentially a sum, the first term of which is

7.309. $$\frac{\partial x^{k_1}}{\partial y^{\alpha_1}} \ldots \frac{\partial x^{k_M}}{\partial y^{k_M}} d_{(1)}y^{\alpha_1} \ldots d_{(M)}y^{\alpha_M}.$$

The other terms in the sum are obtained from this by permuting the subscripts $k_1, k_2, \ldots, k_M$, a minus sign being attached to a term if the permutation is odd. However a permutation of $k_1, \ldots, k_M$ in 7.309 is equivalent to the corresponding permutation of $\alpha_1, \ldots, \alpha_M$ in $d_{(1)}y^{\alpha_1} \ldots d_{(M)}y^{\alpha_M}$. For example,

$$\frac{\partial x^{k_2}}{\partial y^{\alpha_1}} \frac{\partial x^{k_1}}{\partial y^{\alpha_2}} \cdots \frac{\partial x^{k_M}}{\partial y^{\alpha_M}} d_{(1)}y^{\alpha_1} d_{(2)}y^{\alpha_2} \ldots d_{(M)}y^{\alpha_M}$$
$$= \frac{\partial x^{k_1}}{\partial y^{\alpha_2}} \frac{\partial x^{k_2}}{\partial y^{\alpha_1}} \cdots \frac{\partial x^{k_M}}{\partial y^{\alpha_M}} d_{(1)}y^{\alpha_1} d_{(2)}y^{\alpha_2} \ldots d_{(M)}y^{\alpha_M},$$

and this becomes, on interchanging the dummy suffixes α_1, α_2,

$$\frac{\partial x^{k_1}}{\partial y^{\alpha_1}} \frac{\partial x^{k_2}}{\partial y^{\alpha_2}} \cdots \frac{\partial x^{k_M}}{\partial y^{\alpha_M}} d_{(1)}y^{\alpha_2} d_{(2)}y^{\alpha_1} \ldots d_{(M)}y^{\alpha_M}.$$

It follows that 7.308 may be rewritten in the form

7.310. $$d\tau_{(M)}{}^{k_1 \ldots k_M} = \frac{\partial x^{k_1}}{\partial y^{\alpha_1}} \cdots \frac{\partial x^{k_M}}{\partial y^{\alpha_M}} \delta^{\alpha_1 \ldots \alpha_M}_{\gamma_1 \ldots \gamma_M} d_{(1)}y^{\gamma_1} \ldots d_{(M)}y^{\gamma_M}.$$

Introducing the permutation symbols $\epsilon^{\alpha_1 \ldots \alpha_M}$ for the range $1, \ldots, M$, we can, by 7.114, write $\epsilon^{\alpha_1 \ldots \alpha_M} \epsilon_{\gamma_1 \ldots \gamma_M}$ for $\delta^{\alpha_1 \ldots \alpha_M}_{\gamma_1 \ldots \gamma_M}$. But

7.311. $$\epsilon_{\gamma_1 \ldots \gamma_M} d_{(1)}y^{\gamma_1} \ldots d_{(M)}y^{\gamma_M} = \begin{vmatrix} d_{(1)}y^1 & d_{(1)}y^2 & \ldots & d_{(1)}y^M \\ \cdot & \cdot & \cdot & \cdot \\ \cdot & \cdot & \cdot & \cdot \\ d_{(M)}y^1 & d_{(M)}y^2 & \ldots & d_{(M)}y^M \end{vmatrix}$$
$$= \left| d_{(\beta)}y^\gamma \right| ,$$

say. Then 7.310 becomes

7.312. $$d\tau_{(M)}{}^{k_1 \cdots k_M} = \epsilon^{a_1 \cdots a_M} \frac{\partial x^{k_1}}{\partial y^{a_1}} \cdots \frac{\partial x^{k_M}}{\partial y^{a_M}} \left| d_{(\beta)} y^{\gamma} \right| .$$

The most useful expressions for the extension of an infinitesimal M-cell are given by 7.305 and 7.312.

Exercise. Let x^k be rectangular Cartesian coordinates in Euclidean 3-space. Introduce polar coordinates r, θ, ϕ, and consider the surface of the sphere $r = a$. On this sphere form the infinitesimal 2-cell with corners (θ, ϕ), $(\theta + d\theta, \phi)$, $(\theta, \phi + d\phi)$, $(\theta + d\theta, \phi + d\phi)$. Determine the extension of this cell and interpret the rectangular components. In particular, show that the three independent components of the extension are (apart from sign) equal to the areas obtained by normal projection of the cell onto the three rectangular coordinate planes. Does this interpretation remain valid if the sphere is replaced by some other surface?

From the fundamental theorem of multiple integration it follows that, if Φ is a continuous function of the parameters y in R_M, and if the $(M - 1)$-space which bounds R_M is sufficiently smooth, then

7.313. $$\lim \Sigma \, \Phi \left| \Delta_{(\beta)} y^{\gamma} \right|$$

exists and is independent of the manner in which the region R_M is divided into cells. In 7.313 the summation extends over all complete cells in R_M, Φ is evaluated at a point inside each cell, and the limit is that in which the size of each cell tends to zero, that is, $\Delta c^{(\beta)} \rightarrow 0$. The limit in question may be written

7.314. $$\int_{R_M} \Phi \left| d_{(\beta)} y^{\gamma} \right| .$$

This theorem we shall accept without formal proof which properly belongs to the subject of analysis rather than to the tensor calculus. However, we give in Appendix B an intuitive and somewhat incomplete argument which may help to make the theorem plausible.

If we choose the functions $f^{(a)}$ in 7.302 as

7.315. $$f^{(\alpha)}(y) = y^{\alpha},$$

then $d_{(\beta)}y^{\gamma} = 0$ if $\beta \neq \gamma$. Writing simply dy^{γ} for $d_{(\beta)}y^{\gamma}$ when $\beta = \gamma$, 7.314 becomes

7.316. $$\int_{R_M} \Phi \, dy^1 \ldots dy^M,$$

the more familiar form for a multiple integral. The expression 7.314 has the advantage that it exhibits the integral in a form valid for an arbitrary choice of cells.

Let us now take as integrand in 7.314 the expression

$$\Phi = \epsilon^{\alpha_1 \ldots \alpha_M} \frac{\partial x^{k_1}}{\partial y^{\alpha_1}} \cdots \frac{\partial x^{k_M}}{\partial y^{\alpha_M}} \phi,$$

where ϕ is another arbitrary function of position in R_M, that is of the parameters y. Then, by 7.312, we can write the integral 7.314 in the form

7.317. $$\int_{R_M} \phi \, d\tau_{(M)}{}^{k_1 \ldots k_M},$$

which is thus also independent of the choice of cells.

So far our considerations have been confined to a single coordinate system, and no ideas of tensor character have been introduced. The infinitesimal displacement $d_{(\beta)}x^k$ along the βth edge of an M-cell is by definition an absolute contravariant vector. Then, by 7.305, and by virtue of the skew-symmetry and the tensor character of the generalized Kronecker delta, we immediately have: *The extension $d\tau_{(M)}{}^{k_1 \ldots k_M}$ of an infinitesimal M-cell is an absolute covariant tensor of order M which is skew-symmetric in all pairs of suffixes.*

Let us now consider the integral

7.318. $$\int_{R_M} d\tau_{(M)}{}^{k_1 \ldots k_M}.$$

We might call it the extension of the region R_M, but it is of no interest in general as it has no tensorial character. This is, in essence, due to the fact that we are here combining the components of a tensor at different points; the result of such addition is not a tensor. For example

$$(T^k)_A + (T^k)_B$$

is not in general tensorial, A and B being different points, since the transformation coefficients $\partial x'^k/\partial x^r$ at A may differ from those at B. On the other hand, if $T_{k_1 \ldots k_M}$ is a covariant tensor of order M, then

$$T_{k_1 \ldots k_M} d\tau_{(M)}{}^{k_1 \ldots k_M}$$

is an invariant, and so is its integral

7.319. $$\int_{R_M} T_{k_1 \ldots k_M} d\tau_{(M)}{}^{k_1 \ldots k_M}.$$

In the special case when $M = N$, the extension $d\tau_{(N)}{}^{k_1 \ldots k_N}$ of an N-cell is determined by the single component of $d\tau_{(N)}{}^{12 \ldots N}$. In fact it follows immediately from the skew-symmetry of $d\tau_{(N)}{}^{k_1 \ldots k_N}$ that

7.320. $$d\tau_{(N)}{}^{12 \ldots N} = \frac{1}{N!} \epsilon_{k_1 \ldots k_N} d\tau_{(N)}{}^{k_1 \ldots k_N} = d\tau_{(N)},$$

say, and that

7.321. $$d\tau_{(N)}{}^{k_1 \ldots k_N} = \epsilon^{k_1 \ldots k_N} d\tau_{(N)}.$$

We immediately see that $d\tau_{(N)}$ is a relative invariant of weight -1; it is the dual of $d\tau_{(N)}{}^{k_1 \ldots k_N}$. As there is little danger of confusion we shall refer to $d\tau_{(N)}$ also as the extension of the N-cell. It follows from 7.305 and 7.320 that $d\tau_{(N)}$ is the determinant

7.322. $$d\tau_{(N)} = \left| d_{(s)} x^k \right|.$$

If we regard the parameters y as intrinsic coordinates in the subspace V_M, we may define $d\tau_{(M)}$ by the equation

7.323. $$d\tau_{(M)} = \left| d_{(\beta)} y^\alpha \right|.$$

We can now rewrite 7.312 in the form

7.324. $$d\tau_{(M)}{}^{k_1 \ldots k_M} = \nu^{k_1 \ldots k_M} d\tau_{(M)},$$

where

7.325. $$\nu^{k_1 \ldots k_M} = \epsilon^{\alpha_1 \ldots \alpha_M} \frac{\partial x^{k_1}}{\partial y^{\alpha_1}} \cdots \frac{\partial x^{k_M}}{\partial y^{\alpha_M}}.$$

Thus the extension of an infinitesimal M-cell can be written as a product of two factors which we shall now discuss.

The factor $d\tau_{(M)}$ is an intrinsic quantity determined (apart from sign) by the cell. We may call it the *intrinsic extension* of the infinitesimal cell in the subspace V_M. The intrinsic extension is an absolute invariant with respect to transformations of the coordinates x; but it behaves like a relative invariant of weight -1 under transformations of the parameters y, which can be regarded as intrinsic coordinates in V_M.

The factor $v^{k_1 \cdots k_M}$ is quite independent of the cell but is determined by the subspace V_M, including its relation to the parent space V_N, and by the point P in V_M under consideration. It is a generalization of the tangent vector to a curve ($M = 1$) and may therefore be called the *M-direction* of V_M at the point P. The M-direction $v^{k_1 \cdots k_M}$ is a skew-symmetric, absolute, contravariant tensor of order M with respect to transformations of the coordinates x; under transformations of the parameters y it is easily seen to behave like a relative invariant of weight $+1$.

The sign of the intrinsic extension of an infinitesimal M-cell depends on the adopted order of its edges. If we permute the order of the edges this sign does or does not change according as the permutation is odd or even. We thus distinguish between two *orientations* of an M-cell: *two infinitesimal M-cells at the same point P in V_M are said to have the same orientation or opposite orientations according as their intrinsic extensions have the same sign or opposite signs.* Since the M-direction at a point P in V_M is quite independent of any cell under consideration, we see from 7.324 that if $d\tau_{(M)}{}^{k_1 \cdots k_M}$ and $d'\tau_{(M)}{}^{k_1 \cdots k_M}$ are two infinitesimal M-cells at the same point in V_M, then

7.326. $$d'\tau_{(M)}{}^{k_1 \cdots k_M} = \theta \, d\tau_{(M)}{}^{k_1 \cdots k_M},$$

where θ is positive or negative according as the orientations of the cells are the same or opposite.

The intrinsic extension of an infinitesimal M-cell was seen to behave like a relative invariant of weight -1 under transformations of the parameters y. Thus it changes sign under a parameter transformation with negative Jacobian. We can therefore give no invariant meaning to, say, the statement that the intrinsic extension of a cell is positive. However, if two infinitesimal M-cells have intrinsic extensions of equal (or opposite) signs for one system of parameters, this fact remains true for all systems. Thus comparison of the orientation of two M-cells at the same point is of invariant nature.

Comparison of M-cells at two different points A and B in a region R_M of the subspace V_M is achieved as follows: The infinitesimal cell at A, say, is moved in a continuous manner to B, along a path C lying in R_M, such that, at each stage of the continuous motion, the intrinsic extension of the cell is non-zero. The orientations of the two cells are then compared at the point B.

For many types of regions this comparison of orientations is unique. Such regions are called *oriented* or *two-sided*. As an example, consider a region R_M with intrinsic coordinates which are continuous and single valued functions of position in R_M. In the process outlined above, the extension of the M-cell changes continuously during the motion from A to B, and is never zero. Hence it does not change sign. It follows that, in the case considered, the orientations of the M-cells at A and B can be compared by simply comparing the signs of their extensions. This procedure is obviously unique.

There are, however, regions where the orientations of cells at different points cannot be compared because the procedure adopted above is not unique. Such regions are called *unoriented* or *one-sided*. Examples of unoriented regions are the two dimensional polar space of constant curvature (cf. 4.1, Fig. 22) and the well-known Möbius strip.

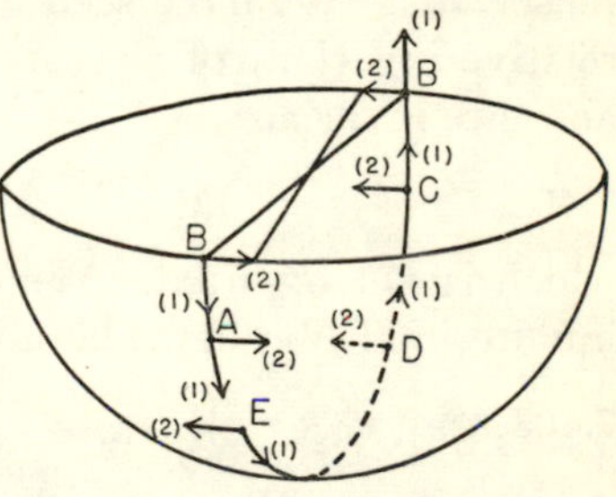

FIG. 22. The pair of vectors, (1) and (2), changes orientation after traversing the closed path $ABCDEA$ of the polar 2-space of constant curvature. This space is therefore unoriented.

In the case of an oriented region, we distinguish between two orientations for the whole region. It should be noted that in formulae such as 7.314 or 7.319, where integrals taken over a region R_M were considered, the region R_M is tacitly assumed to be oriented and all infinitesimal cells are assumed to have the same orientation.

7.4. Volume. In this section we shall introduce a metric (in general not positive-definite), and by means of it pass from the extension tensor to the concept of volume element. The volume is invariant, and can be integrated to give the volume of a finite domain. The extension tensor cannot be used in this way, although, as we have seen, certain integrals involving it have invariant character.

In the previous section we have defined the extension of an N-cell

7.401. $$d\tau_{(N)} = \left| d_{(s)}x^k \right|,$$

and we have seen that it is a relative invariant of weight -1. It follows that $d\tau_{(N)}$ retains or changes sign under a coordinate transformation according as the Jacobian of the transformation is positive or negative. Defining $\epsilon(\tau) = \pm 1$, such that $\epsilon(\tau) d\tau_{(N)}$ is positive, it is immediately seen that $\epsilon(\tau) d\tau_{(N)}$ is an oriented relative invariant of weight -1.

Let us now consider a Riemannian space with the metric tensor a_{mn}. We have seen earlier that $(\epsilon(a)a)^{\frac{1}{2}}$ is an oriented relative invariant of weight $+1$. We can therefore form the absolute invariant

7.402. $$dv_{(N)} = (\epsilon(a)a)^{\frac{1}{2}} \epsilon(\tau) d\tau_{(N)},$$

which is not oriented. We define it to be the *volume* of the infinitesimal N-cell. The integral

7.403. $$v_{(N)} = \int_{R_N} dv_{(N)} = \int_{R_N} (\epsilon(a)a)^{\frac{1}{2}} \epsilon(\tau) \left| d_{(s)}x^k \right|,$$

taken over a region R_N, is an absolute invariant, the volume of the region. Note that 7.403 always gives a *positive* volume.

To show that the above definition of volume agrees with the ordinary definition in the case of Euclidean 3-space, we take rectangular Cartesian coordinates, and define the cells as in 7.315, that is, $d_{(s)}x^k = dx^k$ if $s = k$ and $d_{(s)}x^k = 0$ if $s \neq k$. Then we have $\epsilon(a) = 1$, $a = 1$, $\epsilon(\tau) = 1$, and so the element of volume is

7.404. $$dv_{(3)} = dx^1 dx^2 dx^3,$$

as it should be.

In a general V_N, the square of the element of volume of a cell is

7.405. $$dv_{(N)}{}^2 = \epsilon(a)a \left| d_{(s)}x^k \right|^2 = \epsilon(a) \left| a_{kr} d_{(s)}x^k d_{(t)}x^r \right|.$$

If the edges of the cell are orthogonal, the only surviving elements in this determinant are in the leading diagonal ($s = t$), and they are, to within a sign, the squares of the lengths of the edges. Hence, since we have defined $dv_{(N)}$ to be positive, we have for the volume of an infinitesimal rectangular cell

7.406. $$dv_{(N)} = ds_1\, ds_2 \ldots ds_N,$$

where $ds_1, \ldots, ds_N$ are the lengths of the edges.

Exercise. Using polar coordinates in Euclidean 3-space find the volume of an infinitesimal cell whose edges are tangent to the coordinate curves. Obtain the volume of a sphere by integration.

A subspace V_M in a Riemannian space V_N possesses a metric by virtue of its immersion in V_N. Let V_M have the equation 7.301. Then the distance between two adjacent points of V_M is given by

7.407. $$ds^2 = \epsilon a_{ks} dx^k dx^s,$$

since the points in question are also points of V_N. Here $\epsilon (= \pm 1)$ is the indicator of the direction dx^k. But if we adopt the range 1 to M for Greek suffixes α, β, γ, this is

7.408. $$ds^2 = \epsilon a_{ks} \frac{\partial x^k}{\partial y^\alpha} \frac{\partial x^s}{\partial y^\beta} dy^\alpha\, dy^\beta = \epsilon b_{\alpha\beta}\, dy^\alpha dy^\beta,$$

where

7.409. $$b_{\alpha\beta} = a_{ks} \frac{\partial x^k}{\partial y^\alpha} \frac{\partial x^s}{\partial y^\beta}.$$

Then $b_{\alpha\beta}$ is the metric tensor of V_M for the intrinsic coordinate system y^α. Since V_M is itself a Riemannian space, any portion of it possesses a volume. Thus in a V_N we have to consider N-volume, $(N-1)$-volume, $(N-2)$-volume, etc. A 2-volume is also called an area, a 1-volume a length.

The volume of an infinitesimal M-cell is the invariant

7.410. $$dv_{(M)} = (\epsilon(b)b)^{\frac{1}{2}}\epsilon(\tau)d\tau_{(M)} = (\epsilon(b)b)^{\frac{1}{2}}\,\epsilon(\tau)\left|\,d_{(\beta)}y^\alpha\,\right|.$$

Since $\epsilon^{\alpha_1 \cdots \alpha_M}\epsilon_{\alpha_1 \cdots \alpha_M} = M!$, we obtain

7.411. $$dv_{(M)}{}^2 = \frac{\epsilon(b)}{M!}\,\epsilon^{\alpha_1 \cdots \alpha_M}\,\epsilon_{\alpha_1 \cdots \alpha_M} b\; d\tau_{(M)}{}^2.$$

We easily see that

7.412. $$\epsilon_{\alpha_1 \cdots \alpha_M} b = \epsilon^{\beta_1 \cdots \beta_M}\, b_{\alpha_1\beta_1}\, b_{\alpha_2\beta_2} \cdots b_{\alpha_M\beta_M}$$
$$= \epsilon^{\beta_1 \cdots \beta_M} a_{k_1 s_1} \cdots a_{k_M s_M} \frac{\partial x^{k_1}}{\partial y^{\alpha_1}} \cdots \frac{\partial x^{k_M}}{\partial y^{\alpha_M}} \frac{\partial x^{s_1}}{\partial y^{\beta_1}} \cdots \frac{\partial x^{s_M}}{\partial y^{\beta_M}},$$

by 7.409. Substituting from this into equation 7.411 we immediately have, by use of 7.312,

$$dv_{(M)}^2 = \frac{\epsilon(b)}{M!}\, a_{k_1 s_1} \cdots a_{k_M s_M}\; d\tau_{(M)}{}^{k_1 \cdots k_M}\; d\tau_{(M)}{}^{s_1 \cdots s_M},$$

or, equivalently,

7.413. $$dv_{(M)}^2 = \frac{\epsilon(b)}{M!}\, d\tau_{(M)}{}^{k_1 \cdots k_M}\; d\tau_{(M) k_1 \cdots k_M};$$

$\epsilon(b) = \pm 1$ and must be such as to make the right-hand side of this equation positive.

Given a region R_M in the M-space V_M, the invariant integral

7.414. $$\int_{R_M} dv_{(M)}$$

is defined to be the M-volume of the region. The volume integral of an invariant T,

7.415. $$\int_{R_M} T dv_{(M)}$$

is itself an invariant. But the integral of a vector or tensor of higher order, such as

$$\int_{R_M} T^k dv_{(M)},$$

does not possess tensor character.

Exercise. In the relativistic theory of the finite, expanding universe, the following line element is adopted:

$$ds^2 = R^2[dr^2 + \sin^2 r(d\theta^2 + \sin^2\theta d\phi^2)] - dt^2,$$

where $R = R(t)$ is a function of the "time" t. The ranges of the coordinates may be taken to be $0 \leqslant r \leqslant \pi$, $0 \leqslant \theta \leqslant \pi$, $0 \leqslant \phi < 2\pi$, $-\infty < t < \infty$. Find the total volume of "space," i.e., of the surface t = constant, and show that it varies with the "time" t as $R^3(t)$.

In a space V_N with no metric, the normal to a surface V_{N-1} has no meaning, but we do have an extension tensor associated with every element of V_{N-1}. We shall now show, when there is a metric, the connection between the normal and the extension tensor.

Let us take in V_{N-1} an infinitesimal cell with edges $d_{(1)}x^k, \ldots, d_{(N-1)}x^k$. The extension of the cell is

7.416. $$d\tau_{(N-1)}{}^{k_1 \ldots k_{N-1}} = \delta^{k_1 \ldots k_{N-1}}_{s_1 \ldots s_{N-1}} d_{(1)}x^{s_1} \ldots d_{(N-1)}x^{s_{N-1}}.$$

According to 7.122 the dual of $d\tau_{(N-1)}{}^{k_1 \ldots k_{N-1}}$ is $(-1)^{N-1}\nu_r$, where ν_r is the covariant relative vector defined by

7.417. $$\nu_r = \frac{1}{(N-1)!} \epsilon_{k_1 \ldots k_{N-1} r}\, d\tau_{(N-1)}{}^{k_1 \ldots k_{N-1}}.$$

Then, by 7.416,

$$\begin{aligned} \nu_r\, d_{(1)}x^r &= \frac{1}{(N-1)!} \epsilon_{k_1 \ldots k_{N-1} r}\, \delta^{k_1 \ldots k_{N-1}}_{s_1 \ldots s_{N-1}} \\ &\qquad d_{(1)}x^{s_1} \ldots d_{(N-1)}x^{s_{N-1}}\, d_{(1)}x^r \\ &= \epsilon_{s_1 \ldots s_{N-1} r}\, d_{(1)}x^{s_1} \ldots d_{(N-1)}x^{s_{N-1}}\, d_{(1)}x^r \\ &= 0, \end{aligned}$$

since $d_{(1)}x^r d_{(1)}x^{s_1}$ is symmetric in the suffixes r and s_1, and $\epsilon_{s_1 \ldots s_{N-1} r}$ is skew-symmetric. Similarly, we can show that $\nu_r d_{(2)}x^r = 0$, etc., and thus

7.418. $\nu_r d_{(\beta)} x^r = 0, \qquad \beta = 1, 2, \ldots, N-1.$

This shows that the vector with covariant components ν_r is orthogonal to the vectors $d_{(1)}x^r, \ldots, d_{(N-1)}x^r$, and hence to every infinitesimal displacement in V_{N-1}. In short, ν_r is normal to V_{N-1}.

We now normalize ν_r by introducing the vector* n_r:

7.419.
$$\epsilon(n)n_r = \epsilon(\tau)\,\frac{(\epsilon(a)a)^{\frac{1}{2}}}{dv_{(N-1)}}\,\nu_r$$
$$= \epsilon(\tau)\,\frac{(\epsilon(a)a)^{\frac{1}{2}}}{dv_{(N-1)}}\,\frac{\epsilon_{k_1\ldots k_{N-1}r}}{(N-1)!}\,d\tau_{(N-1)}{}^{k_1\ldots k_{N-1}},$$

$\epsilon(n)$ denoting the indicator of $n_r(\epsilon(n) = \pm 1$, such that $\epsilon(n)n_r n^r$ is positive); $\epsilon(\tau)$ is also ± 1 and determines the sense of n_r; its significance will be discussed shortly. Obviously n_r is also normal to V_{N-1}. We shall now show that n_r has unit magnitude, so that n_r is the *unit normal* to V_{N-1}. Using the metric tensor to move suffixes up and down, we have, by 7.419 and 7.211,

$$n_r n^r = \frac{\epsilon(a)}{[(N-1)!]^2}\cdot\frac{1}{dv_{(N-1)}{}^2}\,\epsilon_{k_1\ldots k_{N-1}r}\,\epsilon^{s_1\ldots s_{N-1}r}$$
$$d\tau_{(N-1)}{}^{k_1\ldots k_{N-1}}\,d\tau_{(N-1)s_1\ldots s_{N-1}}$$
$$= \frac{\epsilon(a)}{[(N-1)!]^2}\cdot\frac{1}{dv_{(N-1)}{}^2}\,\delta^{s_1\ldots s_{N-1}}_{k_1\ldots k_{N-1}}$$
$$d\tau_{(N-1)}{}^{k_1\ldots k_{N-1}}\,d\tau_{(N-1)s_1\ldots s_{N-1}}$$
$$= \frac{\epsilon(a)}{(N-1)!}\cdot\frac{1}{dv_{(N-1)}{}^2}\,d\tau_{(N-1)}{}^{k_1\ldots k_{N-1}}$$
$$d\tau_{(N-1)k_1\ldots k_{N-1}}$$
$$= \epsilon(a)\epsilon(b),$$

the last stage of simplification being justified by 7.413. This establishes that n_r is a unit vector, and also shows that

*The definition of n_r breaks down if $dv_{(N-1)} = 0$ and we therefore exclude this possibility; $dv_{(N-1)}$ vanishes if $d\tau_{(N-1)}{}^{k_1\ldots k_{N-1}} = 0$; however, even if this equation is not satisfied, $dv_{(N-1)}$ can vanish in the case of an indefinite metric—for example, if the cell lies in a null cone (cf. Ex. 7 at end of chapter).

7.420. $$\epsilon(n) = \epsilon(a)\epsilon(b).$$

At any point P of V_{N-1} there are, of course, two unit vectors which differ in sign. Let us assume that V_N is oriented, or, at least, a region R_N of it which includes all points of V_{N-1}, thus, in particular, also P. Let us adopt an orientation in R_N, and let $\epsilon(\tau)$ be $+1$ or -1 according as an elementary N-cell of this orientation has, at the point P, a positive or negative extension $d\tau_{(N)}$. With this convention concerning $\epsilon(\tau)$ we are now in a position to discuss which of the two possible unit normals is given by the n_r of 7.419.

Let

$$dx^r = n^r ds,$$

where ds is a positive infinitesimal; dx^r is codirectional with the unit normal n^r. Thus

$$\epsilon(n) n_r dx^r > 0.$$

Substituting in this relation from 7.419 and 7.416, we obtain on division by positive factors

$$\epsilon(\tau)\epsilon_{s_1 \ldots s_{N-1} r}\, d_{(1)}x^{s_1} \ldots d_{(N-1)}x^{s_{N-1}}\, dx^r > 0.$$

This can be written

7.421. $$\epsilon(\tau) d\tau_{(N)} > 0,$$

where $d\tau_{(N)}$ is the extension of an N-cell at P whose first $N-1$ edges, taken in order, are those of the $(N-1)$-cell whose extension tensor is the $d\tau_{(N-1)}{}^{k_1 \ldots k_{N-1}}$ which occurs in 7.419, and whose last edge has the direction of the normal n^r. Thus, by 7.421, the unit vector n^r must be such that this N-cell has the orientation adopted for the N-space.

7.5. Stokes' theorem. Let x^r be rectangular Cartesian coordinates in Euclidean 3-space, R_2 a finite two dimensional region on a surface V_2 in this 3-space, and R_1 the closed curve bounding R_2. Let n_r be the normal to V_2, and T_r a vector field assigned on R_2. Stokes' well-known theorem* states

*R. Courant, *Differential and Integral Calculus*, New York: Interscience Publishers, Inc., 1936, II, chap. v, sect. 6.

7.501.
$$\iint\limits_{R_2}\left[\left(\frac{\partial T_2}{\partial x^3} - \frac{\partial T_3}{\partial x^2}\right) n_1 + \left(\frac{\partial T_3}{\partial x^1} - \frac{\partial T_1}{\partial x^3}\right) n_2 + \left(\frac{\partial T_1}{\partial x^2} - \frac{\partial T_2}{\partial x^1}\right) n_3\right] dv_{(2)} = \int\limits_{R_1}(T_1\, dx^1 + T_2\, dx^2 + T_3\, dx^3)\,.$$

Here, the integral on the left side is taken over the region R_2, $dv_{(2)}$ denoting an element of surface area (2-volume), and the integral on the right side is taken along the bounding curve R_1. The sense in which the closed curve R_1 is described depends on the choice of the normal n_r, which is determined by V_2 only to within an arbitrary sign. Introducing the summation convention and the permutation symbol ϵ^{rks}, 7.501 simplifies to

7.502.
$$\int\limits_{R_2} \epsilon^{rks} T_{k,s} n_r\, dv_{(2)} = \int\limits_{R_1} T_r\, dx^r,$$

where the comma denotes partial differentiation.

We shall now introduce the extension of V_2 by means of 7.419, which in the special case of Euclidean 3-space and rectangular Cartesian coordinates gives

7.503.
$$n_r dv_{(2)} = \tfrac{1}{2}\, \epsilon_{rmn} d\tau_{(2)}{}^{mn}.$$

Remembering the definition 7.305 of elementary extension, we have, trivially, $dx^r = d\tau_{(1)}{}^r$. Thus we can rewrite Stokes' theorem in the form

$$\int\limits_{R_2} \epsilon^{rks} T_{k,s}\, \tfrac{1}{2}\, \epsilon_{rmn}\, d\tau_{(2)}{}^{mn} = \int\limits_{R_1} T_r\, d\tau_{(1)}{}^r,$$

which immediately simplifies to

7.504.
$$\int\limits_{R_2} T_{k,s}\, d\tau_{(2)}{}^{ks} = \int\limits_{R_1} T_r\, d\tau_{(1)}{}^r.$$

Consider now a transformation to a curvilinear coordinate system. Since $\partial T_k/\partial x^s$ differs from the tensor $\frac{1}{2}(T_{k,s} - T_{s,k})$ by an expression symmetric in k, s and $d\tau_{(2)}{}^{ks}$ is an absolute tensor, skew-symmetric in these suffixes, it follows that the integrand on the left is an invariant. Thus 7.504 is a tensor equation and therefore valid in all coordinate systems. However, this last form of Stokes' theorem has another very important advantage: the metric of the parent space is completely absent in the formulation of 7.504. This reveals the true non-metrical nature of Stokes' theorem.

Our purpose in this section is to generalize Stokes' theorem to arbitrary numbers of dimensions. Possibilities for such generalization are immediately suggested by 7.504. The formulation is as follows:

Let V_M be an M-dimensional subspace of the non-metrical N-space V_N $(M \leqslant N)$. Let R_M be an oriented finite region of V_M, bounded by the closed $(M - 1)$-space R_{M-1}, and let $T_{k_1 \ldots k_{M-1}}$ be a set of functions* of the coordinates. Then the *generalized Stokes' theorem* states

7.505.
$$\int_{R_M} T_{k_1 \ldots k_{M-1}, k_M} \, d\tau_{(M)}^{k_1 \ldots k_M} = \int_{R_{M-1}} T_{k_1 \ldots k_{M-1}} d\tau_{(M-1)}^{k_1 \ldots k_{M-1}},$$

provided the orientations of the M-cell (with extension $d\tau_{(M)}^{k_1 \ldots k_M}$) and the $(M-1)$-cell (with extension $d\tau_{(M-1)}^{k_1 \ldots k_{M-1}}$) are related as follows: If to the edges of the $(M - 1)$-cell there is added as an Mth edge an infinitesimal vector lying in V_M and pointing *out* from R_M, then the orientation of the M-cell so formed is to be the same as that of the M-cell with extension $d\tau_{(M)}^{k_1 \ldots k_M}$ (Fig. 23).

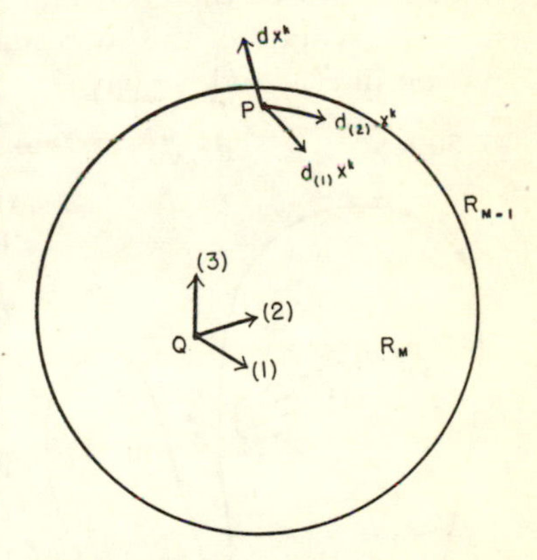

FIG. 23. Cell at P with edges $d_{(1)}x^k$, $d_{(2)}x^k$, dx^k has the same orientation as cell (1), (2), (3) at Q in R_M.

In proving 7.505, it is convenient to establish it first for the particular case $M = N$. Thus we set out to prove

*Since we do not need to transform the coordinates in establishing 7.505, the question of the tensor character of $T_{k \ldots k_{M-1}}$ does not arise.

7.506.
$$\int_{R_N} T_{k_1 \ldots k_{N-1}, k_N} \, d\tau_{(N)}^{k_1 \ldots k_N} = \int_{R_{N-1}} T_{k_1 \ldots k_{N-1}} \, d\tau_{(N-1)}^{k_1 \ldots k_{N-1}}.$$

We recall that the extension tensor is skew-symmetric; hence in any surviving term in 7.506 no two k's can take the same value. Thus we can write

7.507.
$$\int_{R_N} T_{k_1 \ldots k_{N-1}, k_N} \, d\tau_{(N)}^{k_1 \ldots k_N} = I_1 + I_2 + \ldots + I_N,$$

where

7.508.
$$I_1 = \int_{R_N} (T_{k_1 \ldots k_{N-1}, 1} \, d\tau_{(N)}^{k_1 \ldots k_{N-1} 1})_{(k \neq 1)},$$
$$I_2 = \int_{R_N} (T_{k_1 \ldots k_{N-1}, 2} \, d\tau_{(N)}^{k_1 \ldots k_{N-1} 2})_{(k \neq 2)},$$
$$\ldots \ldots \ldots \ldots \ldots,$$

the symbol $k \neq 1$ indicating that no k is to take the value 1, and $k \neq 2, \ldots,$ having similar meanings.

Let us choose the cells in R_N with edges along the parametric lines of the coordinates, in order, in the sense of increasing values of the coordinates. This establishes an orientation in R_N, and we have

7.509.
$$d\tau_{(N)}^{k_1 \ldots k_N} = \epsilon^{k_1 \ldots k_N} dx^1 \ldots dx^N,$$

the infinitesimals being positive. Then, by 7.508, we may write

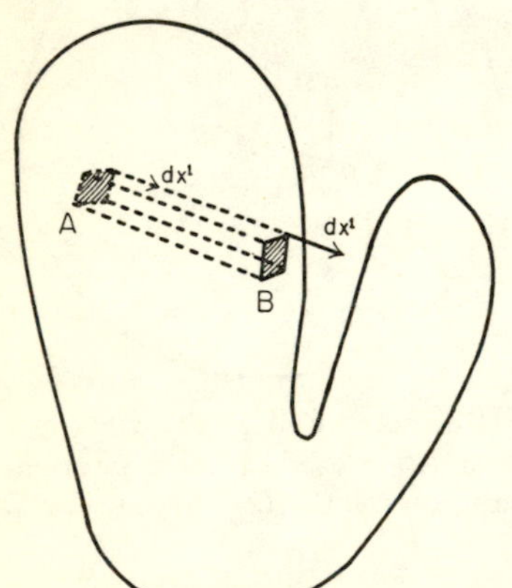

FIG. 24. Thin tube in R_N.

7.510.
$$I_1 = \int_{R_N} T_{\theta_2 \ldots \theta_N, 1} \, \epsilon^{\theta_2 \ldots \theta_N 1} dx^1 \ldots dx^N.$$

Here and throughout the rest of this section, the Greek suffix θ refers to the range of values $2, 3, \ldots, N$. We proceed to evaluate this integral by integrating first along thin tubes made up of parametric lines of x^1 (Fig. 24). Any such tube, taken in the sense of x^1 increasing, cuts R_{N-1} in two $(N - 1)$-cells, the tube entering R_N at a point A and leaving R_N at a point B.* Any one of the cells cut out on R_{N-1} by the tube has the $N-1$ edges

7.511.
$$\begin{array}{llllll} (f_{,2}\,dx^2, & dx^2, & 0, & 0, \ldots. & 0), \\ (f_{,3}\,dx^3, & 0, & dx^3, & 0, \ldots. & 0), \\ & \ldots\ldots & & & \\ (f_{,N}\,dx^N, & 0, & 0, & 0, \ldots. & dx^N), \end{array}$$

where the terms in the first column involve the partial derivatives of the function f by which R_{N-1} is given in the form $x^1 = f(x^2, \ldots, x^N)$. Carrying out the integration with respect to x^1 in 7.510, we have

7.512.
$$I_1 = \int_{R_{N-1}}^{(B)} T_{\theta_2 \ldots \theta_N} \epsilon^{\theta_2 \ldots \theta_N 1}\, dx^2 \ldots dx^N - \int_{R_{N-1}}^{(A)} T_{\theta_2 \ldots \theta_N} \epsilon^{\theta_2 \ldots \theta_N 1}\, dx^2 \ldots dx^N,$$

the integrals being evaluated respectively over the B-points and the A-points.

We now ask an important question: Have the $(N - 1)$-cells 7.511 the prescribed orientations? To test this, we add at a B-point an Mth or final edge with components

7.513.
$$(dx^1, 0, \ldots, 0),$$

with dx^1 positive; this edge points out from R_N. Equation 7.322 gives us, with substitution from 7.511 and 7.513, the following expression for the determinantal extension of the M-cell so obtained:

7.514.
$$\begin{vmatrix} f_{,2}\,dx^2, & dx^2, & 0, & \ldots & 0 \\ f_{,3}\,dx^3, & 0, & dx^3, & \ldots & 0 \\ & \ldots. & & & \\ f_{,N}\,dx^N, & 0, & 0, & \ldots & dx^N \\ dx^1, & 0, & 0, & \ldots & 0 \end{vmatrix}.$$

This is equal to

$$(-1)^{N-1} dx^1 dx^2 \ldots dx^N,$$

which is the determinantal extension of an N-cell of R_N, multi-

*If the tube meets R_{N-1} more than twice, we shall consider each connected portion of the tube, which lies in R_N, as a separate tube.

plied by $(-1)^{N-1}$. Accordingly, *the orientation of the* $(N-1)$-*cell* 7.511 *is correct at* B *if* N *is odd and incorrect if* N *is even.* Similarly, at A, where the final edge to be added is

$$(-dx^1, 0, \ldots, 0),$$

the orientation of 7.511 *is correct if* N *is even and incorrect if* N *is odd.*

Let $(d\tau_{(N-1)}{}^{k_1 \ldots k_{N-1}})_A$ and $(d\tau_{(N-1)}{}^{k_1 \ldots k_{N-1}})_B$ denote the extensions of the $(N-1)$-cells in which the x^1-tube cuts R_{N-1}, with the proper orientations. Introducing the appropriate factors to correct for orientation, where necessary, we have

7.515.
$$\begin{aligned} (d\tau_{(N-1)}{}^{k_1 \ldots k_{N-1}})_B &= (-1)^{N-1} \delta^{k_1 \ldots k_{N-1}}_{s_2 \ldots s_N} \\ &\qquad d_{(2)}x^{s_2} \ldots d_{(N)}x^{s_N}, \\ (d\tau_{(N-1)}{}^{k_1 \ldots k_{N-1}})_A &= (-1)^{N} \delta^{k_1 \ldots k_{N-1}}_{s_1 \ldots s_N} \\ &\qquad d_{(2)}x^{s_2} \ldots d_{(N)}x^{s_N}, \end{aligned}$$

where $d_{(\theta)}x^s (\theta = 2, \ldots, N)$ are the edges 7.511, taken in order. Now if no k in 7.515 is given the value 1, no s can take the value 1 if the term is to survive. So we have

7.516. $$(d\tau_{(N-1)}{}^{\theta_2 \ldots \theta_N})_B = (-1)^{N-1} \delta^{\theta_2 \ldots \theta_N}_{2 \ldots N} dx^2 \ldots dx^N.$$

But

$$\delta^{\theta_2 \ldots \theta_N}_{2 \ldots N} = \delta^{1\, \theta_2 \ldots \theta_N}_{1\, 2 \ldots N} = \epsilon^{1 \theta_2 \ldots \theta_N} = (-1)^{N-1} \epsilon^{\theta_2 \ldots \theta_N 1}$$

and so, with the similar result for the cell at A,

7.517.
$$\begin{aligned} (d\tau_{(N-1)}{}^{\theta_2 \ldots \theta_N})_B &= \epsilon^{\theta_2 \ldots \theta_N 1}\, dx^2 \ldots dx^N, \\ (d\tau_{(N-1)}{}^{\theta_2 \ldots \theta_N})_A &= -\epsilon^{\theta_2 \ldots \theta_N 1}\, dx^2 \ldots dx^N. \end{aligned}$$

Substitution in 7.512 gives

7.518.
$$\begin{aligned} I_1 &= \int_{R_{N-1}} T_{\theta_2 \ldots \theta_N}\, d\tau_{(N-1)}{}^{\theta_2 \ldots \theta_N} \\ &= \int_{R_{N-1}} (T_{k_1 \ldots k_{N-1}}\, d\tau_{(N-1)}{}^{k_1 \ldots k_{N-1}})_{k \neq 1}. \end{aligned}$$

Similarly,

7.519.
$$\begin{aligned} I_2 &= \int_{R_{N-1}} (T_{k_1 \ldots k_{N-1}}\, d\tau_{(N-1)}{}^{k_1 \ldots k_{N-1}})_{k \neq 2}, \\ I_3 &= \int_{R_{N-1}} (T_{k_1 \ldots k_{N-1}}\, d\tau_{(N-1)}{}^{k_1 \ldots k_{N-1}})_{k \neq 3}, \\ &\quad \ldots \ldots \end{aligned}$$

We have now very nearly proved 7.506; by 7.507, 7.518, and 7.519, we have shown that the left-hand side of 7.506 is equal to the sum of a set of terms of the general type of the right-hand side of 7.506, but with the understanding that in the first term no k takes the value 1, in the second term no k takes the value 2, and so on. However, since there are $N - 1$ suffixes in the right-hand side of 7.506, and the extension is skew-symmetric, it is easy to see that we can break down the summation occurring in 7.506 into precisely the restricted sum considered above. Hence Stokes' theorem in the form 7.506 is proved.

The general case for an arbitrary $M (M \leqslant N)$ is easily proved from the above result, as will now be shown. Let the M-dimensional subspace V_M be given by

7.520. $$x^k = x^k(y^1, \ldots, y^M).$$

The parameters y may be regarded as intrinsic coordinates of V_M. Let R_M be a finite region in V_M, bounded by the closed $(M - 1)$-space R_{M-1}, whose parametric equations are

7.521. $$y^\alpha = y^\alpha(z^1, \ldots, z^{M-1}).$$

As in the previous section, the Greek letters α, β, γ range from 1 to M, but the Greek letters ρ, σ, τ range from 1 to $M - 1$. We can disregard the fact that V_M is a subspace of V_N, and apply our previous result 7.506 to the region R_M. In order to avoid confusion, the extensions of the elementary cells will be written in explicit form (cf. equation 7.312). We then have

7.522. $$\int_{R_M} \frac{\partial}{\partial y^{\alpha_M}} t_{\alpha_1 \ldots \alpha_{M-1}} \epsilon^{\alpha_1 \ldots \alpha_M} \left| d_{(\beta)} y^\gamma \right| = \int_{R_{M-1}} t_{\alpha_1 \ldots \alpha_{M-1}} \frac{\partial y^{\alpha_1}}{\partial z^{\rho_1}} \cdots \frac{\partial y^{\alpha_{M-1}}}{\partial z^{\rho_{M-1}}} \epsilon^{\rho_1 \ldots \rho_{M-1}} \left| d_{(\sigma)} z^\tau \right|,$$

where $t_{\alpha_1 \ldots \alpha_{M-1}}$ is a set of quantities defined at each point of R_M.

If $T_{k_1 \ldots k_{M-1}}$ is any set of quantities defined in the full space V_N, then the expression

7.523.
$$T_{k_1 \ldots k_{M-1}} \frac{\partial x^{k_1}}{\partial y^{\alpha_1}} \cdots \frac{\partial x^{k_{M-1}}}{\partial y^{\alpha_{M-1}}}$$

is a function of position in R_M. Substituting this expression for $t_{\alpha_1 \ldots \alpha_{M-1}}$ in 7.522, and simplifying*, we obtain

7.524.
$$\int\limits_{R_M} T_{k_1 \ldots k_{M-1}, k_M} \frac{\partial x^{k_1}}{\partial y^{\alpha_1}} \cdots \frac{\partial x^{k_M}}{\partial y^{\alpha_M}} \epsilon^{\alpha_1 \ldots \alpha_M} \left| d_{(\beta)} y^\gamma \right|$$
$$= \int\limits_{R_{M-1}} T_{k_1 \ldots k_{M-1}} \frac{\partial x^{k_1}}{\partial z^{\rho_1}} \cdots \frac{\partial x^{k_{M-1}}}{\partial z^{\rho_{M-1}}} \epsilon^{\rho_1 \ldots \rho_{M-1}} \left| d_{(\sigma)} z^\tau \right|.$$

This is the generalized Stokes' theorem. By 7.312, it may be written

7.525.
$$\int\limits_{R_M} T_{k_1 \ldots k_{M-1}, k_M} d\tau_{(M)}^{k_1 \ldots k_M}$$
$$= \int\limits_{R_{M-1}} T_{k_1 \ldots k_{M-1}}\, d\tau_{(M-1)}^{k_1 \ldots k_{M-1}},$$

as in 7.505.

So far no assumption has been made regarding the tensor character of $T_{k_1 \ldots k_{M-1}}$ in establishing the generalized Stokes' theorem. If $T_{k_1 \ldots k_{M-1}}$ is an absolute covariant tensor, it is at once evident that the integrand on the right-hand side of 7.525 is an invariant, and so consequently is the integral. Due to the skew-symmetry of $d\tau_{(M)}^{k_1 \ldots k_M}$, the integrand on the left-hand side is also an invariant. This may be proved in a few lines, preferably with use of the compressed notation of 1.7; the proof is left as an exercise for the reader. Thus *both sides of* 7.525 *are invariants*.

Let us add a final remark. Stokes' theorem holds for an arbitrary collection of N^{M-1} functions $T_{k_1 \ldots k_{M-1}}$ of position in R_N. However, the two integrals in Stokes' theorem have tensor character only if $T_{k_1 \ldots k_{M-1}}$ is an absolute covariant tensor of order $M - 1$. Consider, for example, a tensor $T_{k_1 \ldots k_{M-1} r_1 r_2 r_3}$ of order $M + 2$. We still have Stokes' theorem n the form

*The second derivatives of x's with respect to the y's disappear on account of the skew-symmetry of the ϵ-symbol.

7.526.
$$\int_{R_M} T_{k_1 \ldots k_{M-1} r_1 r_2 r_3,\, k_M} d\tau_{(M)}{}^{k_1 \ldots k_M} = \int_{R_{M-1}} T_{k_1 \ldots k_{M-1} r_1 r_2 r_3}\, d\tau_{(M-1)}{}^{k_1 \ldots k_{M-1}}.$$

This equation holds for all values of r_1, r_2, r_3, but neither side of this equation is a tensor. In 7.525 we may use different coordinate systems to evaluate the two sides of the equation; we cannot do this in the case of 7.526.

Exercise. The skew-symmetric part of a tensor $T_{k_1 \ldots k_M}$ is defined to be

$$T_{[k_1 \ldots k_M]} = (M!)^{-1} \delta^{s_1 \ldots s_M}_{k_1 \ldots k_M} T_{s_1 \ldots s_M}.$$

Show that the left-hand side of equation 7.525 is unchanged if $T_{k_1 \ldots k_{M-1}}$ is replaced by its skew-symmetric part. Show that the same is true for the right-hand side.

7.6. Green's theorem. If a metric is given in the N-space V_N, Stokes' theorem, for $M = N$, can be thrown into a different form which is an obvious generalization of *Green's theorem.**

We start with Stokes' theorem in the form

7.601.
$$\int_{R_N} T_{k_1 \ldots k_{N-1},\, k_N}\, d\tau_{(N)}{}^{k_1 \ldots k_N} = \int_{R_{N-1}} T_{k_1 \ldots k_{N-1}}\, d\tau_{(N-1)}{}^{k_1 \ldots k_{N-1}},$$

where $T_{k_1 \ldots k_{N-1}}$ are skew-symmetric functions of position, not necessarily components of a tensor. This, by 7.321, can be written

7.602.
$$\int_{R_N} (N-1)!\, \frac{\partial}{\partial x^{k_N}} \hat{T}^{k_N} d\tau_{(N)} = \int_{R_{N-1}} \hat{T}^{k_N} \epsilon_{k_1 \ldots k_N} d\tau_{(N-1)}{}^{k_1 \ldots k_{N-1}},$$

where $\hat{T}^{k_N}$ is given by

*In the case of an indefinite metric there are exceptional surfaces, such as null cones, for which the transition to Green's theorem breaks down. For such surfaces Stokes' theorem, as formulated in the previous section, must be used.

7.603. $$\hat{T}^{k_N} = \frac{1}{(N-1)!}\epsilon^{k_1 \dots k_N} T_{k_1 \dots k_{N-1}},$$

or, equivalently, by

7.604. $$T_{k_1 \dots k_{N-1}} = \epsilon_{k_1 \dots k_N} \hat{T}^{k_N},$$

since $T_{k_1 \dots k_{N-1}}$ is skew-symmetric in all its suffixes.

If we now define T^r by

7.605. $$T^r = (\epsilon(a)a)^{-\frac{1}{2}} \hat{T}^r,$$

the left-hand side of 7.602 may be written, by 7.402,

7.606. $$(N-1)!\epsilon(\tau)\int_{R_N} \frac{\partial}{\partial x^r}[(\epsilon(a)a)^{\frac{1}{2}} T^r](\epsilon(a)a)^{-\frac{1}{2}}\, dv_{(N)}.$$

By 7.419, the right-hand side of 7.602 may be written

7.607. $$\int_{R_{N-1}} T^r(\epsilon(a)a)^{\frac{1}{2}}\epsilon_{k_1 \dots k_{N-1} r}\, d\tau_{(N-1)}{}^{k_1 \dots k_{N-1}}$$
$$= (N-1)!\epsilon(\tau) \int_{R_{N-1}} \epsilon(n) T^r n_r\, dv_{(N-1)},$$

n_r being the unit normal to R_{N-1}, and $\epsilon(n)$ being its indicator. From our convention, 7.5, linking the orientations of R_N and R_{N-1}, and from the discussion at the end of 7.4, it follows that n_r is the *outward* unit normal to R_{N-1}. From 7.606 and 7.607, we have Green's theorem in the form

7.608. $$\int_{R_N} \frac{\partial}{\partial x^r}[(\epsilon(a)a)^{\frac{1}{2}} T^r]\ (\epsilon(a)a)^{-\frac{1}{2}} dv_{(N)}$$
$$= \int_{R_{N-1}} \epsilon(n) T^r n_r dv_{(N-1)}\,.$$

In this formula, we must in general use the same coordinate system in evaluating the two sides. But if T^r is an absolute contravariant vector, then Green's theorem may be written

7.609. $$\int_{R_N} T^r{}_{|r}\, dv_{(N)} = \int_{R_{N-1}} \epsilon(n) T^r n_r\, dv_{(N-1)}.$$

It is obvious that each side is an invariant.

In Euclidean 3-space, with rectangular Cartesian coordinates x^r, 7.609 reduces to the well-known form of Green's theorem

7.610.
$$\int_{R_3} \frac{\partial T^r}{\partial x^r}\, dv_{(3)} = \int_{R_2} T^r n_r\, dv_{(2)},$$

$\partial T^r/\partial x^r$ being the divergence of the vector T^r and $T^r n_r$ the scalar product of T^r and the unit normal n_r. This formula has already been used earlier in the treatment of hydrodynamics.

SUMMARY VII

Relative tensor of weight W:

$$T'^{r\,\cdots}_{\;s\,\cdots} = J^W T^{m\,\cdots}_{\;n\,\cdots} \frac{\partial x'^r}{\partial x^m} \cdots \frac{\partial x^n}{\partial x'^s} \cdots .$$

Generalized Kronecker delta, permutation symbols:

$\delta^{k_1 \ldots k_M}_{s_1 \ldots s_M}$ relative tensor of weight 0, or absolute tensor,

$\epsilon_{r_1 \ldots r_N}$ relative tensor of weight -1,

$\epsilon^{r_1 \ldots r_N}$ relative tensor of weight $+1$;

$$\delta^{k_1 \ldots k_M}_{s_1 \ldots s_M} \epsilon^{s_1 \ldots s_M r_1 \ldots r_{N-M}} = M!\, \epsilon^{k_1 \ldots k_M\, r_1 \ldots r_{N-M}},$$

$$\epsilon^{k_1 \ldots k_M r_1 \ldots r_{N-M}} \epsilon_{s_1 \ldots s_M r_1 \ldots r_{N-M}} = (N-M)!\, \delta^{k_1 \ldots k_M}_{s_1 \ldots s_M}.$$

Dual tensors:

$$\hat{T}^{r_1 \ldots r_{N-M}} = \frac{1}{M!} \epsilon^{s_1 \ldots s_M r_1 \ldots r_{N-M}} T_{s_1 \ldots s_M},$$

$$T_{s_1 \ldots s_M} = \frac{1}{(N-M!)} \epsilon_{s_1 \ldots s_M r_1 \ldots r_{N-M}} \hat{T}^{r_1 \ldots r_{N-M}}.$$

Determinant of metric tensor:

$a = |a_{mn}|$ relative invariant of weight 2.

Differentiation of relative tensor:

$$\frac{\delta}{\delta u} T^{r\,\cdots}_{s\,\cdots} = (\epsilon(a)a)^{\frac{1}{2}W} \frac{\delta}{\delta u} [(\epsilon(a)a)^{-\frac{1}{2}W}\, T^r{}_{s\,\cdots}^{\,\cdots}],$$

$$T^{r\,\cdots}_{s\,\cdots\,|k} = (\epsilon(a)a)^{\frac{1}{2}W} [(\epsilon(a)a)^{-\frac{1}{2}W}\, T^r{}_{s\,\cdots}^{\,\cdots}]_{|k};$$

$$a_{|k} = 0,\ \epsilon_{r_1 \ldots r_N|k} = 0,\ \epsilon^{r_1 \ldots r_N}{}_{|k} = 0;$$

$$T^n{}_{|n} = \frac{\partial T^n}{\partial x^n} \text{ if } T^n \text{ is of weight 1.}$$

Extension of M-cell:

$$d\tau_{(M)}{}^{k_1\ldots k_M} = \delta^{k_1\ldots k_M}_{s_1\ldots s_M}\, d_{(1)}x^{s_1}\ldots d_{(M)}x^{s_M},$$

$$d\tau_{(M)}{}^{k_1\ldots k_M} = \epsilon^{a_1\ldots a_M}\frac{\partial x^{k_1}}{\partial y^{a_1}}\cdots\frac{\partial x^{k_M}}{\partial y^{a_M}}\, d\tau_{(M)},\quad d\tau_{(M)} = |d_{(\beta)}y^{a}|.$$

Volume:

$$dv_{(N)} = (\epsilon(a)a)^{\frac{1}{2}}\,\epsilon(\tau)d\tau_{(N)},$$

$$dv_{(N)} = ds_1\, ds_2\ldots ds_N \text{ for rectangular cells.}$$

Generalized Stokes' theorem:

$$\int_{R_M} T_{k_1\ldots k_{M-1},k_M} d\tau_{(M)}{}^{k_1\ldots k_M}$$

$$= \int_{R_{M-1}} T_{k_1\ldots k_{M-1}}\, d\tau_{(M-1)}{}^{k_1\ldots k_{M-1}}.$$

Green's theorem:

$$\int_{R_N} T^r{}_{|r}\, dv_{(N)} = \int_{R_{N-1}} \epsilon(n)T^r n_r\, dv_{(N-1)}.$$

EXERCISES VII

1. Show from 7.312 that the number of independent components of the extension of an M-cell in N-space is

$$\frac{N!}{M!\,(N-M)!}.$$

2. Prove that the covariant derivative of a generalized Kronecker delta is zero.

3. Show that

$$\delta^{k_1\ldots k_M}_{s_1\ldots s_M} = \begin{vmatrix} \delta^{k_1}_{s_1} & \delta^{k_1}_{s_2} & \ldots & \delta^{k_1}_{s_M} \\ & \ldots\ldots & & \\ & \ldots\ldots & & \\ \delta^{k_M}_{s_1} & \delta^{k_M}_{s_2} & \ldots & \delta^{k_M}_{s_M} \end{vmatrix}.$$

4. If T_{rs} is a symmetric tensor density and S_{rs} a skew-symmetric tensor density, show that

$$T^r_{s\,|r} = \frac{\partial}{\partial x^r}T^r_s - \tfrac{1}{2}\,T^{rk}\frac{\partial a_{rk}}{\partial x^s},$$

$$S^{rs}{}_{|r} = \frac{\partial}{\partial x^r}S^{rs}.$$

5. Let b_{mn} be an absolute covariant tensor. Show that the cofactors of the elements b_{mn} in the determinant $|b_{mn}|$ are the components of a relative contravariant tensor of weight 2.

6. Determine the tensor character of the cofactors in the determinants formed by the components of (a) a mixed absolute tensor, (b) a relative contravariant tensor of weight 1.

7. In the space-time of relativity, with metric form

$$(dx^1)^2 + (dx^2)^2 + (dx^3)^2 - (dx^4)^2,$$

the 3-space with equation

$$(x^1)^2 + (x^2)^2 + (x^3)^2 - (x^4)^2 = 0$$

is called a null cone. Prove that the 3-volume of any portion of the null cone is zero.

8. Prove that a polar N-space of constant curvature is oriented if N is odd and unoriented if N is even, but that an antipodal N-space of constant curvature is always oriented.

9. Show that the total volume of a polar 3-space of constant curvature R^{-2} is $\pi^2 R^3$, and that the volume is $2\pi^2 R^3$ if the space is antipodal.

10. If r_{mn} is an absolute tensor in 4-space, show that L, defined by

$$L = |r_{mn}|^{\frac{1}{2}},$$

is an invariant density. Discuss the tensor character of f^{mn} and g^{mn}, defined by

$$f^{mn} = \frac{\partial L}{\partial r_{mn}}, \quad g^{mn} = f^{mn} |f^{ks}|^{-\frac{1}{2}},$$

and also of g_{mn}, defined by

$$g_{ms} g^{ns} = \delta^n_m .$$

Prove that $|g_{mn}| = |f^{mn}|$. (Schrödinger)

11. In a non-metrical space V_4 there is a skew-symmetric absolute tensor field F_{mn} satisfying the partial differential equations

$$F_{mn,r} + F_{nr,m} + F_{rm,n} = 0.$$

Show that, if U_2 and W_2 are two closed 2-spaces in V_4, deformable into one another with preservation of orientation, then $$\int_{U_2} F_{mn} d\tau_{(2)}{}^{mn} = \int_{W_2} F_{mn}\, d\tau_{(2)}{}^{mn}.$$

12. In a flat space V_4 there is a skew-symmetric Cartesian tensor field F_{mn}, satisfying the partial differential equations $F_{mn,n} = 0$. Show that if U_3 and W_3 are two closed 3-spaces in V_4, deformable into one another with preservation of the sense of the unit normal n_r, then

$$\int_{U_3} \epsilon(n) F_{rs} n_s dv_{(3)} = \int_{W_3} \epsilon(n) F_{rs} n_s dv_{(3)}.$$

13. In a non-metrical space V_3 there is given an absolute covariant vector field v_r. Then the derived vector field $\omega^r = \epsilon^{rmn} v_{n,m}$ determines an invariant direction at each point of V_3 and hence a congruence of curves in V_3. A tube T is chosen, composed of such curves, and C, C' are two closed curves lying in T, and such that they may be deformed into one another without leaving T. Prove that

$$\int_C v_r dx^r = \int_{C'} v_r dx^r,$$

the integrals being taken in senses which coincide when one curve is deformed into the other. (Note that this is a generalization of Exercises VI, No. 4.)

14. Show that, if V_2 is a closed subspace of a non-metrical space V_3, and v_r any absolute covariant vector field assigned in V_3, then $$\int_{V_2} v_{r,s} d\tau_{(2)}{}^{rs} = 0.$$

15. Generalize Ex. 14 to the case of a closed V_M immersed in a non-metrical V_N in the form

$$\int_{V_M} v_{k_1 k_2 \ldots k_{M-1}, k_M} d\tau_{(M)}{}^{k_1 k_2 \ldots k_M} = 0.$$

16. In a flat space of N dimensions, with positive definite metric form, the volume V of the interior of a sphere of radius R is $V = \int \ldots \int dz_1 \ldots dz_N$, where z_p are rectangular Cartesian coordinates, and the integration is taken over all values of the coordinates satisfying $z_p z_p \leqslant R^2$. Show at once that $V = KR^N$, where K is independent of R. Further show that

$$V = 2^N R^N \int_0^{F_N} dz_N \int_0^{F_{N-1}} dz_{N-1} \ldots \int_0^{F_2} dz_2 \int_0^{F_1} dz_1$$

where

$$F_P = (1 - z_{P+1}^2 - \cdots - z_N^2)^{\frac{1}{2}},$$

and that

$$\int_0^{F_{P+1}} (F_P)^P \, dz_{P+1} = \frac{\pi^{\frac{1}{2}}}{2} \frac{\Gamma(\frac{1}{2}P + 1)}{\Gamma(\frac{1}{2}P + \frac{3}{2})} (F_{P+1})^{P+1};$$

hence prove that the volume of a sphere in N-space is

$$V = \frac{\pi^{\frac{1}{2}N} R^N}{\Gamma(\frac{1}{2}N + 1)}.$$

Check this result for $N = 2$ and $N = 3$.

17. In a flat N-space, as in Ex. 16, if we put $r^2 = z_p z_p$, then $z_p = \frac{\partial}{\partial z_p}(\frac{1}{2} r^2)$, and the outward-drawn unit normal to the sphere $z_p z_p = R^2$ is $n_p = z_p/R$. Use these facts and Green's theorem to prove that

$$\int r^2 dV = \frac{N}{N+2} R^2 V,$$

where the integral is taken throughout the interior of the sphere, and V is the volume of the sphere.

18. Using Ex. 17, or otherwise, prove that the fourth-order moment of inertia tensor (cf. equation 5.330) for a sphere of uniform density ρ and radius R, calculated with its centre as origin, is

$$J_{nprq} = \frac{\rho V R^2}{N+2} \delta_{np}^{|rq}.$$

$$\left(\text{Prove first that} \int z_p z_q dV = \frac{\delta_{pq} V R^2}{N+2} \right).$$

Show also that the tensor I_{st}, if defined by 5.335, has components

$$I_{st} = \rho V R^2 \frac{N-1}{N+2} \delta_{st}.$$

(Note that the fraction $(N - 1)/(N + 2)$ gives for $N = 2, 3$, respectively, the values of 1/4, 2/5, familiar in the evaluation of the moments of inertia of a circular disc and a sphere, about diameters).

CHAPTER VIII

NON-RIEMANNIAN SPACES

8.1. Absolute derivative. Spaces with a linear connection. Paths. This concluding chapter lays no claim to completeness. It is rather of an introductory nature and its only object is to give the reader an idea of some of the more modern developments of differential geometry. It may be added that the generalizations from Riemannian geometry, considered in this chapter, have found application in the many unified relativistic theories of gravitation and electrodynamics.

In Chapter II we introduced the Riemannian N-space with metric tensor a_{mn}. Perhaps the most important use which we made of the metric tensor was to define by means of it absolute and covariant differentiation. It is our purpose here to show that the ideas of absolute and covariant differentiation can be introduced without going as far as to require a metric, i.e., without asserting the possibility of associating a length with an infinitesimal displacement dx^r or an angle with two vectors. Thus we are led to consider *non-Riemannian spaces*, which are more general than Riemannian spaces but more specialized than the "amorphous" space of Chapter I and of parts of Chapter VII.

In order to simplify our discussion, we shall restrict ourselves throughout this chapter to *absolute tensors* and refer to them as *tensors* without any qualifying adjective. Generalization to the case of relative tensors presents no difficulty (see exercises 17 to 21 at the end of this chapter).

Given a contravariant vector field T^r, defined along a curve $x^r = x^r(u)$, the ordinary derivative of T^r is easily seen to transform according to the equation

8.101.
$$\frac{dT^p}{du} = \frac{dT^r}{du} X^p_r + X^p_{rs} T^r \frac{dx^s}{du},$$

where the compressed notation of 1.7 is used. Thus, as we already know, dT^r/du is not tensorial. We now search for a *vector* $\delta T^r/\delta u$ associated with the vector field T^r; this we shall call the *absolute derivative* of T^r. Similarly, we require absolute derivatives of covariant vector fields and of other tensor fields defined along a curve.

We shall assume that the absolute derivative of a tensor satisfies a certain minimum number of requirements. The smaller the number of these properties which the absolute derivative is required to possess, the more general is the resulting space. As more conditions are added, the space becomes specialized—until it reduces to Riemannian space or, if the commutativity of absolute differentiation is added, to flat space.

We shall impose the following basic postulates on the absolute derivative of a tensor field, defined along the curve $x^r = x^r(u)$:

A. *The absolute derivative of a tensor is a tensor of the same order and type.*

B. *The absolute derivative of an outer product of tensors is given, in terms of the factors, by the usual rule for differentiating a product.* Symbolically:

8.102.
$$\frac{\delta}{\delta u}(UV) = \frac{\delta U}{\delta u} V + U \frac{\delta V}{\delta u}.$$

C. *The absolute derivative of the sum of tensors of the same type is equal to the sum of the absolute derivatives of the tensors.* Symbolically:

8.103.
$$\frac{\delta}{\delta u}(U + V) = \frac{\delta U}{\delta u} + \frac{\delta V}{\delta u}.$$

D. *The absolute derivatives of contravariant and covariant vectors are respectively:*

8.104.
$$\frac{\delta T^r}{\delta u} = \frac{dT^r}{du} + \Gamma^r_{mn} T^m \frac{dx^n}{du},$$

8.105.
$$\frac{\delta T_r}{\delta u} = \frac{dT_r}{du} - \bar{\Gamma}^m_{rn} T_m \frac{dx^n}{du},$$

where the coefficients Γ^r_{mn}, $\bar{\Gamma}^r_{mn}$ *are functions of the coordinates.*

We know that these postulates are consistent, because they are all satisfied in Riemannian geometry, where $\Gamma^r_{mn} = \bar{\Gamma}^r_{mn} = \left\{ \begin{matrix} r \\ mn \end{matrix} \right\}$. Riemannian geometry was constructed out of a symmetric tensor a_{mn} with $\frac{1}{2}N(N+1)$ components. Now we have $2N^3$ quantities Γ^r_{mn}, $\bar{\Gamma}^r_{mn}$ — a larger number. Thus, in a sense, our new geometry is more complicated than Riemannian geometry; but, in another sense, it is simpler, because Riemannian geometry contains results which are not necessarily true in the new geometry.

The four postulates A, B, C, D define explicitly the absolute derivatives of vectors only. We shall now show that they also determine uniquely the absolute derivative of a tensor of arbitrary order.

Let T be an invariant and S^r an arbitrary contravariant vector. Applying 8.104 to S^r and to the vector TS^r, we immediately find

$$\frac{\delta}{\delta u}(TS^r) = \frac{dT}{du} S^r + T \frac{\delta S^r}{\delta u}.$$

But, by postulate B, we have

$$\frac{\delta}{\delta u}(TS^r) = \frac{\delta T}{\delta u} S^r + T \frac{\delta S^r}{\delta u}.$$

Thus

$$\frac{\delta T}{\delta u} S^r = \frac{dT}{du} S^r$$

and, since S^r is arbitrary,

8.106.
$$\frac{\delta T}{\delta u} = \frac{dT}{du}.$$

This proves that the *absolute derivative of an invariant is its ordinary derivative.*

Before we consider tensors of higher orders, we must establish the following lemma:

Any tensor can be expressed as a sum of outer products of vectors.

For the sake of concreteness we shall prove this lemma for the case of a third-order tensor $T_{mn}^{\cdot\cdot r}$. The extension of our proof to a tensor of arbitrary order and type will then be obvious. We introduce one set of N^2 vectors and two sets of N vectors each, and we denote these vectors by $X_{(p,q)m}$, $Y_{(p)m}$, $Z_{(p)}{}^m$. Here the letters p and q assume all values from 1 to N and are used to label the individual vectors of the set; m is a suffix and denotes the component of a vector (e.g., $X_{(2,3)m}$ is a covariant vector and $Z_{(1)}{}^m$ is a contravariant vector). In some fixed coordinate system we define those $N^2 + 2N$ vectors as follows:

$$\textbf{8.107.} \qquad X_{(p,q)m} = T_{mp}^{\cdot\cdot q}, \; Y_{(p)m} = \delta_m^p, \; Z_{(p)}{}^m = \delta_p^m,$$

where $T_{mp}^{\cdot\cdot q}$ is the given tensor. Then, with summation understood for repeated labels p and q, the equation

$$\textbf{8.108.} \qquad T_{mn}^{\cdot\cdot r} = X_{(p,q)m} Y_{(p)n} Z_{(q)}{}^r$$

is obviously true in the coordinate system considered. But 8.108 is a tensor equation and therefore true generally. This establishes our lemma.

Exercise. Show that any tensor T^{mn} may be written in the form

$$T^{mn} = X_{(p)}{}^m Y_{(p)}{}^n,$$

and use this result to prove problem 11, Exercises III.

Taking the absolute derivative on both sides of 8.108, we have, by postulates B and C,

$$\textbf{8.109.} \qquad \frac{\delta}{\delta u} T_{mn}^{\cdot\cdot r} = \left(\frac{\delta}{\delta u} X_{(p,q)m}\right) Y_{(p)n} Z_{(q)}{}^r$$
$$+ X_{(p,q)m} \left(\frac{\delta}{\delta u} Y_{(p)n}\right) Z_{(q)}{}^r$$
$$+ X_{(p,q)m} Y_{(p)n} \frac{\delta}{\delta u} Z_{(q)}{}^r.$$

Using 8.104 and 8.105, this easily reduces to

8.110. $$\frac{\delta}{\delta u} T_{mn}^{\cdot\cdot r} = \frac{d}{du} T_{mn}^{\cdot\cdot r} - \overline{\Gamma}^s_{mt} T_{sn}^{\cdot\cdot r} \frac{dx^t}{du} - \overline{\Gamma}^s_{nt} T_{ms}^{\cdot\cdot r} \frac{dx^t}{du} + \overline{\Gamma}^r_{st} T_{mn}^{\cdot\cdot s} \frac{dx^t}{du}.$$

Thus the absolute derivative of the tensor $T_{mn}^{\cdot\cdot r}$ is uniquely determined by our postulates. It is clear that this method may be extended to establish a unique absolute derivative of an arbitrary tensor.

Exercise. Show that, if the parameter along the curve $x^r = x^r(u)$ is changed from u to v, then the absolute derivative of a tensor field with respect to v is du/dv times the absolute derivative with respect to u. Symbolically:

8.111. $$\frac{\delta}{\delta v} = \frac{du}{dv} \frac{\delta}{\delta u}.$$

So far we have not considered the consequences of the tensor character of the absolute derivative, which is postulated in A. It is easy to see that the vector character of $\delta T^r/\delta u$ and $\delta T_r/\delta u$ determines the transformation properties of the quantities Γ^r_{mn}, $\overline{\Gamma}^r_{mn}$; in the compressed notation of 1.7, the transformations are as follows:

8.112. $$\Gamma^\rho_{\mu\nu} = \Gamma^r_{mn} X^\rho_r X^m_\mu X^n_\nu + X^r_{\mu\nu} X^\rho_r,$$

8.113. $$\overline{\Gamma}^\rho_{\mu\nu} = \overline{\Gamma}^r_{mn} X^\rho_r X^m_\mu X^n_\nu + X^r_{\mu\nu} X^\rho_r.$$

Note that Γ^r_{mn}, $\overline{\Gamma}^r_{mn}$ transform exactly like Christoffel symbols of the second kind (cf. 2.508).

Exercise. Prove that C^r_{mn}, defined by

8.114. $$C^r_{mn} = \Gamma^r_{mn} - \overline{\Gamma}^r_{mn},$$

is a tensor.

Exercise. Show that the right-hand side of 8.110 may also be obtained formally by the method of 2.516, i.e., by differentiating the invariant $T_{mn}^{\cdot\cdot r} X^m Y^n Z_r$, and using $\delta X^m/\delta u = 0$, $\delta Y^n/\delta u = 0$, $\delta Z_r/\delta u = 0$.

Consider an N-space in which two sets of quantities Γ^r_{mn}, $\bar{\Gamma}^r_{mn}$ are assigned as functions of the coordinates in some coordinate system x^r. When the coordinate system is changed, the functions Γ^r_{mn}, $\bar{\Gamma}^r_{mn}$ are to be transformed according to 8.112 and 8.113. Thus, having assigned Γ^r_{mn}, $\bar{\Gamma}^r_{mn}$ in one system of coordinates they are defined in all. We then say that the space has a *linear connection* and that Γ^r_{mn}, $\bar{\Gamma}^r_{mn}$ are the *coefficients of linear connection*. In a space with a linear connection, the absolute derivative of a tensor field, assigned along some curve $x^r = x^r(u)$, is defined by an expression of the form 8.110; the absolute derivative of an invariant is given by 8.106.

Having set up this technique of absolute differentiation, guided by the postulates A to D, it remains to show that these postulates are in fact satisfied without imposing on the coefficients of linear connection any conditions other than the formulae of transformation 8.112, 8.113.

As regards postulate A, it is clear from 8.106 that the absolute derivative of an invariant is an invariant. The formulae 8.112, 8.113 have been developed expressly to secure the tensor character of $\delta T^r/\delta u$ and $\delta T_r/\delta u$. As for the tensor character of 8.110, this is established by inspection of 8.109; this argument may be extended to cover any absolute derivative, and so we see that postulate A is completely satisfied.

As regards postulate B, the method (cf. 8.109) which we have used to define the absolute derivative of a tensor ensures that postulate B is satisfied for any outer product of vectors, and hence for any outer product of tensors.

The linear character of 8.104, 8.105, and 8.110 ensures the satisfaction of postulate C. As for postulate D, it has been incorporated in the technique. Hence all the postulates are satisfied.

Following the analogy of the Riemannian case, we define the *co-variant derivative* of a tensor field $T^{r}_{\cdot\, s\, \cdot\,\cdot}{}^{\cdot\,\cdot}$ by the relation

8.115. $$T^{r}_{\cdot\, s\, \cdot\,\cdot}{}^{\cdot\,\cdot}{}_{|n} \frac{dx^n}{du} = \frac{\delta}{\delta u} T^{r}_{\cdot\, s\, \cdot\,\cdot}{}^{\cdot\,\cdot},$$

this relation being assumed to hold for all curves $x^r = x^r(u)$.

It is easily seen that $T^{r}_{.\,s}{}^{\cdot\,\cdot}_{\cdot\,\cdot}{}_{|n}$ is a tensor field given by

8.116. $$T^{r}_{.\,s}{}^{\cdot\,\cdot}_{\cdot\,\cdot}{}_{|n} = T^{r}_{.\,s}{}^{\cdot\,\cdot}_{\cdot\,\cdot}{}_{,n} + \underline{\Gamma}^{r}_{mn} T^{m}_{.\,s}{}^{\cdot\,\cdot}_{\cdot\,\cdot} + \ldots - \Gamma^{m}_{sn} T^{r}_{.\,m}{}^{\cdot\,\cdot}_{\cdot\,\cdot} - \ldots$$

where the comma in $T^{r}_{.\,s}{}^{\cdot\,\cdot}_{\cdot\,\cdot}{}_{,n}$ denotes partial differentiation.

Exercise. Prove that $\delta^{r}_{s\,|\,n} = C^{r}_{sn}$, where C^{r}_{sn} is defined by 8.114.

We are now in a position to introduce ideas of curvature in a manner analogous to the method of Chapter III (cf. Exercises 2, 3 at end). However, we shall not do this at present; to obtain simpler formulae, we shall wait until we have specialized our space to the case of a symmetric connection.

We say that *a tensor $T^{r}_{.\,s}{}^{\cdot\,\cdot}_{\cdot\,\cdot}$ is propagated parallelly along a curve C if its absolute derivative along C vanishes:*

8.117. $$\frac{\delta}{\delta u} T^{r}_{.\,s}{}^{\cdot\,\cdot}_{\cdot\,\cdot} = 0.$$

The property 8.111 insures that this definition of parallel propagation is independent of the choice of parameter along the curve.

Exercise. Prove that an invariant remains constant under parallel propagation.

We are now ready to examine the possibility of defining curves analogous to the geodesics in Riemannian space. Obviously the property of stationary length cannot be applied since no length is defined in our space. Let us, however, recall the important property of a geodesic, that a vector, initially tangent to the curve and propagated parallelly along it, remains tangent to the curve at all points. This is the content of equation 2.427 and of the argument following it. The property just quoted is, in fact, sufficient to define geodesics in Riemannian space. This definition is immediately generalized to spaces with a linear connection. We shall call these curves "paths":

A path in a space with a linear connection is a curve such that a vector, initially tangent to the curve and propagated parallelly along it, remains tangent to the curve at all points.

Thus a path $x^r = x^r(u)$ must be such that the equation $\delta T^r/\delta u = 0$ has a solution $T^r = \theta\lambda^r$, where

8.118. $$\lambda^r = \frac{dx^r}{du},$$

and θ is an invariant function of u. Hence $\delta(\theta\lambda^r)/\delta u = 0$, or

8.119. $$\frac{\delta\lambda^r}{\delta u} = \mu\lambda^r,$$

where μ is an invariant function of u.

Eliminating μ from the system of equations 8.119, we easily see that a necessary and sufficient condition for a curve to be a path is that it satisfy the differential equations

8.120. $$\frac{\delta\lambda^r}{\delta u}\lambda^s = \lambda^r\frac{\delta\lambda^s}{\delta u},$$

or, more explicitly,

8.121. $$\left(\frac{d^2x^r}{du^2} + \Gamma^r_{mn}\frac{dx^m}{du}\frac{dx^n}{du}\right)\frac{dx^s}{du} = \frac{dx^r}{du}\left(\frac{d^2x^s}{du^2} + \Gamma^s_{mn}\frac{dx^m}{du}\frac{dx^n}{du}\right).$$

Exercise. Show that by a suitable choice of the parameter u along a path, the differential equation 8.119 simplifies to $\delta\lambda^r/\delta u = 0$.

Attention must be drawn to the fact that postulate B concerned the *outer* products of tensors, i.e., expressions such as $U^{rs}V_{mn}$ and not contracted (inner) expressions such as $U^{rm}V_{mn}$. In fact, the postulate is not in general true for inner products, and in this respect our new geometry differs from Riemannian geometry. Let us consider the invariant U^rV_r, where U^r and V_r are two vectors defined along a curve with parameter u. Since U^rV_r is an invariant, 8.106 gives

$$\frac{\delta}{\delta u}(U^r V_r) = \frac{d}{du}(U^r V_r)$$
$$= \frac{dU^r}{du} V_r + U^r \frac{dV^r}{du}.$$

Hence, by 8.104 and 8.105, with $\lambda^n = dx^n/du$,

$$\frac{\delta}{\delta u}(U^r V_r) = \frac{\delta U^r}{\delta u} V_r + U^r \frac{\delta V_r}{\delta u} - (\Gamma^r_{mn} - \bar{\Gamma}^r_{mn}) U^m V_r \lambda^n.$$

Thus

$$\text{8.122.} \quad \frac{\delta}{\delta u}(U^r V_r) - \left(\frac{\delta U^r}{\delta u} V_r + U^r \frac{\delta V_r}{\delta u} \right) = - C^r_{mn} U^m V_r \lambda^n.$$

The right-hand side of this equation will vanish only under special circumstances. Hence we may say that, *if U^r and V_r are both propagated parallelly along a curve, the inner product $U^r V_r$ will not be propagated parallelly; in fact we shall have*

$$\text{8.123.} \quad \frac{\delta}{\delta u}(U^r V_r) = - C^r_{mn} U^m V_r \lambda^n.$$

If C^r_{mn} are of the form

$$\text{8.124.} \quad C^r_{mn} = \delta^r_m C_n,$$

where C_n are functions of the coordinates, then 8.123 may be written

$$\frac{d}{du}(U^r V_r) = - C_n \lambda^n (U^r V_r).$$

Under these circumstances the relation $U^r V_r = 0$ is satisfied along a curve if it is satisfied at one point and the vectors U^r, V_r are propagated parallelly. Although we have no metric, the invariant condition $U^r V_r = 0$ may be said to imply the *orthogonality* of the vectors U^r and V_r. Since, under condition 8.124, this orthogonality is conserved under parallel propagation of the two vectors, we call a linear connection *ortho-invariant* if it satisfies 8.124.

Exercise. Show that, in a space with ortho-invariant linear connection, the Kronecker delta is propagated parallelly along curves satisfying $C_n \lambda^n = 0$.

What we have been considering in 8.122 is actually the question of the commuting of the operations of contraction and absolute differentiation, and we have seen that in general these operations do not commute. To simplify our geometry, we shall now add a fifth postulate:

E. *The operations of contraction and absolute differentiation commute.*

Let us examine the consequences of this postulate. Starting with the outer product $U^r V_s$, we first contract to $U^r V_r$ and then take the absolute derivative $\delta(U^r V_r)/\delta u$. Next, we first take the absolute derivative

$$\frac{\delta}{\delta u}(U^r V_s) = \frac{\delta U^r}{\delta u} V_s + U^r \frac{\delta V_s}{\delta u},$$

and then contract, obtaining

$$\frac{\delta U^r}{\delta u} V_r + U^r \frac{\delta V_r}{\delta u}.$$

The difference between the two results is given by 8.122. If this is to vanish, for all vectors U^r, V_r and for all tangent vectors λ^n, we must have

8.125. $$C^r_{mn} = 0,$$

or, equivalently,

8.126. $$\bar{\Gamma}^r_{mn} = \Gamma^r_{mn}.$$

By the method used above (expressing a tensor as a sum of outer products) it can be shown that 8.126 is sufficient to ensure that contraction and absolute differentiation commute for any tensor.

If a connection satisfies postulate E, or equivalently 8.126, it may be called *contraction-invariant*. However, since the relation 8.126 shows that we have to deal with only one set of coefficients of connection (N^3 of them instead of $2N^3$ in the general case), we shall call a space for which 8.126 is satisfied a space with a *single connection*.* A single connection is of course ortho-invariant ($C_n = 0$). In the following sections only single connections will be considered.

*If 8.126 is not satisfied, we may speak of a *double connection*, i.e., one connection for contravariant components and the other for covariant.

Exercise. Show that in the case of a single connection,

$$\frac{\delta}{\delta u}\delta^r_s = 0. \tag{8.127}$$

8.2. Spaces with symmetric connection. Curvature. To the five postulates A to E of the previous Section we now add a sixth:

F. *The coefficients of the single linear connection satisfy*

$$\Gamma^r_{mn} = \Gamma^r_{nm}. \tag{8.201}$$

Such a connection is called *symmetric* or *affine.*

The significance of this requirement becomes clearer by considering two statements each of which is completely equivalent to F, namely:

F_1. There exist infinitesimal parallelograms.

F_2. Given any point O, there exists a coordinate system such that, at O, the absolute derivative of a tensor reduces to its ordinary derivative.*

The statement F_1 is rather loosely formulated, but its precise meaning will become clear in proving the equivalence of F and F_1.

Let us remember the condition for the parallel propagation of a vector T^r along a curve $x^r = x^r(u)$:

$$\frac{\delta T^r}{\delta u} = \frac{dT^r}{du} + \Gamma^r_{mn}T^m\frac{dx^n}{du} = 0. \tag{8.202}$$

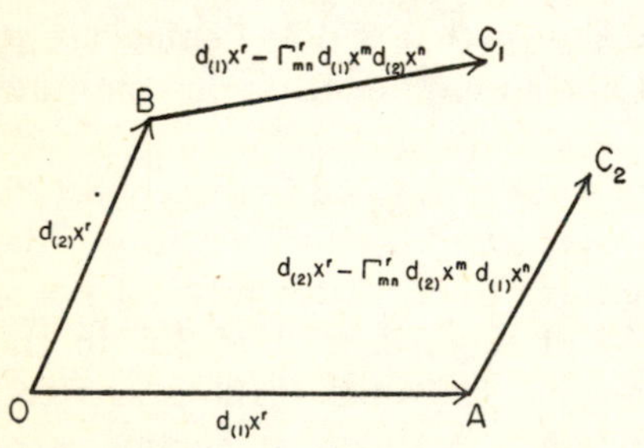

FIG. 25. Infinitesimal parallelogram.

For a small displacement dx^r, the increment of T^r under parallel propagation is given by

$$dT^r = -\Gamma^r_{mn}T^m dx^n, \tag{8.203}$$

to the first order in the small quantities dx^r.

Let OAB (Fig. 25) be an infinitesimal triangle, the displacement $\overrightarrow{OA}$ being given by $d_{(1)}x^r$ and $\overrightarrow{OB}$ by $d_{(2)}x^r$. If we

*Such a coordinate system is often called *geodesic.*

propagate $\overrightarrow{OA}$ parallelly from O to B we obtain an infinitesimal vector $\overrightarrow{BC_1}$, given by

8.204. $$d_{(1)}x^r - \Gamma^r_{mn} d_{(1)}x^m d_{(2)}x^n.$$

Similarly, if we propagate $\overrightarrow{OB}$ parallelly from O to A we obtain an infinitesimal vector $\overrightarrow{AC_2}$, given by

8.205. $$d_{(2)}x^r - \Gamma^r_{mn} d_{(2)}x^m d_{(1)}x^n.$$

The expressions 8.204 and 8.205 are correct to the second order in the differentials $d_{(1)}x^r$ and $d_{(2)}x^r$. This remains true if, in 8.204 and 8.205, Γ^r_{mn} is evaluated at O. The infinitesimal displacement $\overrightarrow{C_1C_2}$ is now easily seen to be

8.206. $$(\Gamma^r_{mn} - \Gamma^r_{nm}) d_{(1)}x^m d_{(2)}x^n,$$

correct to the second order. This always vanishes if and only if 8.201 holds. Thus, neglecting third order terms, C_1 and C_2 coincide in a single point C and $OACB$ is an infinitesimal parallelogram. This establishes the equivalence of F and F_1.

Let us now consider the statement F_2. Assuming F_2 to hold and given an arbitrary point O, we have a coordinate system x^r, say, in which the absolute derivative of any tensor field reduces to its ordinary derivative at O. This immediately implies the vanishing at O of all the components of Γ^r_{mn} in the coordinates x^r. If x^ρ is any other coordinate system*, the components of the linear connection at O are given in the new coordinate system by the transformation equation 8.112. These, by virtue of the vanishing of Γ^r_{mn}, reduce to

8.207. $$\Gamma^\rho_{\mu\nu} = X^r_{\mu\nu} X^\rho_r.$$

Since $X^r_{\mu\nu} = X^r_{\nu\mu}$, the $\Gamma^\rho_{\mu\nu}$ are symmetric in the two lower suffixes. But O being an arbitrary point and the coordinate system x^ρ being arbitrary also, this symmetry property holds generally. Thus we see that F_2 implies F.

We shall now prove the converse, namely, that F implies F_2. Let x^r be an arbitrary coordinate system and O an arbi-

*The compressed notation of 1.7 is used.

trary point in space. The coordinates of O are denoted by x_0^r and the coefficients of linear connection at O by $(\Gamma^r_{mn})_0$. We introduce new coordinates x^ρ by the equations

8.208. $$x^\rho = \delta^\rho_r(x^r - x^r_0) - \tfrac{1}{2}\,\delta^\rho_r(\Gamma^r_{mn})_0(x^m - x^m_0)(x^n - x^n_0),$$

where δ^ρ_r is unity if ρ and r have the same numerical value and 0 otherwise. We immediately have

8.209. $$(X^\rho_r)_0 = \delta^\rho_r,\ X^\rho_{mn} = \delta^\rho_r(\Gamma^r_{mn})_0,$$

where, in the second expression, use is made of the symmetry of Γ^r_{mn}. Using the relation

$$X^\sigma_r X^r_\rho = \delta^\sigma_\rho,$$

we find that

8.210. $$(X^r_\rho)_0 = \delta^r_\rho.$$

By virtue of the identity 1.706, the transformation equation 8.112 becomes

8.211. $$\Gamma^\rho_{\mu\nu} = \Gamma^r_{mn}X^\rho_r X^m_\mu X^n_\nu - X^\rho_{mn}X^m_\mu X^n_\nu.$$

Using 8.209 and 8.210, we see that at O, 8.211 reduces to

8.212. $$(\Gamma^\rho_{\mu\nu})_0 = \delta^\rho_r\delta^m_\mu\delta^n_\nu\,[(\Gamma^r_{mn})_0 - (\Gamma^r_{mn})_0] = 0.$$

Thus the coordinate system x^ρ satisfies the requirements of F_2. This completes the proof of the equivalence of statements F and F_2.

Exercise. Deduce immediately from F_2 that

$$T_{|mn} = T_{|nm},$$

where T is an invariant.

We shall now investigate the *curvature* of a space with symmetric connection. If we carry out the operations which led to 3.105, using the coefficients of connection instead of Christoffel symbols, we obtain, for any covariant vector field T_r,

8.213. $$T_{r|mn} - T_{r|nm} = T_s R^s_{.rmn},$$

where

8.214. $$R^s_{.rmn} = \frac{\partial\Gamma^s_{rn}}{\partial x^m} - \frac{\partial\Gamma^s_{rm}}{\partial x^n} + \Gamma^p_{rn}\Gamma^s_{pm} - \Gamma^p_{rm}\Gamma^s_{pn}.$$

By 8.213, $R^{s}_{.rmn}$ is a tensor; we call it the *curvature tensor* of the space with symmetric connection. The curvature tensor satisfies the following identities, analogous to those of Chapter III:

8.215. $$R^{s}_{.rmn} = - R^{s}_{.rnm},$$

8.216. $$R^{s}_{.rmn} + R^{s}_{.mnr} + R^{s}_{.nrm} = 0,$$

8.217. $$R^{s}_{.rmn|k} + R^{s}_{.rnk|m} + R^{s}_{.rkm|n} = 0.$$

Exercise. Prove the above identites by using a coordinate system of the type considered in F_2.

Note that we can not define a covariant curvature tensor as in 3.112 since there is now no metric which enables us to lower a suffix. Thus the curvature tensor $R^{s}_{.rmn}$ does not possess any symmetry property involving the contravariant suffix s. In fact, by contracting the curvature tensor, we can, in contrast to the Riemannian case, form two distinct tensors of second order, as follows:

8.218. $$R_{rm} = R^{s}_{.rms} = \frac{\partial \Gamma^{s}_{rs}}{\partial x^{m}} - \frac{\partial \Gamma^{s}_{rm}}{\partial x^{s}} + \Gamma^{n}_{rs}\Gamma^{s}_{nm} - \Gamma^{n}_{rm}\Gamma^{s}_{ns},$$

8.219. $$F_{mn} = \tfrac{1}{2} R^{s}_{.smn} = \tfrac{1}{2}\left(\frac{\partial \Gamma^{s}_{ns}}{\partial x^{m}} - \frac{\partial \Gamma^{s}_{ms}}{\partial x^{n}}\right).$$

Exercise. Verify that F_{mn} is skew-symmetric and that $R_{mn} + F_{mn}$ is symmetric. Show also directly from 8.219 that F_{mn} vanishes in a Riemannian space.

It is obvious from 8.214 that if Γ^{r}_{mn} all vanish identically in some coordinate system then the curvature tensor is identically zero:

8.220. $$R^{s}_{.rmn} = 0.$$

This last equation, being tensorial, must hold in all coordinate systems; but the Γ^{r}_{mn} will, of course, not vanish in a general coordinate system.

A space in which the curvature tensor vanishes identically is called *flat*.

By methods closely analogous to those employed in 3.5, which, however, we shall not repeat here, it can be shown that *the vanishing of* $R^{s}_{.rmn}$ *is sufficient* (*as well as necessary*) *for the existence of a coordinate system in which the coefficients of connection are all identically zero.*

In such a coordinate system the equations of a path are linear in the coordinates, i.e., a path is given by $N - 1$ independent equations of the form

8.221. $$A_{\sigma r}x^{r} + B_{\sigma} = 0, \qquad \sigma = 1, 2, \ldots, N - 1,$$

where N is the dimension of the space and $A_{\sigma r}$, B_{σ} are constants. The proof of this statement is quite simple and, since we shall not require the result in our later work, it is left as an exercise.

In a coordinate system in which the Γ^{r}_{mn} are all identically zero, the parallel propagation of a tensor along an arbitrary curve leaves each of the components of the tensor unchanged. (This is immediately seen from the definition 8.117 of parallel propagation and the fact that absolute differentiation coincides here with ordinary differentiation.) It follows that *parallel propagation of a tensor from a point A to another point B is independent of the path, joining A and B, along which the tensor is propagated.* This last statement is of an invariant nature and thus holds for flat space whatever the coordinate system used. In fact, using the arguments of 3.5, it can be shown that this statement is equivalent to the vanishing of the curvature tensor and may thus be used as an alternative definition of flat space with symmetric connection.

8.3. Weyl spaces. Riemannian spaces. Projective spaces. In this final Section we shall start with a space with symmetric connection. First, by two specializations, we shall arrive at Weyl's geometry and Riemannian geometry in this order. Next, we shall consider briefly the generalization to projective space. This generalization does not consist in loosening the restrictions imposed on the coefficients of connection (which would merely lead back to unsymmetric connections); it is obtained by introducing a new type of transformation affecting the coefficients of connection but not the coordinates of the space.

A *Weyl space* is obtained by imposing on *a space with symmetric connection* Γ^r_{mn} the following requirement:

W. *There exists a covariant vector* ϕ_r *and a symmetric tensor* a_{mn} *whose determinant is non-zero, such that*

8.301. $$a_{mn\,|\,r} + a_{mn}\phi_r = 0,$$

8.302. $$a = \left| a_{mn} \right| \neq 0.$$

We shall treat a_{mn} as a metric tensor and define a^{mn} as in 2.203. The tensors a_{mn} and a^{mn} will be used to lower or raise suffixes in the usual manner. Thus we can associate with the coefficients of connection Γ^r_{mn} the new quantities defined by

8.303. $$\Gamma_{mnr} = a_{rs}\Gamma^s_{mn}, \quad \Gamma^r_{mn} = a^{rs}\Gamma_{mns}.$$

The Γ_{mnr} are close analogues of the Christoffel symbols of the first kind and share their transformation properties, while the Γ^r_{mn} are analogues of the Christoffel symbols of the second kind, as was remarked earlier.

Writing 8.301 out in full, we have

8.304. $$a_{mn,\,r} - \Gamma^s_{mr}a_{sn} - \Gamma^s_{nr}a_{ms} + a_{mn}\phi_r = 0,$$

where, as usual, the comma in $a_{mn,r}$ denotes partial differentiation. By 8.303, this becomes

8.305. $$\Gamma_{mrn} + \Gamma_{nrm} = a_{mn,r} + a_{mn}\phi_r.$$

If, in this equation, we permute the suffixes $m\ n\ r$ cyclicly we obtain, in addition to 8.305, two further equations. Adding those two equations and subtracting 8.305, we obtain

8.306. $$\Gamma_{mnr} = [mn,r] + \tfrac{1}{2}\left(a_{mr}\phi_n + a_{nr}\phi_m - a_{mn}\phi_r\right),$$

where $[mn,r]$ is the Christoffel symbol of the first kind:

8.307. $$[mn,r] = \tfrac{1}{2}\left(a_{mr,n} + a_{nr,m} - a_{mn,r}\right).$$

If we raise the suffix r by means of the contravariant metric tensor a^{rs}, 8.306 becomes

8.308. $$\Gamma^r_{mn} = \left\{ {r \atop mn} \right\} + \tfrac{1}{2}\left(\delta^r_m\phi_n + \delta^r_n\phi_m - a_{mn}\phi^r\right),$$

where

8.309. $$\left\{ {r \atop mn} \right\} = a^{rs}\,[mn,s], \quad \phi^r = a^{rs}\phi_s.$$

Exercise. Prove that 8.301 implies

8.310. $$a^{mn}{}_{|r} - a^{mn}\phi_r = 0.$$

If we are given a vector ϕ_r and a symmetric covariant tensor a_{mn}, then 8.308 gives us a symmetric connection for which 8.301 is satisfied. However, we can look at the question the other way round: Given an arbitrary symmetric connection, do a_{mn} and ϕ_r exist to satisfy 8.301 and 8.302 ? The answer is "No", but we shall not attempt to establish this (cf. Ex. 22 at end). (If the answer were "Yes", then postulate W would not restrict the connection; every space with symmetric connection would be a Weyl space.)

We shall now show that, given a symmetric connection satisfying W, the choice of a_{mn} and ϕ_r is far from unique.

Consider a'_{mn}, ϕ'_r, defined by

8.311. $$a'_{mn} = \lambda a_{mn}, \quad \phi'_r = \phi_r - (\ln\lambda)_{,r} = \phi_r - \frac{\lambda_{,r}}{\lambda},$$

where λ is an arbitrary invariant function of position. Then

8.312.
$$\begin{aligned} a'_{mn|r} &+ a'_{mn}\phi'_r \\ &= \lambda a_{mn|r} + a_{mn}\lambda_{,r} + \lambda a_{mn}\phi_r - a_{mn}\lambda_{,r} \\ &= \lambda(a_{mn|r} + a_{mn}\phi_r) \\ &= 0, \end{aligned}$$

by virtue of 8.301. Thus if a_{mn}, ϕ_r satisfy 8.301, then a'_{mn}, ϕ'_r defined by 8.311, satisfy a relation of the same form.

A Weyl space therefore contains an infinite set of tensor pairs (a_{mn}, ϕ_r), (a'_{mn}, ϕ'_r), . . . , mutually related by equations of the form 8.311. There is no *a priori* reason why one such tensor pair should be preferred over all others. Thus the fundamental relations and quantities in Weyl's geometry are those which do not depend on the particular choice of the tensor pair (a_{mn}, ϕ_r).

The process of replacing the tensor pair (a_{mn}, ϕ_r) by a new pair (a'_{mn}, ϕ'_r), defined by 8.311, is called a *gauge transformation.* We can now restate the content of the previous paragraph by saying that in Weyl's geometry we are concerned with *gauge-invariant* relations and *gauge-invariant* quantities, i.e., with relations and quantities invariant under all gauge transformations.

We shall conclude our brief survey of Weyl's geometry by an enumeration of the most important gauge-invariant relations and quantities.

The squared magnitude of a contravariant vector

8.313. $$X^2 = \epsilon a_{mn} X^m X^n$$

is not gauge-invariant since it gains a factor λ under a gauge transformation. However, the property of having zero magnitude, i.e., to be a null vector, is gauge-invariant. Moreover, the ratio of the magnitudes of two contravariant vectors at the same point is a gauge-invariant quantity. The angle θ between two contravariant vectors X^r and Y^r, defined in the usual way by

8.314. $$\cos\theta = \frac{a_{mn} X^m Y^n}{(a_{pq} X^p X^q a_{rs} Y^r Y^s)^{\frac{1}{2}}},$$

is gauge-invariant. Since a gauge transformation preserves angles it is *conformal.*

Exercise. Is δ^r_s a gauge-invariant tensor?

Exercise. Show that, under the gauge transformation 8.311, a^{mn} and a transform as follows:

8.315. $$a'^{mn} = \frac{1}{\lambda} a^{mn}, \quad a' = \lambda^N a.$$

Any geometrical object either involves (a_{mn}, ϕ_r), or does not involve them. In the former case, it may, or may not be gauge-invariant. In the latter case, it is certainly gauge-invariant, because it is unaltered by the transformation 8.311. The connection Γ^r_{mn} is in a peculiar position; it is supposed given *a priori,* subject only to the condition that (a_{mn}, ϕ_r) exist to satisfy 8.301 and 8.302; but it is expressible in terms of (a_{mn}, ϕ_r) by 8.308. Any doubt regarding its gauge-invariance can be removed by substituting 8.311 in 8.308, and showing that it remains unchanged.

Any quantity defined in terms of the Γ^r_{mn} exclusively must of course be gauge-invariant. Thus the curvature tensor, defined by

$$8.316. \quad R^{s}_{.rmn} = \frac{\partial \Gamma^{s}_{rn}}{\partial x^{m}} - \frac{\partial \Gamma^{s}_{rm}}{\partial x^{n}} + \Gamma^{p}_{rn}\Gamma^{s}_{pm} - \Gamma^{p}_{rm}\Gamma^{s}_{pn},$$

is a gauge-invariant tensor.

Exercise. The covariant curvature tensor is defined by

$$R_{srmn} = a_{sp}R^{p}_{.rmn}.$$

How does it behave under gauge transformations?

Contracting the curvature tensor $R^{s}_{.rmn}$ with respect to s and n we obtain the gauge-invariant tensor

$$8.317. \quad R_{rm} = R^{s}_{.rms}.$$

Contracting $R^{s}_{.rmn}$ with respect to s and r, we obtain the gauge-invariant tensor

$$F_{mn} = \tfrac{1}{2} R^{s}_{.smn} = \tfrac{1}{2}\left(\frac{\partial \Gamma^{s}_{sn}}{\partial x^{m}} - \frac{\partial \Gamma^{s}_{sm}}{\partial x^{n}}\right).$$

By 8.308, we have

$$\Gamma^{s}_{sn} = \left\{ {s \atop sn} \right\} + \tfrac{1}{2}N\,\phi_{n},$$

where N is the dimension of the Weyl space. Substituting this into the expression for F_{mn} and using 2.542, we find

$$8.318. \quad F_{mn} = \frac{N}{4}(\phi_{n,m} - \phi_{m,n}).$$

It is easy to show that the gauge-invariant tensor G_{mn}, defined by

$$8.319. \quad R_{mn} = G_{mn} - F_{mn},$$

is symmetric (cf. Exercise following 8.219).

In Weyl's unified relativistic field theory an important role is played by relative invariants of weight 1 which are gauge-invariant. It follows immediately from 8.315 that for $N = 4$, the following relative invariants of weight 1 are gauge-invariant*:

*We limit ourselves to gauge transformations with positive λ.

8.320.
$$(\epsilon(a)a)^{\frac{1}{2}}R^2,\ (\epsilon(a)a)^{\frac{1}{2}}F_{mn}F^{mn},$$
$$(\epsilon(a)a)^{\frac{1}{2}}G_{mn}G^{mn},\ (\epsilon(a)a)^{\frac{1}{2}}R_{srmn}R^{srmn},$$

where $R = a^{mn}R_{mn}$ and $\epsilon(a) = \pm 1$ such as to make $\epsilon(a)a$ positive. Note that the tensor densities 8.320 are gauge-invariant only if the Weyl space is four dimensional. As a consequence Weyl's unified field theory gives special significance to the fact that space-time has four dimensions.

We shall now make the transition from a Weyl space to a *Riemannian space.*

Consider a Weyl space in which the tensor F_{mn}, given by 8.318, vanishes identically:

8.321.
$$F_{mn} = \frac{N}{4}(\phi_{n,m} - \phi_{m,n}) = 0.$$

A necessary and sufficient condition for the vanishing of F_{mn} is that ϕ_r be the gradient of some invariant ϕ:

8.322.
$$\phi_r = \phi_{,r}.$$

The sufficiency of the condition is immediately obvious. The necessity will now be proved by actual construction of the function ϕ. Let O be a fixed point and P a variable one. We put

8.323.
$$\phi(P, C) = \int_{O(C)}^{P} \phi_r dx^r,$$

where the integral is taken over some curve C joining O and P. Then as indicated in 8.323 ϕ is a function of the coordinates of P and also depends on the choice of the curve C. Consider any other curve C_1 which joins O to P. The two curves C and C_1 together form a closed curve $OCPC_1O$ which we denote by Γ. Let S be any 2-space which is bounded by Γ (Fig. 26). Then

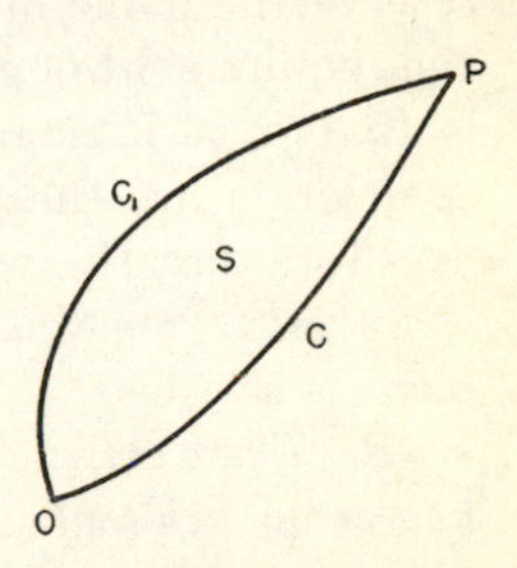

FIG. 26.

$$\phi(P, C) - \phi(P, C_1) = \int_{O(C)}^{P} \phi_r dx^r - \int_{O(C_1)}^{P} \phi_r dx^r$$
$$= \int_{O(C)}^{P} \phi_r dx^r + \int_{P(C_1)}^{O} \phi_r dx^r = \int_{\Gamma} \phi_r dx^r.$$

By Stokes' theorem 7.505, this becomes

$$\phi(P, C) - \phi(P, C_1) = \int_S \phi_{r,s} d\tau_{(2)}{}^{rs}$$
$$= \tfrac{1}{2} \int_S (\phi_{r,s} - \phi_{s,r})\, d\tau_{(2)}{}^{rs},$$

and the last integral vanishes by 8.321. Thus

$$\phi(P, C) = \phi(P, C_1),$$

and this shows that ϕ is independent of the path of integration so that ϕ is a function of position only. We may now write 8.323 in the differential form

$$d\phi = \phi_r dx^r,$$

and 8.322 follows immediately. This establishes the equivalence of 8.321 and 8.322.

Under a gauge transformation ϕ_r goes into

8.324. $$\phi'_r = \phi_r - (\ln \lambda)_{,r}.$$

It follows that if and only if ϕ_r is the gradient of an invariant, 8.322, there exists a gauge transformation such that ϕ'_r vanishes identically. This is the case if F_{mn} is zero everywhere. Then, with $\phi'_r = 0$, equation 8.312 reads

8.325. $$a'_{mn\,|\,r} = 0.$$

The tensor pair $(a'_{mn}, 0)$ is distinguished from all other pairs (a_{mn}, ϕ_r) by its greater simplicity. It is therefore natural to reserve the name of metric tensor for a'_{mn} alone and to drop the requirement of gauge-invariance. The resulting geometry is that of a Riemannian space; by 8.308 the coefficients of connection are the Christoffel symbols.

More directly, we can introduce a *Riemannian space* as a *space with symmetric connection* in which the following requirement is satisfied:

R. *There exists a symmetric tensor a_{mn}, whose determinant is non-zero, such that*

8.326. $$a_{mn\,|\,r} = 0.$$

Repeating the process which led from 8.301 to 8.306 and 8.308, we obtain

8.327. $$\Gamma_{mnr} = [mn, r], \quad \Gamma^r_{mn} = \left\{ {r \atop mn} \right\}.$$

The whole structure of Riemannian geometry, as discussed in the earlier chapters, follows from these equations.

We shall now give a very brief account of the generalization of a space with symmetric connection to a *projective space*. This generalization is similar to that by which we pass from a Riemannian space to a Weyl space.

In classical projective geometry the basic elements are the points and straight lines and the basic relation between them is that of incidence. Concepts such as parallelism, length, or magnitude of an angle are outside the domain of projective geometry. In differential geometry the obvious analogues of straight lines are the paths. While retaining the significance of paths we must attempt to exclude the concept of parallelism or, more precisely, of parallel propagation along a curve, a concept well defined in a space with symmetric connection. The result of this generalization will be a projective space within the framework of differential geometry.

The equations 8.120 of a path are explicitly as follows:

8.328. $$\lambda^s\left(\frac{d\lambda^r}{du} + \Gamma^r_{mn}\lambda^m\lambda^n\right) = \lambda^r\left(\frac{d\lambda^s}{du} + \Gamma^s_{mn}\lambda^m\lambda^n\right),$$

where

8.329. $$\lambda^r = \frac{dx^r}{du}.$$

While retaining the coordinate system used, we consider all possible changes in the coefficients of connection Γ^r_{mn} which leave the equations 8.328 of a path invariant. We shall regard all symmetric connections obtained in this manner as equally fundamental. All quantities or relations which are common to the spaces with these different symmetric connections will be called *projective* quantities or relations. Let Γ'^r_{mn} be the coefficients of another such symmetric connection. Then the equations

8.330. $$\lambda^s\left(\frac{d\lambda^r}{du} + \Gamma'^r_{mn}\lambda^m\lambda^n\right) = \lambda^r\left(\frac{d\lambda^s}{du} + \Gamma'^s_{mn}\lambda^m\lambda^n\right)$$

must be identical with 8.328. Subtracting 8.328 from 8.330 we obtain

8.331 $$(\delta^s_p A^r_{mn} - \delta^r_p A^s_{mn}) \lambda^p \lambda^m \lambda^n = 0,$$

where

8.332. $$A^r_{mn} = A^r_{nm} = \Gamma'^r_{mn} - \Gamma^r_{mn}.$$

Since Γ^r_{mn} and Γ'^r_{mn} transform according to 8.112 under changes of coordinates, it is easily seen that A^r_{mn} is a tensor. With a little manipulation of dummy suffixes, 8.331 can be brought into the form

8.333. $$(\delta^s_p A^r_{mn} + \delta^s_n A^r_{pm} + \delta^s_m A^r_{np} - \delta^r_p A^s_{mn} - \delta^r_n A^s_{pm} - \delta^r_m A^s_{np}) \lambda^p \lambda^m \lambda^n = 0.$$

The bracket on the left-hand side of this expression is symmetric in the suffixes p, m, n. Since λ^r is an arbitrary vector it follows that this bracket must be zero:

8.334. $$\delta^s_p A^r_{mn} + \delta^s_n A^r_{pm} + \delta^s_m A^r_{np} - \delta^r_p A^s_{mn} - \delta^r_n A^s_{pm} - \delta^r_m A^s_{np} = 0.$$

Contracting with respect to s and p, we obtain

8.335. $$A^r_{mn} = \delta^r_n \psi_m + \delta^r_m \psi_n,$$

where ψ_r is a vector given by

8.336. $$\psi_r = \frac{1}{N+1} A^s_{rs}.$$

Conversely, if 8.335 is substituted in 8.331, the equation is satisfied identically in λ^r and ψ_r. By 8.332, the coefficients of the new symmetric connection must be of the form

8.337. $$\Gamma'^r_{mn} = \Gamma^r_{mn} + \delta^r_n \psi_m + \delta^r_m \psi_n.$$

It is now clear that the paths are the same for two symmetric connections if and only if their coefficients are related by 8.337, ψ_r being an arbitrary vector. The changes 8.337 of the connection are called *projective transformations* of the connection. We may now say that projective geometry consists of the quantities and relations which are *projective-invariant*, i.e., invariant under projective transformations.

In a projective space the notion of parallel propagation is lost since there is an infinite set of equivalent symmetric connections with conflicting equations for parallel propagation.

We conclude with a short statement on the curvature properties of projective space, omitting all proofs.*

By direct computation of its behaviour under projective transformations, the following tensor can be shown to be a projective-invariant:

8.338. $$W^{s}_{.rmn} = R^{s}_{.rmn} - \frac{2}{N+1}\delta^{s}_{r}F_{mn} + \frac{1}{N-1}(\delta^{s}_{m}R_{rn} - \delta^{s}_{n}R_{rm}) - \frac{2}{N^2-1}(\delta^{s}_{n}F_{rm} - \delta^{s}_{m}F_{rn}),$$

where N is the number of dimensions of the space, and $R^{s}_{.rmn}$, R_{rm}, F_{mn} are given by 8.214, 8.218, 8.219. The tensor $W^{s}_{.rmn}$, discovered by H. Weyl, is called the *projective curvature tensor*.

In two dimensions ($N = 2$) the projective curvative tensor vanishes identically.

In higher dimensional spaces ($N \geqslant 3$), it can be shown that the vanishing of the projective curvature tensor is necessary and sufficient for the existence of a projective transformation which makes the space flat. In more mathematical language: If $W^{s}_{.rmn} = 0$ and if $N \geqslant 3$, then there exists a vector ψ_r, such that the projective transformation induced by it transforms $R^{s}_{.rmn}$ into $R'^{s}_{.rmn} = 0$. A projective space of 3 or more dimensions with vanishing projective curvature tensor is called *projectively flat*.

It can also be shown that if and only if a space is projectively flat, there exists a coordinate system such that the equations of all paths are linear in the coordinates, i.e., such that paths are given by equations of the form 8.221.

*For proofs see the books by Eisenhart (Non-Riemannian Geometry), Schouten (Der Ricci-Kalkül) and Thomas (Differential Invariants of Generalized Spaces) listed in the bibliography.

SUMMARY VIII

Coefficients of linear connection:

$$\Gamma^r_{mn},\ \overline{\Gamma}^r_{mn},$$

$$\Gamma^\rho_{\mu\nu} = \Gamma^r_{mn} X^\rho_r X^m_\mu X^n_\nu + X^r_{\mu\nu} X^\rho_r,\quad \overline{\Gamma}^\rho_{\mu\nu} = \overline{\Gamma}^r_{mn} X^\rho_r X^m_\mu X^n_\nu + X^r_{\mu\nu} X^\rho_r.$$

Absolute derivative:

$$\frac{\delta}{\delta u} T^{r}_{.s}{}^{::}_{::} = \frac{d}{du} T^{r}_{.s}{}^{::}_{::} + \Gamma^r_{mn} T^{m}_{.s}{}^{::}_{::} \frac{dx^n}{du} + \ldots$$

$$- \overline{\Gamma}^m_{sn} T^{r}_{.m}{}^{::}_{::} \frac{dx^n}{du} - \ldots.$$

Covariant derivative:

$$T^{r}_{.s}{}^{::}_{::|n} = T^{r}_{.s}{}^{::}_{::,n} + \Gamma^r_{mn} T^{m}_{.s}{}^{::}_{::} - \overline{\Gamma}^m_{sn} T^{r}_{.m}{}^{::}_{::} + \ldots.$$

Parallel propagation:

$$\frac{\delta}{\delta u} T^{r}_{.s}{}^{::}_{::} = 0.$$

Path:

$$\frac{\delta\lambda^r}{\delta u}\lambda^s = \lambda^r \frac{\delta\lambda^s}{\delta u},\quad \lambda^r = \frac{dx^r}{du}.$$

Ortho-invariant connections:

$$\Gamma^r_{mn} - \overline{\Gamma}^r_{mn} = \delta^r_m C_n.$$

Single (contraction-invariant) connection:

$$\Gamma^r_{mn} = \overline{\Gamma}^r_{mn},$$

$$\frac{\delta}{\delta u}(T^{r}{}^{::}_{::}\ S_{r}{}^{::}_{::}) = \frac{\delta T^{r}{}^{::}_{::}}{\delta u}\ S_{r}{}^{::}_{::} + T^{r}{}^{::}_{::}\ \frac{\delta S_{r}{}^{::}_{::}}{\delta u},$$

$$\frac{\delta}{\delta u}\delta^r_s = 0.$$

Symmetric connection: $\Gamma^r_{mn} = \Gamma^r_{nm}$;

given a point O, there exists a coordinate system such that

$$\Gamma^r_{mn} = 0 \text{ at } O.$$

Curvature tensor:

$$R^s_{.rmn} = \frac{\partial\Gamma^s_{rn}}{\partial x^m} - \frac{\partial\Gamma^s_{rm}}{\partial x^n} + \Gamma^p_{rn}\Gamma^s_{pm} - \Gamma^p_{rm}\Gamma^s_{pn},$$

$$R^s_{.rmn} = -R^s_{.rnm}, \quad R^s_{.rmn} + R^s_{.mnr} + R^s_{.nrm} = 0,$$

$$R^s_{.rmn|k} + R^s_{.rnk|m} + R^s_{.rkm|n} = 0,$$

$$R_{rm} = R^s_{.rms}, \quad F_{mn} = \tfrac{1}{2} R^s_{.smn} = \tfrac{1}{2}\left(\frac{\partial \Gamma^s_{sn}}{\partial x^m} - \frac{\partial \Gamma^s_{sm}}{\partial x^n}\right).$$

Flat space:

$$R^s_{.rmn} = 0.$$

Weyl space:

$$a_{mn|r} + a_{mn}\phi_r = 0, \quad a_{mn} = a_{nm}, \quad a \neq 0,$$

$$\Gamma^r_{mn} = \left\{ {r \atop mn} \right\} + \tfrac{1}{2}\left(\delta^r_m \phi_n + \delta^r_n \phi_m - a_{mn}\phi^r\right).$$

Gauge transformation:

$$a'_{mn} = \lambda a_{mn}, \quad \phi'_r = \phi_r - (\ln\lambda)_{,r}, \quad \Gamma'^r_{mn} = \Gamma^r_{mn}.$$

Riemannian space:

$$a_{mn|r} = 0, \quad \Gamma^r_{mn} = \left\{ {r \atop mn} \right\}.$$

Projective transformation:

$$\Gamma'^r_{mn} = \Gamma^r_{mn} + \delta^r_n \psi_m + \delta^r_m \psi_n.$$

Projective curvature tensor:

$$W^s_{.rmn} = R^s_{.rmn} - \frac{2}{N+1}\delta^s_r F_{mn} + \frac{1}{N-1}\left(\delta^s_m R_{rn} - \delta^s_n R_{rm}\right)$$

$$- \frac{2}{N^2-1}\left(\delta^s_n F_{rm} - \delta^s_m F_{rn}\right).$$

Projectively flat space ($N \geqslant 3$) :

$$W^s_{.rmn} = 0.$$

EXERCISES VIII

1. In a space with a general linear connection, show that the expressions

$$\Gamma^r_{mn} - \Gamma^r_{nm}, \quad \bar{\Gamma}^r_{mn} - \bar{\Gamma}^r_{nm}$$

are tensors of the type indicated by the position of the suffixes.

2. If T is an invariant in a space with a single linear connection, show that

$$T_{|mn} - T_{|nm} = -2\,T_{|r}\,L^r_{mn},$$

where

$$L^r_{mn} = -L^r_{nm} = \tfrac{1}{2}\,(\Gamma^r_{mn} - \Gamma^r_{nm}).$$

3. If T_r is a covariant vector field in a space with a single linear connection, show that

$$T_{r|mn} - T_{r|nm} = T_s R^s_{.rmn} - 2\,T_{r|s}\,L^s_{mn},$$

where L^s_{mn} is defined in Ex. 2, and where

$$R^s_{.rmn} = \frac{\partial \Gamma^s_{rn}}{\partial x^m} - \frac{\partial \Gamma^s_{rm}}{\partial x^n} + \Gamma^p_{rn}\Gamma^s_{pm} - \Gamma^p_{rm}\Gamma^s_{pn}.$$

4. In a space with a single linear connection, show that *if* there exists a coordinate system for each point O of space such that, at O and for all curves through O, the absolute and ordinary derivatives of any tensor differ by a multiple of the tensor, *then* the coefficients of linear connection satisfy a relation of the form

$$\Gamma^r_{mn} - \Gamma^r_{nm} = \delta^r_m A_n - \delta^r_n A_m,$$

where A_r is some covariant vector. (Such a single linear connection is said to be *semi-symmetric*).

5. Prove the converse of Ex. 4.

[Hint: Consider the coordinate transformation (in the notation of 8.208):

$$x^\rho = \delta^\rho_r(x^r - x^r_0) + \tfrac{1}{2}\left\{ \tfrac{1}{2}\delta^\rho_r(\Gamma^r_{mn} + \Gamma^r_{nm})_0 - \frac{1}{N-1}\,\delta^\rho_n(\Gamma^p_{pm} - \Gamma^p_{mp})_0 \right\}(x^m - x^m_0)(x^n - x^n_0).]$$

6. Show that

$$T_{r|s} - T_{s|r} = T_{r,s} - T_{s,r},$$

T_r being a covariant vector, if the connection is symmetric but not, in general, if the connection is unsymmetric.

7. Show that, in the generalized Stokes' theorem 7.505, the partial differentiation in the integrand on the left-hand side

can be replaced by covariant differentiation if the space has a symmetric connection.

8. In a Weyl space, the rate of change of the squared magnitude X^2 of a vector X^r under parallel propagation of X^r along some curve $x^r = x^r(u)$ is given by

$$\frac{\Delta}{\Delta u} X^2 = \frac{d}{du}(\epsilon a_{mn} X^m X^n) = \epsilon \frac{da_{mn}}{du} X^m X^n + 2\,\epsilon a_{mn} X^m \frac{dX^n}{du},$$

where da_{mn}/du is the ordinary derivative along the curve of a_{mn} (which is defined throughout the space), whereas dX^n/du is obtained by parallel propagation, i.e.,

$$\frac{dX^n}{du} = -\,\Gamma^n_{sp}\, X^s \frac{dx^p}{du}.$$

Show that

$$\frac{\Delta}{\Delta u} X^2 = -\,X^2 \phi_r \frac{dx^r}{du}.$$

Hence prove that the change in X^2 under parallel propagation of X^r around an infinitesimal circuit, bounding a 2-element of extension $d\tau_{(2)}{}^{mn}$, is given by

$$\Delta X^2 = \frac{2}{N} X^2 F_{mn} d\tau_{(2)}{}^{mn}.$$

(Hint: Use Stokes' theorem).

9. Verify that the projective curvature tensor 8.338 is invariant under all projective transformations.

10. In a projective space, the *coefficients of projective connection* P^r_{mn} are defined as follows:

$$P^r_{mn} = \Gamma^r_{mn} - \frac{1}{N+1}(\delta^r_m \Gamma^p_{pn} + \delta^r_n \Gamma^p_{pm}).$$

Show that P^r_{mn} is invariant under projective transformations of Γ^r_{mn}. Verify that $P^r_{mn} = P^r_{nm}$, $P^r_{rn} = 0$. Find the transformation properties of P^r_{mn} under changes of the coordinate system. (T. Y. Thomas.)

11. Show that the differential equation of a path can be written in the form

$$\lambda^s\left(\frac{d\lambda^r}{du} + P^r_{mn}\lambda^m\lambda^n\right) = \lambda^r\left(\frac{d\lambda^s}{du} + P^s_{mn}\lambda^m\lambda^n\right),$$

where P^r_{mn} are the coefficients of projective connection defined in Ex. 10. Deduce that no change in the Γ^r_{mn} other than a projective transformation leaves the P^r_{mn} invariant.

12. Defining

$$P^s_{.rmn} = \frac{\partial P^s_{rn}}{\partial x^m} - \frac{\partial P^s_{rm}}{\partial x^n} + P^p_{rn}P^s_{pm} - P^p_{rm}P^s_{pn},$$

$$P_{rm} = P^s_{.rms},$$

where P^r_{mn} are the coefficients of projective connection of Ex. 10, show that

$$P^s_{.smn} = 0, \quad P_{rm} = -\frac{\partial P^s_{rm}}{\partial x^s} + P^p_{rs}P^s_{pm}.$$

Prove that

$$W^s_{.rmn} = P^s_{.rmn} + \frac{1}{N-1}(\delta^s_m P_{rn} - \delta^s_n P_{rm}),$$

where $W^s_{.rmn}$ is the projective curvature tensor.

13. Show that

$$W^s_{.smn} = 0, \quad W^s_{.rsn} = 0, \quad W^s_{.rms} = 0,$$

$$W^s_{.rmn} = -W^s_{.rnm}, \quad W^s_{.rmn} + W^s_{.mnr} + W^s_{.nrm} = 0.$$

14. In a space with a linear connection, we say that the *directions* of two vectors, X^r at a point A and Y^r at B, are parallel with respect to a curve C which joins A and B if the vector obtained by parallel propagation of X^r along C from A to B is a multiple of Y^r. Prove that the most general change of linear connection which preserves parallelism of directions (with respect to all curves) is given by

$$\Gamma'^r_{mn} = \Gamma^r_{mn} + 2\,\delta^r_m\psi_n,$$

where ψ_n is an arbitrary vector. If Γ^r_{mn} are the coefficients of a symmetric connection, show that Γ'^r_{mn} are semi-symmetric (cf. Exercise 4).

15. In a space with symetric connection, show that

$$T^r_{\ |mn} - T^r_{\ |nm} = -T^s R^r_{.smn}.$$

16. By use of the lemma of 8.1, or otherwise, show that in a space with symmetric connection

$$T^{r}_{.s}{}^{\cdots}_{\cdots|mn} - T^{r}_{.s}{}^{\cdots}_{\cdots|nm} =$$
$$- T^{p}_{.s}{}^{\cdots}_{\cdots} R^{r}_{.pmn} + T^{r}_{.p}{}^{\cdots}_{\cdots} R^{p}_{.smn} + \dots \quad .$$

17. Using the compressed notation of 1.7, show that, in a space with a double linear connection, the contractions of the connections Γ^{r}_{rn}, $\bar{\Gamma}^{r}_{rn}$ transform as follows:

$$\Gamma^{\rho}_{\rho\nu} = X^{n}_{\nu}\Gamma^{r}_{rn} + \frac{1}{J}\frac{\partial J}{\partial x^{\nu}},$$

$$\bar{\Gamma}^{\rho}_{\rho\nu} = X^{n}_{\nu}\bar{\Gamma}^{r}_{rn} + \frac{1}{J}\frac{\partial J}{\partial x^{\nu}},$$

where $J = |X^{r}_{\rho}|$ is the Jacobian of the transformation.

18. Let $T^{r}_{.s}{}^{\cdot}_{\cdot}$ be a relative tensor field of weight W, so that its transformation character is given by 7.102. In a space with a double linear connection, we now define the absolute derivative of $T^{r}_{.s}{}^{\cdot}_{\cdot}$ along a curve $x^{r} = x^{r}(u)$ as follows:

$$\frac{\delta}{\delta u} T^{r}_{.s}{}^{\cdot}_{\cdot} = \frac{d}{du} T^{r}_{.s}{}^{\cdot}_{\cdot} + \Gamma^{r}_{mt} T^{m}_{.s}{}^{\cdot}_{\cdot} \frac{dx^{t}}{du} + \cdots$$
$$- \bar{\Gamma}^{m}_{st} T^{r}_{.m}{}^{\cdot}_{\cdot} \frac{dx^{t}}{du} - \cdots - W\Gamma^{m}_{mt} T^{r}_{.s}{}^{\cdot}_{\cdot} \frac{dx^{t}}{du}.$$

Prove that $\delta T^{r}_{.s}{}^{\cdot}_{\cdot}/\delta u$ is a relative tensor of the same type and weight as $T^{r}_{.s}{}^{\cdot}_{\cdot}$. Verify that the absolute derivatives of outer products and sums of relative tensors obey the usual rules for differentiating products and sums (cf. 8.102, 8.103).

19. Are the statements of Ex. 18 correct if, in the definition of the absolute derivative, Γ^{m}_{mt} is replaced by (a) $\bar{\Gamma}^{m}_{mt}$, (b) $\frac{1}{2}(\Gamma^{m}_{mt} + \bar{\Gamma}^{m}_{mt})$? Show that, in the special case of a single connection, postulate E applies to relative tensors.

20. From the definition of Ex. 18 and from 8.115, deduce an expression for the covariant derivative of a general relative tensor.

21. In the special case of a Riemannian space

$$\left(\Gamma^r_{mn} = \bar{\Gamma}^r_{mn} = \begin{Bmatrix} r \\ mn \end{Bmatrix}\right),$$

verify that the definition of Ex. 18 agrees with 7.212.

22. Substituting into the equations $a_{mn\cdot rs} - a_{mn\cdot sr} = 0$ from 8.304, and using 8.318, 8.219, show that, in a Weyl space, the metric tensor a_{mn} satisfies the algebraic equations

(a) $$a_{pq}G^{pq}_{\cdot\cdot mnrs} = 0,$$

where $G^{pq}_{\cdot\cdot mnrs}$ is a function of the Γ^r_{mn} and their derivatives, given by

$$G^{pq}_{\cdot\cdot mnrs} = \tfrac{1}{2}(\delta^p_m R^q_{\cdot nrs} + \delta^q_m R^p_{\cdot nrs} + \delta^p_n R^q_{\cdot mrs} + \delta^q_n R^p_{\cdot mrs})$$
$$- \frac{1}{N}(\delta^p_m \delta^q_n + \delta^q_m \delta^p_n)\, R^t_{\cdot trs}.$$

By repeated covariant differentiation of (a) and by use of 8.301, deduce the following set of algebraic equations for a_{mn}:

(b) $$a_{pq}G^{pq}_{\cdot\cdot mnrs\,|\,t_1} = a_{pq}G^{pq}_{\cdot\cdot mnrs\,|\,t_1t_2} = \ldots = 0.$$

(Note: A symmetric connection can always be chosen such that, in some fixed system of coordinates and at a fixed point, the quantities Γ^r_{mn}, $\Gamma^r_{mn\cdot s}$, $\Gamma^r_{mn\cdot st}$, etc., assume arbitrarily assigned values subject only to certain symmetry requirements, such as $\Gamma^r_{mn} = \Gamma^r_{nm}$, $\Gamma^r_{mn\cdot st} = \Gamma^r_{nm\cdot ts}$, etc. Hence, in a space with a general symmetric connection, the infinite set (a), (b) of linear algebraic equations for a_{mn} ($= a_{nm}$) has, in general, only the trivial solution $a_{mn} = 0$. It follows that postulate W does not restrict the connection, as was stated on p. 298.)

APPENDIX A

REDUCTION OF A QUADRATIC FORM

(Reference to p. 58)

In the reduction of a metric form to a sum of squares, 2.6, a number of pertinent considerations have been omitted from the text so as not to detract from the essential simplicity of the basic arguments. Some of the subsidiary questions will be answered here.

1. The process by which 2.601 was obtained breaks down if $a_{11} = 0$. Even if $a_{11} \neq 0$, the same difficulty may arise at some of the following stages which lead to 2.603.

If $a_{11} = 0$, and if any one of the remaining "diagonal" coefficients of the metric form, i.e., $a_{22}, a_{33}, \ldots, a_{NN}$, does not vanish, then we renumber the coordinates so as to make this coefficient the first and proceed as in the text. The only remaining possibility is that $a_{11} = a_{22} = \ldots = a_{NN} = 0$. Then, since Φ cannot vanish identically (we assume $a \neq 0$), at least one of the non-diagonal coefficients must be different from zero, $a_{12} \neq 0$, say. In this case we first perform the simple coordinate transformation

A 1.1 $\quad x^2 = x'^1 + x'^2, \; x^m = x'^m \quad m = 1, 3, 4, \ldots, N.$

Then $a'_{11} = 2a_{12} \neq 0$, and we can proceed as in the text.

If the difficulty considered here arises at any other stage of the process which leads to 2.603, the device outlined above can be used again. An alternative procedure is indicated in Exercises II, No. 15.

2. We shall now consider another question. It appears that when we apply the process described in the text to Φ_r and

obtain a Ψ_{r+1} and a Φ_{r+1}, more than one differential may conceivably be eliminated so that we end up with fewer than N forms Ψ_m in 2.603. We shall now show that this is not possible.

We can always write Φ in the form

A 2.1. $$\Phi = \epsilon_1\Psi_1^2 + \epsilon_2\Psi_2^2 + \ldots + \epsilon_N\Psi_N^2,$$

if we temporarily permit ϵ_m to assume the value 0 in addition to the usual values ± 1. Put

A 2.2. $$\begin{aligned} E_{rs} &= \epsilon_r \text{ if } r = s \\ &= 0 \text{ if } r \neq s. \end{aligned}$$

Then, by 2.604,

A 2.3. $$\Phi = E_{rs}\Psi_r\Psi_s = E_{rs}b_{rm}b_{sn}dx^m dx^n = a_{mn}dx^m dx^n,$$

and therefore, since $E_{rs}b_{rm}b_{sn}$ is symmetric in m and n,

A 2.4. $$a_{mn} = E_{rs}b_{rm}b_{sn}.$$

Taking the determinant of both sides, we have

A 2.5. $$a = |E_{rs}|\,|b_{mn}|^2 = \epsilon_1\epsilon_2 \ldots \epsilon_N |b_{mn}|^2.$$

Since $a \neq 0$, we deduce that no ϵ_m vanishes and that $|b_{mn}| \neq 0$. The first conclusion proves our assertion. The non-vanishing of $|b_{mn}|$ shows that $\Psi_1, \Psi_2, \ldots, \Psi_N$ are independent linear combinations of $dx^1, dx^2, \ldots, dx^N$, i.e., the set of equations 2.604 can be solved and the dx^m expressed linearly in terms of the Ψ_m.

3. The last question we shall consider here concerns the signs of the ϵ_m in A 2.1. Sylvester's famous *theorem of inertia* states: If a quadratic form $\Phi = a_{mn}dx^m dx^n$ (with $a \neq 0$) is reduced to a sum of squares, A 2.1, then the number R of ϵ's which have the value $+1$ and the number $N - R$ of ϵ's which have the value -1 are invariants of the quadratic form; this means that these numbers do not depend on the manner in which the reduction to a sum of squares is accomplished. The difference between invariants R and $N - R$ is called the *signature* of the quadratic form.

We proceed to prove the theorem of inertia. By renumbering the Ψ's in A 2.1 we can make the first R ϵ's positive and the remaining ones negative. Then

A 3.1.
$$\Phi = \Psi_1^2 + \Psi_2^2 + \ldots + \Psi_R^2 - \Psi_{R+1}^2 - \Psi_{R+2}^2 - \ldots - \Psi_N^2,$$
$$\Psi_m = b_{mn}dx^n, \qquad |b_{mn}| \neq 0.$$

Let us consider any other reduction of Φ to a sum of squares, such as

A 3.2.
$$\Phi = X_1^2 + X_2^2 + \ldots + X_S^2 - X_{S+1}^2 - X_{S+2}^2 - \ldots - X_N^2,$$
$$X_m = c_{mn}dx^n, \qquad |c_{mn}| \neq 0.$$

We must show that $R = S$. We first assume that this is not so, i.e., that $R \neq S$. Then we can, without loss of generality, take

A 3.3.
$$S > R.$$

Combining A 3.1 and A 3.2 we have

A 3.4.
$$\Psi_1^2 + \ldots + \Psi_R^2 + X_{S+1}^2 + \ldots + X_N^2 = X_1^2 + \ldots + X_S^2 + \Psi_{R+1}^2 + \ldots + \Psi_N^2.$$

Now consider the set of linear homogeneous equations

A 3.5.
$$\Psi_m = b_{mn}dx^n = 0, \quad m = 1, 2, \ldots, R,$$
$$X_m = c_{mn}dx^n = 0, \quad m = S+1, S+2, \ldots, N.$$

By A 3.3 the number of equations $(N - S + R)$ in this system is less than the number of unknowns $dx^m (N)$. Therefore equations A 3.5 have a non-trivial solution where the dx^n are not all zero.* Obviously the left-hand side of A 3.4 vanishes for this solution. Since $|b_{mn}| \neq 0$, the set of equations

A 3.6.
$$\Psi_m = b_{mn}dx^n = 0, \quad m = 1, 2, \ldots, N,$$

has only the trivial solution $dx^n = 0$. Hence for the non-trivial solution of A 3.5, at least one of $\Psi_{R+1}, \ldots, \Psi_N$ must be different from zero. Thus the right-hand side of A 3.4 is positive while the left-hand side vanishes. This contradiction shows that the assumption A 3.3 is untenable and thus establishes the theorem of inertia.

*M. Bôcher, *Introduction to Higher Algebra*, New York, The Macmillan Company, 1907, sect. 17, Thm. 3, Cor. 1.

APPENDIX B

MULTIPLE INTEGRATION

(Reference to p. 257)

In this appendix we present a plausibility argument in favour of the theorem on multiple integration stated in 7.3.

Consider the edge $AB_{(1)}$ of the M-cell $AB_{(1)} \dots B_{(M)}$ introduced earlier in 7.3. Along this edge $c^{(1)}$ alone changes. Subdivide this edge by points $B'_{(1)}, B''_{(1)}, \dots$, lying on it and let $c^{(1)\prime}, c^{(1)\prime\prime}, \dots$, be the corresponding values of $c^{(1)}$. Then the $(M-1)$ dimensional spaces

$$f^{(1)}(y) = c^{(1)\prime}, f^{(1)}(y) = c^{(1)\prime\prime}, \dots,$$

divide the cell into subcells. If the original cell, and therefore also the subcells, are infinitesimal, then

B 1. $$d_{(1)}y^a = d'_{(1)}y^a + d''_{(1)}y^a + \dots,$$

where $d_{(1)}y^a, d'_{(1)}y^a, d''_{(1)}y^a, \dots$, denote the increments in the parameter y^a in passing from A to $B_{(1)}$, from A to $B'_{(1)}$, from $B'_{(1)}$ to $B''_{(1)}, \dots$, respectively. Let $\Delta = |d_{(\beta)}y^a|$ and let Δ', Δ'', ..., denote the corresponding determinants for the subcells. Then

$$\begin{aligned} \Delta &= \epsilon_{a_1 \dots a_M}(d'_{(1)}y^{a_1} + d''_{(1)}y^{a_1} + \dots) d_{(2)}y^{a_2} \dots d_{(M)}y^{a_M} \\ &= \epsilon_{a_1 \dots a_M} d'_{(1)}y^{a_1} d_{(2)}y^{a_2} \dots d_{(M)}y^{a_M} \\ &\qquad + \epsilon_{a_1 \dots a_M} d''_{(1)}y^{a_1} d_{(2)}y^{a_2} \dots d_{(M)}y^{a_M} + \dots, \end{aligned}$$

or, equivalently,

B 2. $$\Delta = \Delta' + \Delta'' + \dots.$$

Thus the determinant Δ is the sum of the corresponding determinants for the subcells.

We may proceed in the manner described above to choose points $B'_{(2)}, B''_{(2)}, \dots, B'_{(3)}, B''_{(3)}, \dots$, on the other edges of our original cell, and so subdivide it further. Obviously our result still holds, namely, that Δ is the sum of the corresponding determinants for the subcells.

If the region R_M is divided into sufficiently small cells by the M families of surfaces

B 3. $$f^{(\alpha)}(y) = c^{(\alpha)},$$

then for any further subdivision of these cells, of the type considered above, equation B 2 may be assumed to hold. Also, if Φ is a continuous function in the region R_M (including the boundary R_{M-1}), then, neglecting terms of the order of the Δc's, Φ is a constant in each cell. Thus any finer subdivision of R_M, obtained by adding new members of the families B 3 to those already considered, changes the expression

B 4. $$\Sigma\Phi|\Delta_{(\beta)}y^\alpha|, \quad \text{or } \Sigma\Phi\Delta,$$

by a small quantity of the order of the Δc's. This indicates that if a definite subdivision of R_M is continually refined in the manner just considered, such that all the Δc's tend to zero, then the sum B 4 tends to a finite limit.

Equation B 2 also indicates that lim $\Sigma\Phi\Delta$ is independent of the manner in which R_M is divided into cells—provided these cells have edges that are tangent to the curves of intersection of the $(M - 1)$-spaces B 3. For, given two such divisions of R_M into infinitesimal cells, we can subdivide each further to obtain a common division into smaller cells; this process of subdivision changes the sum B 4 by an infinitesimal only, as was seen above.

Let us again consider an infinitesimal M-cell $AB_{(1)} \ldots B_{(M)}$ in V_M, the edges being characterized by $d_{(\beta)}y^\alpha$. We say that a displacement dy^α is *coplanar* with some of these edges, say with $AB_{(1)}$, $AB_{(2)}$, $AB_{(3)}$, if numbers a_1, a_2, a_3 exist such that

B 5. $$dy^\alpha = a_1 d_{(1)}y^\alpha + a_2 d_{(2)}y^\alpha + a_3 d_{(3)}y^\alpha.$$

Now if we give to the further extremity of an edge, say $B_{(1)}$, a displacement coplanar with some or all of the other edges, say with $AB_{(2)}$, $AB_{(3)}$, $AB_{(4)}$, we do not alter the value of the determinant $|d_{(\beta)}y^\alpha|$. For we merely add to the elements of one row of the determinant quantities proportional to the elements in the other rows. Obviously any cell may be deformed in this way into another cell whose edges assume M prescribed (non-coplanar) directions in V_M.

Our last result indicates that the limit of B 4 is completely independent of the manner in which R_M is divided into cells. For we can start with an arbitrary division and deform each cell, in the manner just indicated, such that its edges are tangent to the curves of intersection of the $(M - 1)$-spaces, B 3.

BIBLIOGRAPHY

1. G. Ricci and T. Levi–Civita, *Méthodes de calcul differentiel absolu et leurs applications* (Paris, 1923). (Reprinted from *Mathematische Annalen*, tome 54, 1900.)
2. J. A. Schouten, *Der Ricci-Kalkül* (Berlin, 1924).
3. P. Appell, *Traité de mecanique rationelle*, Tome 5 (Paris, 1926).
4. L. P. Eisenhart, *Riemannian Geometry* (Princeton, 1926).
5. T. Levi-Civita, *The Absolute Differential Calculus* (Calculus of Tensors) (London, 1927).
6. L. P. Eisenhart, *Non-Riemannian Geometry* (American Mathematical Society, Colloquium Publications, Vol. VIII, 1927).
7. A. Duschek and W. Mayer, *Lehrbuch der Differentialgeometrie* (Leipzig and Berlin, 1930).
8. H. Jeffreys, *Cartesian Tensors* (Cambridge, 1931).
9. A. J. McConnell, *Applications of the Absolute Differential Calculus* (London, 1931).
10. O. Veblen, *Invariants of Quadratic Differential Forms* (Cambridge, 1933).
11. T. Y. Thomas, *Differential Invariants of Generalized Spaces* (Cambridge, 1934).
12. J. A. Schouten and D. J. Struik, *Einführung in die neueren Methoden der Differentialgeometrie*, 2 Vols. (Groningen, 1935 and 1938).
13. L. Brillouin, *Les Tenseurs en mécanique et en élasticité* (Paris, 1938, New York, 1946).
14. C. E. Weatherburn, *Riemannian Geometry and the Tensor Calculus* (Cambridge, 1938).
15. P. G. Bergmann, *Introduction to the Theory of Relativity* (New York 1942).
16. H. V. Craig, *Vector and Tensor Analysis* (New York, 1943).
17. A. D. Michal, *Matrix and Tensor Calculus* (New York, 1947).
18. L. Brand, *Vector and Tensor Analysis* (New York, 1947).

INDEX